Foreword

You have in your hand the second edition of *Preparing for Your ACS Examination in General Chemistry, The Official Guide* designed to help students prepare for examinations produced under the auspices of the Division of Chemical Education (DivCHED) of the American Chemical Society (ACS). The first edition of this guide was published in 1998. In 2002, a study guide for organic chemistry students was published followed by the physical chemistry study guide.

As is common for materials produced by ACS Exams, we called on colleagues in the chemistry education community to help us put this second edition together. The first edition of this guide was written and produced under previous Director, I. Dwaine Eubanks and previous Associate Director, Lucy T. Eubanks. Their time and effort was utilized by many, many students who used the first edition since 1998. In this second edition, we have utilized many of their underlying principles while updating a substantial portion of the guide to provide a guide aligned to current textbooks and curriculum as well as a better representation of what may be expected on an ACS Exam. This effort was in no way accomplished solely by the three authors, rather was a collective effort with special thanks to our expert editing team and student volunteers. Specific acknowledgments noting each contributor is listed in the *Acknowledgments*.

As a discipline, chemistry is surely unique in the extent to which its practitioners provide beneficial volunteer service to the teaching community. ACS exams have been produced by volunteer teacher-experts for more than eighty-five years. Other projects of the Examinations Institute benefit from the abundance of donated time, effort, and talent. The result is that invariably high-quality chemistry assessment materials are made available to the teaching and learning community at a fraction of their real value.

The three *Official Guides* that have been released so far are intended to be ancillary student materials, particularly in courses that use ACS exams. The care that goes into producing ACS exams may be lost on students who view the exams as foreign and unfamiliar. The purpose of this series of guides is to remove any barriers that might stand in the way of students demonstrating their knowledge of chemistry. The extent to which this goal is achieved in this new edition of the general chemistry study guide will become known only as future generations of chemistry students sit for an ACS exam in general chemistry.

We wish them the best.

Thomas Pentecost
Jeffrey Raker
Kristen Murphy

Milwaukee, Wisconsin
September, 2018

Acknowledgements

The unselfish dedication of hundreds of volunteers who contribute their time and expertise make ACS Exams possible. It is from the reservoir of their work that we have drawn inspiration and examples to produce this book to help students who will be taking an ACS exam. We gratefully acknowledge the efforts of all past General Chemistry Committee members.

This *Official Guide* also benefited from the careful proofreading by several colleagues. We extend our special thanks to these faculty members.

William J. Donovan	The University of Akron
Christine Gaudinski	Aims Community College
Barbara L. Gonzalez	California State University, Fullerton
Daniel Groh	Grand Valley State University
Thomas A. Holme	Iowa State University
Cynthia J. Luxford	Texas State University
Eric Malina	University of Nebraska
Matthew Stoltzfus	The Ohio State University
Melonie A. Teichert	United States Naval Academy

Students also participated in the development of this *Official Guide* through using this guide in draft form and sharing their experiences back to us to aid in providing the student-user perspective. We extend our special thanks to these 45 students.

General Chemistry I students:	University of South Florida
General Chemistry II students:	Grand Valley State University
	The University of Akron
	University of Wisconsin-Milwaukee

The personnel of the ACS Division of Chemical Education Examinations Institute played a central role in helping us to produce ***Preparing for Your ACS Examination in General Chemistry: The Official Guide.*** A very special thank you for all of the work involved is owed to our staff members.

Julie Adams	Cherie Mayes
Jessica Reed	Shalini Srinivasan
Jaclyn Trate	

While all of these reviewers have been very helpful in finding problems large and small, any remaining errors are solely our responsibility. You can assist us in the preparation of an even better product by notifying ACS Exams of any errors you may find.

Thomas Pentecost
Jeffrey Raker
Kristen Murphy

Milwaukee, WI
September, 2018

Table of Contents

Toolbox (Foundational Concepts)

Chapter Summary:

This chapter will focus on foundational concepts and skills you will use throughout this study guide and likely on your exam. This should not be considered a comprehensive list of preparatory material, but does include important concepts that will be referenced in the content chapters.

Specific topics covered in this chapter are:

- Unit conversations
- Significant figures
- Scientific notation
- Nomenclature
- Density
- Classification of matter
- Properties and representations of matter

Additionally, this chapter will include supplemental information in sections at the end of this chapter for preparation for your exam including:

- Sample instructions
- Sample datasheet
- How to use this book

Common representations used in questions related to this material:

Name	Example	Used in questions related to
Particulate representations		Classification of matter
Compound units	$g \cdot cm^{-3}$	Density or unit conversion

Where to find this in your textbook:

The material in this chapter typically aligns to an introductory chapter or chapters (could be labeled as "Matter and Measurement" or "Atoms, Molecules, and Compounds") in your textbook. The name of your chapter may vary.

Practice exam:

There are practice exam questions aligned to the material in this chapter. Because there are a limited number of questions on the practice exam, a review of the breadth of the material in this chapter is advised in preparation for your exam.

How this fits into the big picture:

The material in this chapter aligns to the Big Idea of Experiments, Measurement and Data (9) as listed on page 12 of this study guide.

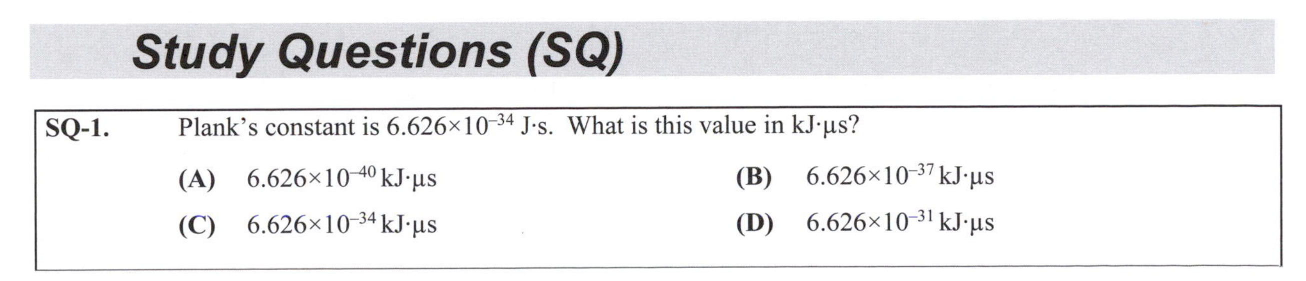

Study Questions (SQ)

SQ-1. Plank's constant is 6.626×10^{-34} J·s. What is this value in kJ·µs?

(A) 6.626×10^{-40} kJ·µs **(B)** 6.626×10^{-37} kJ·µs

(C) 6.626×10^{-34} kJ·µs **(D)** 6.626×10^{-31} kJ·µs

Knowledge Required: (1) Definitions of metric prefixes.

Thinking it Through: As you read this question, you see that the units for the value are given with no prefixes: "J" and "s". The units to which you are converting are prefixed: "kJ" and "μs". Therefore, you could start this question by reviewing your metric prefixes (a good idea during studying as well):

Prefix	Name	Value
M	mega	10^6 or 1,000,000
k	kilo	10^3 or 1,000

Prefix	Name	Value
c	centi	10^{-2} or 0.01
m	milli	10^{-3} or 0.001
μ	micro	10^{-6} or 0.000001
n	nano	10^{-9} or 0.000000001
p	pico	10^{-12} or 0.000000000001

You can then see that you will use the equivalencies of 1 kJ = 1000 J and 1 μs = 10^{-6} s and then use this to convert the value:

$$\left(6.626\times10^{-34}\ \text{J}\cdot\text{s}\right)\left(\frac{1\ \text{kJ}}{1000\ \text{J}}\right)\left(\frac{1\ \mu\text{s}}{10^{-6}\ \text{s}}\right)=6.626\times10^{-31}\ \text{kJ}\cdot\mu\text{s}\ \text{or choice}\ \textbf{(D)}$$

Choice **(A)** incorrectly defines micro as 10^{-3} and inverts the conversion. Choice **(B)** inverts the conversion of seconds only and Choice **(C)** incorrectly defines micro as 10^{-3}.

Practice Questions Related to This: **PQ-1** and **PQ-2**

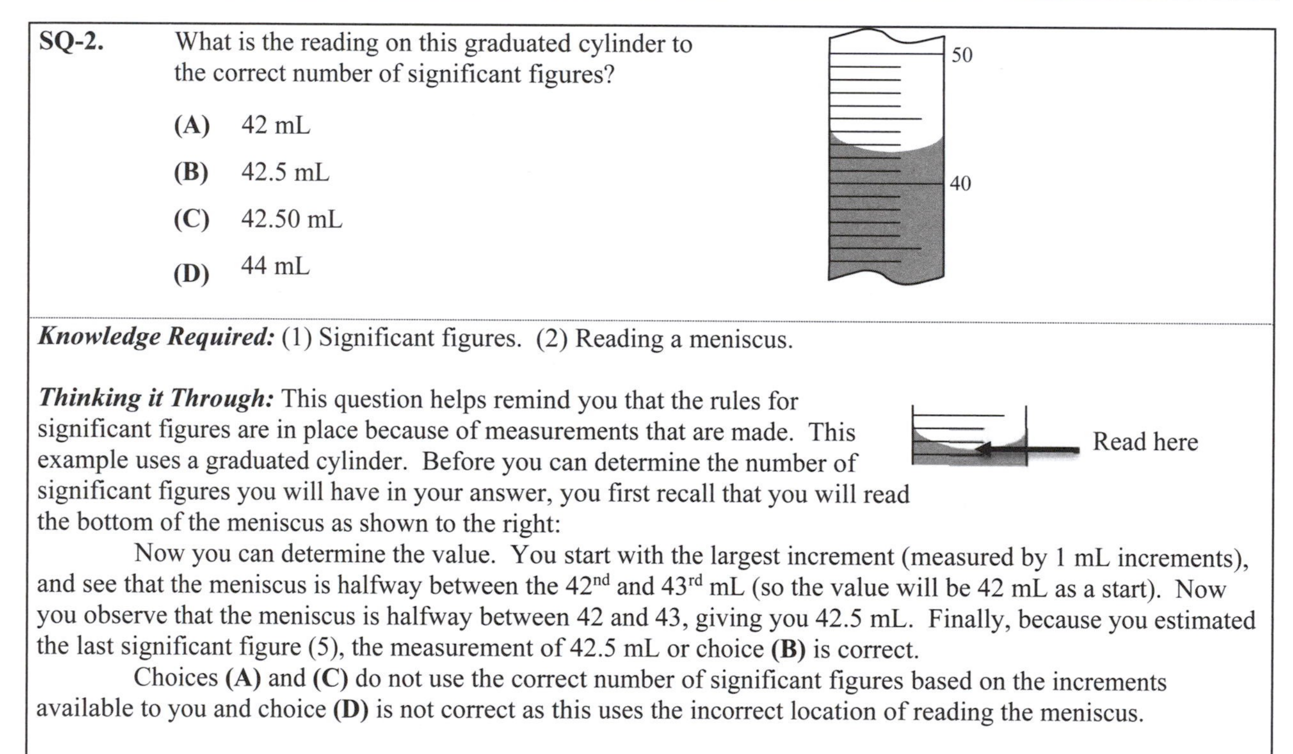

SQ-2. What is the reading on this graduated cylinder to the correct number of significant figures?

(A) 42 mL

(B) 42.5 mL

(C) 42.50 mL

(D) 44 mL

Knowledge Required: (1) Significant figures. (2) Reading a meniscus.

Thinking it Through: This question helps remind you that the rules for significant figures are in place because of measurements that are made. This example uses a graduated cylinder. Before you can determine the number of significant figures you will have in your answer, you first recall that you will read the bottom of the meniscus as shown to the right:

Now you can determine the value. You start with the largest increment (measured by 1 mL increments), and see that the meniscus is halfway between the 42nd and 43rd mL (so the value will be 42 mL as a start). Now you observe that the meniscus is halfway between 42 and 43, giving you 42.5 mL. Finally, because you estimated the last significant figure (5), the measurement of 42.5 mL or choice **(B)** is correct.

Choices **(A)** and **(C)** do not use the correct number of significant figures based on the increments available to you and choice **(D)** is not correct as this uses the incorrect location of reading the meniscus.

Practice Questions Related to This: **PQ-3** and **PQ-4**

SQ-3. Which compound is named correctly?

(A) $FeCO_3$, iron carbonate

(B) K_2SO_3, potassium sulfite

(C) $Sr(NO_3)_2$, strontium nitrite

(D) $Co_2(SO_4)_3$, cobalt(II) sulfate

Knowledge Required: (1) Rules for nomenclature. (2) Formula and names of polyatomic ions.

Thinking it Through: The question tells you that you are going to be using the rules of nomenclature (or naming) to answer this. When you skim the choices, you can see that this question is about ionic nomenclature. Therefore, you know the general rules for ionic nomenclature are:

1. Name the metal or polyatomic cation first (first word for simple compounds).
   a. For those metals with only one cationic charge (such as most of the main group metals), you use just the name of the metal. For example, Na^+ would be "sodium" and Zn^{2+} would be "zinc".
   b. For those metals with more than one cationic charge (such as most of the transition metals), you would use the name of the metal followed by the charge in Roman numerals in parentheses. For example, Fe^{3+} would be "iron(III)" and Pb^{2+} would be "lead(II)".
   c. Polyatomic cations are named with the name of the cation.
2. Name the nonmetal or polyatomic anion second (second word for simple compounds).
   a. For monatomic anions, this is the stem of the element plus the suffix "-ide". For example, S^{2-} is "sulfide" and O^{2-} is "oxide".
   b. Polyatomic anions are named with the name of the anion.

Additionally, for this question, you also need to know the names and formula of polyatomic anions:

CO_3^{2-} = carbonate SO_3^{2-} = sulfite NO_3^- = nitrate SO_4^{2-} = sulfate

Finally, to give the charge for the metals that require them, you will also need to know the charge on the metal. You do this using charge balance of the compound. Algorithmically, this is commonly done through the "cross down" method or the subscript in the formula is the opposite charge on the ion (making sure to still check once you use this that your formula is represented correctly; i.e. it is an empirical formula with the correct ratio of cation:anion; also make sure you understand *why* this method works):

$$Fe^{3+} \quad O^{2-}$$
$$Fe_2O_3$$

Therefore, combining all of these, choice **(B)** does not need a charge on the metal (potassium has only one charge, +1), sulfite is named correctly and the ratio for charge balance is correct.

Choice **(A)** is missing the charge (should be II). Choice **(D)** has a charge, but it is the wrong charge (should be III not II). Finally, choice **(C)** has the wrong name of the polyatomic anion.

Practice Questions Related to This: **PQ-5, PQ-6,** and **PQ-7**

SQ-4. *Conceptual*

In the name arsenic trichloride, the *tri-* prefix specifies

(A) arsenic has an oxidation state of +3.

(B) the geometry of the molecule is trigonal pyramidal.

(C) arsenic is covalently bonded to three chlorine atoms.

(D) arsenic(III) is ionically bonded to three chloride ions.

Knowledge Required: (1) Rules for nomenclature.

Thinking it Through: The question tells you that you are going to be using the rules of nomenclature again to answer this. From the question, you can see this is about molecular nomenclature (or naming molecules or covalent compounds). Therefore, you know the general rules for covalent nomenclature are (for simple compounds):

1. Name the first element shown in the formula first (first word in the name).
   This name is the name of the element with the prefix for the number of atoms of the element using Greek prefixes:

number	prefix	number	prefix
1	mono-	6	hexa-
2	di-	7	hepta-

number	prefix	number	prefix
3	tri-	8	octa-
4	tetra-	9	nona-
5	penta-	10	deca-

When there is only one atom of the element, the prefix of "mono-" is not used. For example, **N**$_2$O$_4$ is "dinitrogen" and **C**O is simply "carbon".

2. Name the second element shown in the formula second (second word in the name). This name is the name of the stem of the element with the prefix for the number of atoms of the element using Greek prefixes for all numbers of atoms including one and the suffix "-ide". When two vowels are adjacent that are either o or a, the letter of the prefix is omitted. For example, N$_2$**O$_4$** is "tetroxide" and C**O** is "monoxide".

Choice **(C)** reflects the correct usage of the Greek prefix (and the formula of arsenic trichloride, $AsCl_3$, reflects this).

Choice **(A)** incorrectly links the prefix of "tri" to oxidation state. Choice **(D)** implies this is an ionic compound, which is not correct. No structural information is provided with this name, so choice **(B)** is not correct either.

Practice Questions Related to This: **PQ-8** and **PQ-9**

SQ-5. What is the formula of iodic acid?

(A) HIO **(B)** HIO_2 **(C)** HIO_3 **(D)** HIO_4

Knowledge Required: (1) Rules for nomenclature. (2) Formula and names of polyatomic ions.

Thinking it Through: The question tells you that you are going to be using the rules of nomenclature (or naming) for acids to answer this. Therefore, you know the general rules for naming acids are based on the type of acid:

Binary acids – those with hydrogen and a monatomic anion:

First word: hydro + the stem of the anion + ic
Second word: acid

Ex: HI = hydro + iod + ic acid = hydroioidic acid

Oxoacids – those with hydrogen and a polyatomic anion:

First word: if the anion has the suffix "-ite" then remove the "-ite" and add "-ous"

Ex: the nitri**te** ion, NO_2^-, becomes nitr**ous**

if the anion has the suffix "-ate" then remove the "-ate" and add "-ic"

Ex: the nitr**ate** ion, NO_3^-, becomes nitr**ic**

Second word: acid

Ex: HNO_2 = nitrous acid HNO_3 = nitric acid

Working backwards, iodic acid means this is the acid with iodate ion or IO_3^-. Therefore, the acid formula is HIO_3 or choice **(C)**. Choice **(A)** is hypoiodous acid; choice **(B)** is iodous acid; and choice **(D)** is periodic acid.

Practice Questions Related to This: **PQ-10** and **PQ-11**

SQ-6. Twenty irregular pieces of an unknown metal have a collective mass of 28.225 g. When carefully placed in a graduated cylinder that originally contained 7.75 mL of water, the final volume read 11.70 mL. What is the density of the metal (in $g{\cdot}cm^{-3}$)?

(A) 2.41 $g{\cdot}cm^{-3}$ **(B)** 3.64 $g{\cdot}cm^{-3}$ **(C)** 7.15 $g{\cdot}cm^{-3}$ **(D)** 112 $g{\cdot}cm^{-3}$

Knowledge Required: (1) Definition of density. (2) Displacement.

Thinking it Through: As you read this question, you recognize that you will need to calculate the density of the metal which you know is the mass of the substance divided by the volume it occupies. You are given the mass in the problem, so you will need to use the information provided to determine the volume. The volume of the metal is determined by displacement of water:

$$V_{\text{f, water + metal}} - V_{\text{i, water}} = V_{\text{metal}} \text{ or } 11.70 \text{ mL} - 7.75 \text{ mL} = 3.95 \text{ mL}$$

Therefore, you can calculate the density of the metal (and convert from mL to cm^3):

$$\left(\frac{28.225 \text{ g}}{3.95 \text{ mL}}\right)\left(\frac{1 \text{ mL}}{1 \text{ cm}^3}\right) = 7.15 \text{ g}\cdot\text{cm}^{-3} \text{ which is choice } \textbf{(C)}$$

Choice **(A)** uses the final volume of water and metal and choice **(B)** uses the initial volume of only the water. Choice **(D)** uses the correct volume of the metal but then calculates the density incorrectly by multiplying the mass and volume.

Practice Questions Related to This: **PQ-12** and **PQ-13**

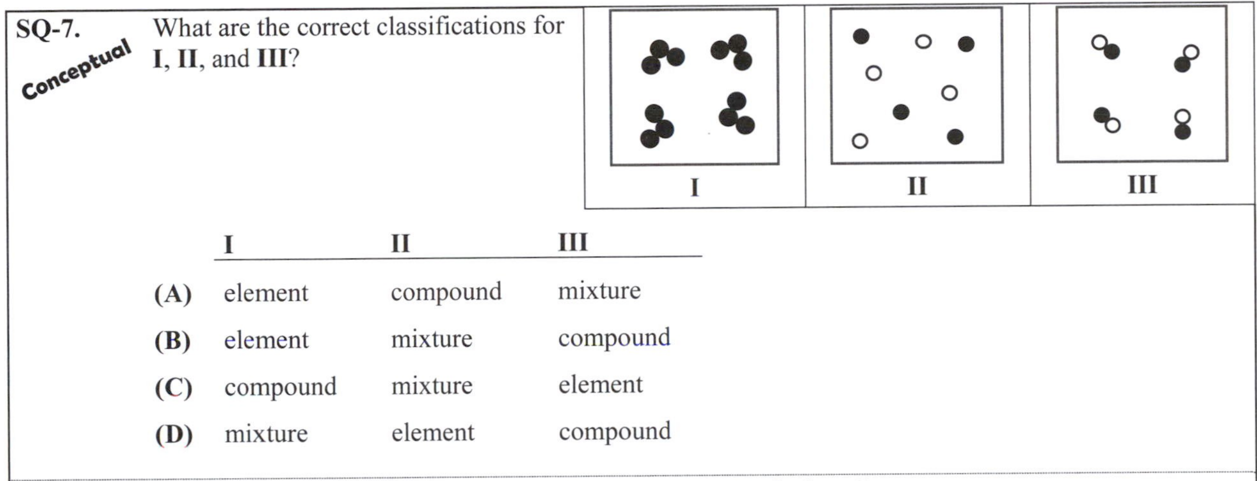

SQ-7. *Conceptual* What are the correct classifications for **I**, **II**, and **III**?

	I	**II**	**III**
(A)	element	compound	mixture
(B)	element	mixture	compound
(C)	compound	mixture	element
(D)	mixture	element	compound

Knowledge Required: (1) Classification of matter/definitions. (2) Particulate diagrams.

Thinking it Through: As you read through this and evaluate the diagrams, you realize that you need to use your definitions of element, compound and mixtures.

First, you know that an element is a group of atoms with the same or similar properties. Merging this with the diagrams you are provided, you see that you are looking for one atom type only which would be sample **I**.

Second, you know that a compound is a pure substance which contains two or more elements chemically combined (or bonded together). Bonded atoms are shown in diagrams **I** and **III**, where the bonded atoms in diagram I are all the same type (so showing a molecular *element*) where diagram **III** is showing bonded atoms of different types (so showing a molecular *compound*); **III** is the compound.

Finally, a mixture is two or more pure substances physically combined. Sample II shows two different atom types mixed together and not bonded together, so this is the mixture of two elements.

These combine to the correct choice of **(B)**.

Practice Questions Related to This: **PQ-14** and **PQ-15**

Practice Questions (PQ)

PQ-1. The speed of light is 3.0×10^8 m·s^{-1}. What is this speed in nm·ms^{-1}?

(A) 3.0×10^{-4} nm·ms^{-1} **(B)** 3.0×10^{2} nm·ms^{-1} **(C)** 3.0×10^{14} nm·ms^{-1} **(D)** 3.0×10^{20} nm·ms^{-1}

PQ-2. The density of helium is 0.164 kg·m^{-3}. What is this density in lb·ft^{-3}? 1 kg = 2.20 lb and 1 m = 3.28 ft

(A) 0.0102 lb·ft^{-3} **(B)** 0.110 lb·ft^{-3} **(C)** 1.18 lb·ft^{-3} **(D)** 12.7 lb·ft^{-3}

PQ-3. Based on the figure, what volume should be reported for the liquid?

21 mL

20 mL

(A) 20.8 mL

(B) 20.68 mL

(C) 20.6 mL

(D) 20.57 mL

PQ-4. What is the correctly reported mass of water based on the data in the table?

Mass of beaker and water	29.62 g
Mass of beaker only	28.3220 g

(A) 1.3 g **(B)** 1.30 g **(C)** 1.298 g **(D)** 1.2980 g

PQ-5. What is the name of $Ti_3(PO_4)_4$?

(A) titanium phosphate **(B)** titanium(III) phosphate

(C) titanium(IV) phosphate **(D)** titanium tetraphosphate

PQ-6. The formula of strontium hexafluorosilicate is $SrSiF_6$. What is the formula of aluminum hexafluorosilicate?

(A) $AlSiF_6$ **(B)** $Al_2(SiF_6)_3$ **(C)** Al_3SiF_6 **(D)** $Al_3(SiF_6)_2$

PQ-7. Which is **NOT** named correctly?

(A) MnO_2 manganese(II) oxide **(B)** $CuSO_4$ copper(II) sulfate

(C) Na_3PO_4 sodium phosphate **(D)** CaF_2 calcium fluoride

PQ-8. What is the correct name for N_2O_5?

(A) nitrogen(II) oxide **(B)** nitrogen(V) oxide

(C) dinitrogen oxide **(D)** dinitrogen pentoxide

PQ-9. What is the formula of sulfur trioxide?

(A) SO_2 **(B)** SO_3 **(C)** SO_3^{2-} **(D)** SO_4^{2-}

PQ-10. What is the name of $HBrO_4$(aq)?

(A) bromic acid **(B)** bromous acid **(C)** hydrobromic acid **(D)** perbromic acid

PQ-11. What is the formula of hydroiodic acid?

(A) HI(aq) **(B)** HIO(aq) **(C)** HIO_2(aq) **(D)** HIO_3(aq)

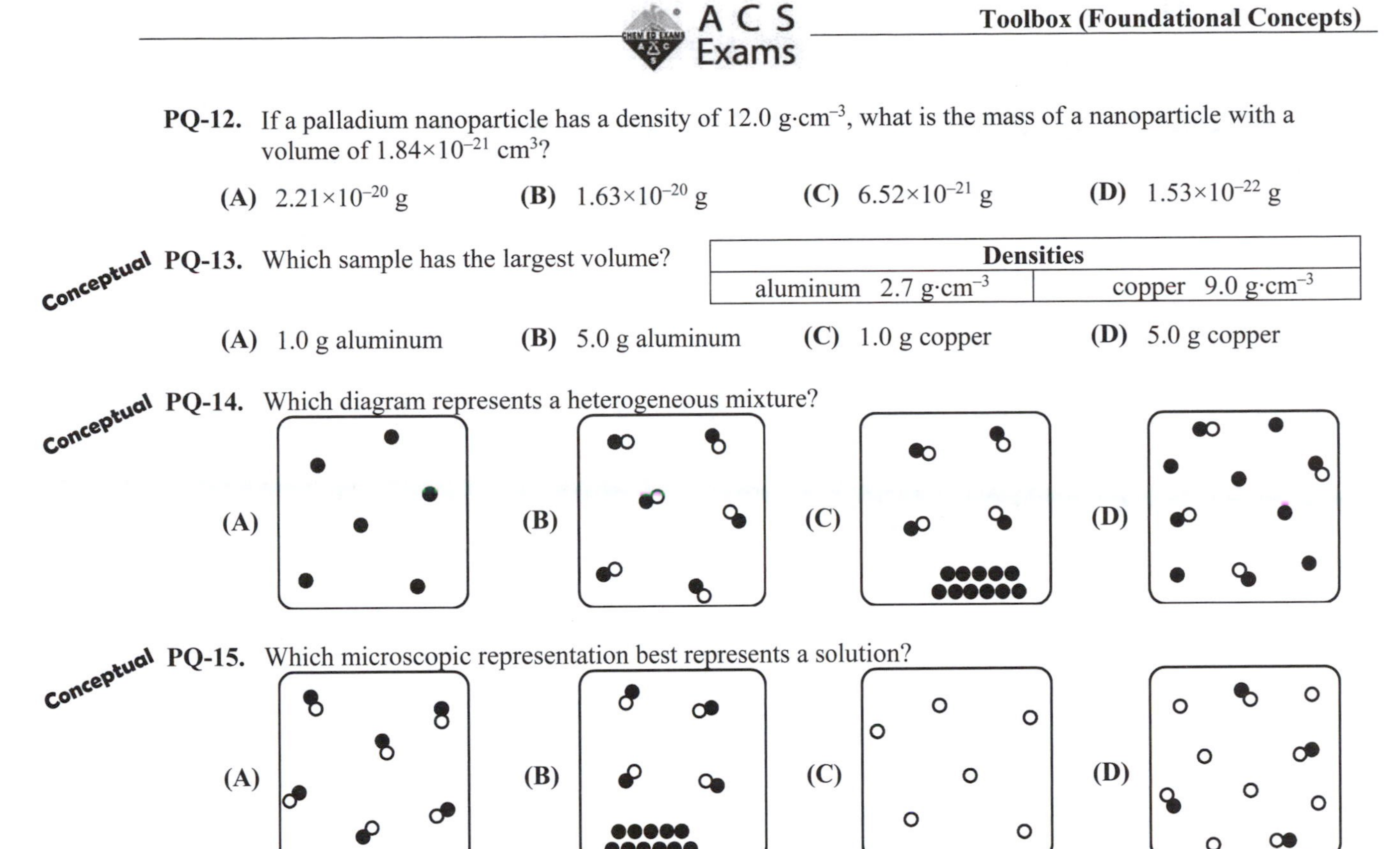

PQ-12. If a palladium nanoparticle has a density of 12.0 g·cm^{-3}, what is the mass of a nanoparticle with a volume of 1.84×10^{-21} cm^3?

(A) 2.21×10^{-20} g **(B)** 1.63×10^{-20} g **(C)** 6.52×10^{-21} g **(D)** 1.53×10^{-22} g

Conceptual **PQ-13.** Which sample has the largest volume?

Densities	
aluminum 2.7 g·cm^{-3}	copper 9.0 g·cm^{-3}

(A) 1.0 g aluminum **(B)** 5.0 g aluminum **(C)** 1.0 g copper **(D)** 5.0 g copper

Conceptual **PQ-14.** Which diagram represents a heterogeneous mixture?

(A) (B) (C) (D)

Conceptual **PQ-15.** Which microscopic representation best represents a solution?

(A) (B) (C) (D)

Answers to Study Questions

1. D
2. B
3. B
4. C
5. C
6. C
7. B

Answers to Practice Questions

1. C
2. A
3. D
4. B
5. C
6. B
7. A
8. D
9. B
10. D
11. A
12. A
13. B
14. C
15. D

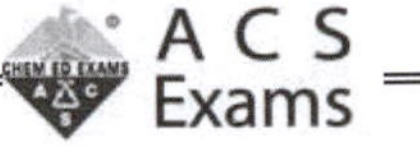

How to use this book

You have an ACS exam coming up and might be feeling a bit overwhelmed by it. You might be worried that the exam will be very different from the ones you are used to and will be on material you haven't covered. The first thing to do is to R-E-L-A-X. The ACS exam you are going to take was written by chemistry faculty from around the country who all teach general chemistry courses. The truth is that most general chemistry courses, like most general chemistry textbooks, are very similar. There is a set of common topics that almost every course covers. What makes courses different are the level of detail of coverage of the topics. Some instructors have favorite topics that they cover in depth and some that they only briefly cover due to time. The good news is that the exam you are going to take was written to be applicable to this wide variety of topic coverage and depth of topic coverage. The committees that write these exams do not look for the most minute details; instead the topics covered are the foundational ones that are common to most courses, and you should recognize them. Due to the variations in courses, described above, you might see a few questions about topics that are not as familiar to you. That is fine, relax!

As you use this study guide, and study chemistry in general, you should focus on the underlying concepts present in a question. Remember that there are many ways to ask a question about stoichiometry (conservation of matter). If you focus on identifying the underlying concept/topic being asked about in the question, you will be able to handle any "surface" differences in the way the question is asked. After you complete a problem, and before you go onto the next question, take a minute to think about other ways the problem could have been asked. For example, what if you had been given the information in the answer and had been asked for some information that was given to you in the original question? When you stop and think like this, you have moved beyond working problems to actual studying.

The chapters in this study guide are arranged in the same order as many common chemistry textbooks. Topics typically covered in the first semester are in chapters (1–8) and typical second semester topics are in chapters (9–16). Each chapter begins with a brief summary and a list of topics covered. This list of topics should correspond to the topics covered in your course. Because chemistry relies on many different types of representations, the types of representations/notations used in the chapter are previewed. Following this you will be told where the material might be found in your chemistry textbook. There is not much variation in chapter titles between books, so you should be able to find the relevant material no matter what book you have. This is an important point; this study guide is not intended to be a replacement for your textbook. It is a supplement specifically designed to assist you to prepare for the ACS exam. The introductory material ends with a mention of online practice exam problems and how the material in the chapter fits into the bigger picture of chemistry concepts.

Following this introductory material, you will find worked Study Questions. The best way to use these would be to cover up the detailed solution and try to answer the question/work the problem. Only after you have tried the question should you move to the discussion of the solution. The worked-out solution is written to model the thinking that could be used to answer the question correctly. Sometimes, this involves solving the problem without looking at the multiple-choice options (i.e., numerical problems), other times this involves an analysis of each choice (i.e., selecting a correct statement). At the end of each study question, you will find an explanation of why each incorrect choice is wrong and which practice problems at the end of the chapter correspond to the study question. There are as many as 30 practice problems at the end of each chapter.

Some study questions and practice problems are marked as **Conceptual** using a special notation next to the problem number. We specifically included this because part, half or possibly all of your ACS Exam will have conceptual problems. These types of questions can also test your deeper understanding of a concept and can be useful gauges of how well you understand the material. Often, these questions do not require detailed calculations, instead these can involve interpreting particulate nature of matter representations and/or graphical information, among other things.

Remember that this study guide was written to give you practice with ACS exam-type questions as you review the chemistry content. As you review, you are likely to find topics that you need to go back to your textbook and notes to review and you will find some topics that you have a very good grasp of. Don't let the thought of a BIG ACS exam scare you; if you have been working throughout the course and put in some quality study time, you will be fine!

Sample Instructions

You will find that the front cover of an ACS Exam will have a set of instructions very similar to this. This initial set of instructions is meant for both the faculty member who administers the exam and the student taking the exam. You will be well advised to read the entire set of instructions while waiting for the exam to begin. This sample is from a general chemistry exam released in 2018.

TO THE EXAMINER:
This test is designed to be taken with a special answer sheet on which the student records his or her responses. All answers are to be marked on this answer sheet, not in the test booklet. Each student should be provided with a test booklet, one answer sheet, and scratch paper; all of which must be turned in at the end of the examination period. The test is to be available to the students only during the examination period. You must collect and account for all exam booklets at the end of the examination. For complete instructions refer to the Directions for Administering Examinations. Only nonprogrammable calculators are permitted. ***All electronic devices with photo-taking capability are prohibited.*** Norms are based on:

Score = Number of right answers
70 items - 110 minutes

TO THE STUDENT:
DO NOT WRITE ANYTHING IN THIS BOOKLET! Do not turn this page until your instructor gives the signal to begin. A periodic table and other useful information is on page 2. When you are told to begin work, open the booklet and read the directions on page 3.

Note the **restriction** on the type of **calculators** that you may use and the **time** for administering the exam. This restriction applies to allow your results to be compared to national norms, ensuring that all students have had the same tools and time to display their knowledge. Your instructor may choose not to follow the time restriction, particularly if they do not plan to submit your data as part of the national process for calculating norms.

Be sure to notice that scoring is based **only** on the number of right answers. There is no penalty for making a reasonable guess even if you are not completely sure of the correct answer. Often you will be able to narrow the choice to two possibilities, improving your odds at success. You will need to keep moving throughout the examination period; it is to your advantage to attempt every question. Do not assume that the questions become harder as you progress through an ACS Exam. Questions are generally grouped by topic rather than difficulty.

Note for the exam in the example, the data sheet and periodic table are in the exam. For other exams it could be on a separate datasheet or occasionally be on the last page of the exam.

Next, here is a sample of the directions you will find at the beginning of an ACS exam.

DIRECTIONS

- When you have selected your answer, blacken the corresponding space on the answer sheet with a soft, black #2 pencil. Make a heavy, full mark, but no stray marks. If you decide to change an answer, erase the unwanted mark very carefully.
- Make no marks in the test booklet. Do all calculations on scratch paper provided by your instructor.
- There is only one correct answer to each question. Any questions for which more than one response has been blackened **will not be counted**.
- Your score is based solely on the number of questions you answer correctly. **It is to your advantage to answer every question.**

Pay close attention to the mechanical aspects of these directions. Marking your answers without erasures helps to create a very clean answer sheet that can be read without error. As you look at your answer sheet before the end of the exam period, be sure that you check that every question has been attempted, and that only one choice has been made per question. As was the case with the cover instructions, note that your attention is again directed to the fact that the score is based on the total number of questions that you answer correctly. You also can expect a reasonable distribution of **A**, **B**, **C**, and **D** responses, something that is not necessarily true for the distribution of questions in *The Official Guide*.

ABBREVIATIONS AND SYMBOLS

amount of substance	n	gas constant	R	molar mass	M
atmosphere	atm	gram	g	mole	mol
atomic mass unit	u	hour	h	Planck's constant	h
atomic molar mass	A	joule	J	pressure	P
Avogadro constant	N_A	kelvin	K	second	s
Celsius temperature	°C	kilo– prefix	k	speed of light	c
centi– prefix	c	liter	L	temperature, K	T
energy of activation	E_a	measure of pressure	mmHg	time	t
enthalpy	H	milli– prefix	m	volume	V
frequency	ν	molar	M		

CONSTANTS & CONVERSIONS

$R = 8.314\ \text{J·mol}^{-1}\text{·K}^{-1}$

$R = 0.0821\ \text{L·atm·mol}^{-1}\text{·K}^{-1}$

$N_A = 6.022 \times 10^{23}\ \text{mol}^{-1}$

$h = 6.626 \times 10^{-34}\ \text{J·s}$

$c = 2.998 \times 10^{8}\ \text{m·s}^{-1}$

$0\ °\text{C} = 273.15\ \text{K}$

1 atm = 760 mmHg

1 atm = 760 torr

EQUATIONS

Equations may be included in this position of your exam.

These vary depending on the type of exam you are taking.

For example, if you are expected to do a calculation involving an integrated rate law, you may be provided with a series of equations in this position.

PERIODIC TABLE OF THE ELEMENTS

1	2	3	4	5	6	7	8	9	10	11	12	13	14	15	16	17	18
1 **H** 1.008																	2 **He** 4.003
3 **Li** 6.941	4 **Be** 9.012											5 **B** 10.81	6 **C** 12.01	7 **N** 14.01	8 **O** 16.00	9 **F** 19.00	10 **Ne** 20.18
11 **Na** 22.99	12 **Mg** 24.31											13 **Al** 26.98	14 **Si** 28.09	15 **P** 30.97	16 **S** 32.07	17 **Cl** 35.45	18 **Ar** 39.95
19 **K** 39.10	20 **Ca** 40.08	21 **Sc** 44.96	22 **Ti** 47.88	23 **V** 50.94	24 **Cr** 52.00	25 **Mn** 54.94	26 **Fe** 55.85	27 **Co** 58.93	28 **Ni** 58.69	29 **Cu** 63.55	30 **Zn** 65.39	31 **Ga** 69.72	32 **Ge** 72.61	33 **As** 74.92	34 **Se** 78.96	35 **Br** 79.90	36 **Kr** 83.80
37 **Rb** 85.47	38 **Sr** 87.62	39 **Y** 88.91	40 **Zr** 91.22	41 **Nb** 92.91	42 **Mo** 95.94	43 **Tc**	44 **Ru** 101.1	45 **Rh** 102.9	46 **Pd** 106.4	47 **Ag** 107.9	48 **Cd** 112.4	49 **In** 114.8	50 **Sn** 118.7	51 **Sb** 121.8	52 **Te** 127.6	53 **I** 126.9	54 **Xe** 131.3
55 **Cs** 132.9	56 **Ba** 137.3	57 **La** 138.9	72 **Hf** 178.5	73 **Ta** 180.9	74 **W** 183.8	75 **Re** 186.2	76 **Os** 190.2	77 **Ir** 192.2	78 **Pt** 195.1	79 **Au** 197.0	80 **Hg** 200.6	81 **Tl** 204.4	82 **Pb** 207.2	83 **Bi** 209.0	84 **Po**	85 **At**	86 **Rn**
87 **Fr**	88 **Ra**	89 **Ac**	104 **Rf**	105 **Db**	106 **Sg**	107 **Bh**	108 **Hs**	109 **Mt**	110 **Ds**	111 **Rg**	112 **Cn**	113 **Nh**	114 **Fl**	115 **Mc**	116 **Lv**	117 **Ts**	118 **Og**

58 **Ce** 140.1	59 **Pr** 140.9	60 **Nd** 144.2	61 **Pm**	62 **Sm** 150.4	63 **Eu** 152.0	64 **Gd** 157.3	65 **Tb** 158.9	66 **Dy** 162.5	67 **Ho** 164.9	68 **Er** 167.3	69 **Tm** 168.9	70 **Yb** 173.0	71 **Lu** 175.0
90 **Th** 232.0	91 **Pa** 231.0	92 **U** 238.0	93 **Np**	94 **Pu**	95 **Am**	96 **Cm**	97 **Bk**	98 **Cf**	99 **Es**	100 **Fm**	101 **Md**	102 **No**	103 **Lr**

Please note that the periodic table changes to keep current with recent discoveries. The periodic table you use may vary from the table shown here.

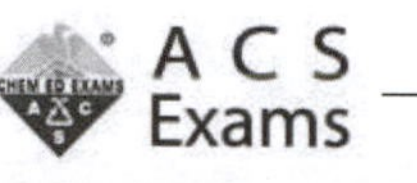

Big Ideas

Big Ideas – ACS Exams Anchoring Concepts Content Map (ACCM)

Studies of past ACS Exams have shown that while individual tests vary somewhat in their content, overall the coverage of big ideas in chemistry is fairly broad when looked at with the lens of the ACCM. What is listed below are anchoring concepts or "big ideas" (in **bold**). These are followed by statements which provide more detail on how these big ideas are explained or included in the undergraduate curriculum. These are included in this guide as these big ideas apply to all levels of chemistry in the undergraduate program, and it can be useful to see how the specific chapter or content area you are studying fits into the bigger picture of your undergraduate program. Towards that end, at the beginning of each chapter in this guide, you are provided with the specific big idea covered in that chapter.

I. **Atoms**
   A. Atoms have unique chemical identities based on the number of protons in the nucleus.
   B. Electrons play the key role for atoms to bond with other atoms.
   C. Atoms display a periodicity in their structures and observable phenomena that depend on that structure.
   D. Most information about atoms is inferred from studies on collections of atoms often involving an interaction with electromagnetic radiation.
   E. Macroscopic samples of matter contain so many atoms that they are counted in moles.
   F. Atoms maintain their identity, except in nuclear reactions.
   G. Ions arise when the number of electrons and protons are not equal, and can be formed from atoms.

II. **Bonding**
   A. Because protons and electrons are charged, physical models of bonding are based on electrostatic forces.
   B. Because chemical bonds arise from sharing of negatively charged valence electrons between positively charged nuclei, the overall electrostatic interaction is attractive.
   C. When chemical bonds form, the overall energy of the bonding atoms is lowered relative to free atoms, and therefore energy is released.
   D. To break a chemical bond requires an input of energy.
   E. A theoretical construct that describes chemical bonding utilizes the construction of molecular orbitals for the bond based on overlap of atomic orbitals on the constituent atoms.
   F. Covalent bonds can be categorized based on the number of electrons (pairs) shared. The most common categories are single, double, and triple bonds.
   G. Metallic bonding arises in many solids and fundamentally involves the sharing of valence electrons among many positively charged "cores" over extended distances.

III. **Structure and Function**
   A. Atoms combine to form new compounds that have new properties based on structural and electronic features.
   B. Models exist that allow the prediction of the shape of chemicals about any bonding atom in a molecule.
   C. Theoretical models are capable of providing detail structure for whole molecules based on energy minimization methods.
   D. Symmetry, based on geometry, plays an important role in how atoms interact within molecules and how molecules are observed in many experiments.
   E. Three-dimensional structures may give rise to chirality, which can play an important role in observed chemical and physical properties.
   F. Reactions of molecules can often be understood in terms of subsets of atoms, called functional groups.
   G. Periodic trends among elements can be used to organize the understanding of structure and function for related chemical compounds.
   H. Many solid state, extended systems exist, and geometric structures play an important role in understanding the properties of these systems.

IV. Intermolecular Interactions

A. Intermolecular forces are generally weaker, on an individual basis, than chemical bonds, but the presence of many such interactions may lead to overall strong interactions.
B. For large molecules, intermolecular forces may occur between different regions of the molecule. In these cases, they are sometimes termed noncovalent forces.
C. Intermolecular forces can be categorized based on the permanence and structural details of the dipoles involved.
D. For condensed phases that are not structures of extended chemical bonds, the physical properties of the state are strongly influenced by the nature of the intermolecular forces.
E. The energy consequences of chemical reactions that take place in condensed phases (solution) usually must include intermolecular forces to be correctly/completely explained.

V. Chemical Reactions

A. In chemical changes, matter is conserved and this is the basis behind the ability to represent chemical change via a balanced chemical equation.
B. Chemical change involves the breaking or forming of chemical bonds, or typically both.
C. Chemical change can be observed at both the particulate and macroscopic levels, and models exist that allow the translation between these two levels of observation.
D. There are a large number of possible chemical reactions, and categories have been devised to organize understanding of these reaction types.
E. Many chemical properties of elements follow periodic trends that can be used to strategically design reactions to achieve desired outcomes.
F. Chemical change can be controlled by choices of reactants, reaction conditions, or use of catalysts.
G. Controlling chemical reactions is a key requirement in the synthesis of new materials.

VI. Energy and Thermodynamics

A. Most chemical changes are accompanied by a net change of energy of the system.
B. Many chemical reactions require an energy input to be initiated.
C. The type of energy associated with chemical change may be heat, light, or electrical energy.
D. Breaking chemical bonds requires energy; formation of chemical bonds releases energy.
E. The forces that are associated with energy change in chemical processes are electrostatic forces.
F. In accord with thermodynamics, energy is conserved in chemical changes, but the change of form in which the energy is present may be harnessed via natural or human-made devices.
G. Thermodynamics provides a detailed capacity to understand energy change at the macroscopic level.
H. The tendency of nature to disperse, particularly in terms of energy distribution, is embodied in the state function called entropy.
I. Energy changes associated with nuclear chemistry are many orders of magnitude larger than those of classical chemical changes.

VII. Kinetics

A. Chemical change can be measured as a function of time and occurs over a wide range of time scales.
B. Empirically derived rate laws summarize the dependence of reaction rates on concentrations of reactants and temperature.
C. Most chemical reactions take place by a series of more elementary reactions, called the reaction mechanism.
D. An elementary reaction requires that the reactants collide (interact) and have both enough energy and appropriate orientation of colliding particles for the reaction to occur.
E. Catalysis increases the rate of reaction and has important applications in a number of subdisciplines of chemistry.
F. Reaction products can be influenced by controlling whether reaction rate or reaction energy plays the key role in the mechanism.

VIII. Equilibrium

A. Both physical and chemical changes may occur in either direction (e.g., from reactants to products or products to reactants).
B. When opposing processes both occur at the same rate the net change is zero.

C. For chemical reactions, the equilibrium state can be characterized via the equilibrium constant.
D. When the equilibrium constant is very large or small, products or reactants, respectively, are primarily present at equilibrium. Systems with K near 1 have significant amounts of both reactants and products present.
E. If perturbed, a system at equilibrium will respond in the direction that tends to offset the perturbation.
F. Thermodynamics provides mathematical tools to understand equilibrium systems quantitatively.
G. Equilibrium concepts have important applications in several subdisciplines of chemistry.

IX. Experiments, Measurement, and Data

A. Quantitative observation of matter can be made at a wide range of distance scales and/or time scales.
B. Because there are a large number of compounds, a system of naming these compounds is used.
C. Experimental control of reactions plays a key role in the synthesis of new materials and analysis of composition.
D. Chemical measurements are based on mass, charge, or interaction with electrons or photons.
E. Observations are verifiable, so experimental conditions, including considerations of the representativeness of samples, must be considered for experiments.
F. Fidelity of inferences made from data requires appropriate experimental design.
G. Chemistry experiments have risks associated with them, so chemical safety is a key consideration in the design of any experiment.

X. Visualization

A. Many theoretical constructs are constructed at the particulate level, while many empirical observations are made at the macroscopic level.
B. The mole represents the key factor for translating between the macroscopic and particulate levels.
C. Macroscopic properties result from large numbers of particles, so statistical methods provide a useful model for understanding the connections between these levels.
D. Quantitative reasoning within chemistry is often visualized and interpreted graphically.

Chapter 1 – Atoms, Molecules, and Ions

Chapter Summary:

This chapter will focus on the structure of the atom, including subatomic particles and how changing these particles for an atom will result in a new species. Also included in this chapter are questions related the structure of the periodic table.

Specific topics covered in this chapter are:

- Subatomic particles
- Ions
- Isotopes
- Average atomic mass
- Metals, nonmetals, metalloids (or semi-metals)
- Groups and periods

Previous material that is relevant to your understanding of questions in this chapter include:

- Significant figures (***Toolbox***)
- Scientific notation (***Toolbox***)

Common representations used in questions related to this material:

Name	Example	Used in questions related to
Nuclear or atomic symbols	$^{2}_{1}H$	Atomic structure, isotopes
Particulate representations		Atomic structure, isotopes

Where to find this in your textbook:

The material in this chapter typically aligns to "Atoms, Molecules, and Ions" or "Atoms and Molecules" in your textbook. The name of your chapter may vary.

Practice exam:

There are practice exam questions aligned to the material in this chapter. Because there are a limited number of questions on the practice exam, a review of the breadth of the material in this chapter is advised in preparation for your exam.

How this fits into the big picture:

The material in this chapter aligns to the Big Idea of Atoms (1) as listed on page 12 of this study guide.

Study Questions (SQ)

SQ-1. *Conceptual*

Using the models of particles below, which are isotopes of the same element?

Key: • = electron ● = neutron ○ = proton

particle 1 | particle 2 | particle 3 | particle 4

(A) particle 1 and particle 2 **(B)** particle 1 and particle 3

(C) particle 2 and particle 4 **(D)** particle 3 and particle 4

Knowledge Required: (1) Definition of an isotope and ion. (2) Interpreting particulate representations of subatomic particles.

Thinking it Through: As you read the question, one of the key things you note is to find "isotopes" in the figures. This means that you are looking for the same number of protons and differing numbers of neutrons. You then look at the key and see that the open circle represents a proton. This means that particles 1, 2 and 3 are all the same element and could be isotopes. You then see that the gray circles are neutrons and particles 1 and 2 have 3 neutrons but are different from particle 3 which has 4 neutrons. That means that particles 1 and 2 have the same mass (6 amu), which means they cannot be isotopes. Therefore, the correct answer combination would be either 1 and 3 or 2 and 3. Because only 1 and 3 is an option, this is the correct answer **(B)**.

Choice **(A)** of 1 and 2 is not correct because these have the same mass.

Choice **(C)** of 2 and 4 or choice **(D)** 3 and 4 is not correct because these are different elements (different number of protons).

It is important to notice that you did not consider the number of electrons in the question because as mentioned above, isotopes have the same number of protons but differing numbers of neutrons. Particles 1 and 2 differ by the number of electrons. Particle 1 is an atom (where the number of protons = number of electrons); particles 2 and 4 are ions (where the number of protons > number of electrons).

Practice Questions Related to This: **PQ-1, PQ-2, PQ-3,** and **PQ-4**

SQ-2. An atom of strontium-90, $^{90}_{38}\text{Sr}$, contains

(A) 38 electrons, 38 protons, 52 neutrons. **(B)** 38 electrons, 38 protons, 90 neutrons.

(C) 52 electrons, 52 protons, 38 neutrons. **(D)** 52 electrons, 38 protons, 38 neutrons.

Knowledge Required: (1) Interpretation of nuclear symbols. (2) Definition of mass number.

Thinking it Through: You see that you are given the name "strontium-90", which tells you that the mass of the atom is 90 amu. You also see that you are given the nuclear symbol of $^{90}_{38}\text{Sr}$. The nuclear symbol gives both the number of protons or atomic number (Z) and the number of protons + number of neutrons or mass number (A):

$$^{A}_{Z}\text{X} \qquad ^{\text{mass number}}_{\text{atomic number}}\text{X} \qquad ^{\text{\# of protons + neutrons}}_{\text{\# of protons}}\text{X}$$

From the symbol, you see that the atomic number is 38 which means that it has 38 protons. The mass number is 90 which means 90 – 38 = 52 neutrons. Finally, because it is an atom, it has no charge or is neutral. Therefore, the number of electrons = the number of protons or also 38. The correct choice is **(A)**.

Choice **(B)** is not correct because the mass number (90) does not equal the number of neutrons.

Choice **(C)** is not correct because the atomic number (Z) is not the number of neutrons.

Choice **(D)** is not correct because it is an atom with the number of protons equals the number of electrons and the number of neutrons + number of protons must equal the mass number.

Practice Questions Related to This: **PQ-5, PQ-6,** and **PQ-7**

SQ-3. What is the atomic number for an ion that has a charge of +1 and contains 18 electrons?

(A) 17 **(B)** 18 **(C)** 19 **(D)** 20

Knowledge Required: (1) Relationship between the charge of an ion and the number of electrons and protons. (2) Definition of atomic number.

Thinking it Through: You recall that the charge of an ion is related to the relative numbers of electrons and protons. Because the ion in the question has a positive one charge (+1), you know that there is one less electron than proton in the ion. Therefore, the number of protons is 18 + 1 = 19. Because there are 19 protons in the nucleus, the atomic number is 19. The correct choice is **(C)**.

Choice **(A)** is not correct because the atomic number (the number of protons) must be greater than the number of electrons in a positive ion. Choice **(B)** is not correct because if the number of protons equals the number of electrons, there is no net charge. Choice **(D)** is not correct because the species with an atomic number of 20 and 18 electrons would have a charge of +2.

Practice Questions Related to This: **PQ-8** and **PQ-9**

SQ-4. How many protons (p^+), neutrons (n^o), and electrons (e^-) are present in $^{47}_{22}Ti^{3+}$?

(A) 22 p^+, 25 n^o, 19 e^- **(B)** 22 p^+, 25 n^o, 25 e^-

(C) 25 p^+, 22 n^o, 19 e^- **(D)** 25 p^+, 22 n^o, 25 e^-

Knowledge Required: (1) Interpretation of nuclear symbols. (2) Definition of mass number.

Thinking it Through: You see that you are given the nuclear symbol of $^{47}_{22}Ti^{3+}$. The nuclear symbol gives both the number of protons or atomic number (Z) and the number of protons + number of neutrons or mass number (A):

$$^{A}_{Z}X \qquad ^{\text{mass number}}_{\text{atomic number}}X \qquad ^{\#\text{ of protons + neutrons}}_{\#\text{ of protons}}X$$

From the symbol, you see that the atomic number is 22, which means that it has 22 protons. The mass number is 47, which means 47 – 22 = 25 neutrons. You see that the charge of the ion is a positive three (3+). This means that the atom has three more protons than electrons. This happens when electrons are removed from the atom, forming the ion. Therefore, the number of electrons is 22 – 3 = 19. The correct choice is **(A)**.

Choice **(B)** is not correct because the number of electrons (25) represents adding three electrons, not removing them to form the ion.

Choice **(C)** is not correct because the atomic number (Z) is not the number of neutrons.

Choice **(D)** is not correct because the atomic number (Z) is not the number of neutrons and the number of electrons represents adding three electrons.

Practice Questions Related to This: **PQ-10, PQ-11, PQ-12,** and **PQ-13**

SQ-5. Chlorine occurs naturally as a mixture of two isotopes with atomic masses of 34.97 amu and 36.97 amu. What are the relative abundances of these isotopes?

(A) 24% and 76% **(B)** 50% and 50% **(C)** 76% and 24% **(D)** 95% and 5%

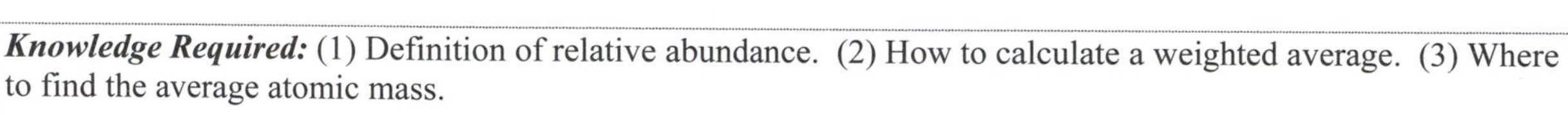

Knowledge Required: (1) Definition of relative abundance. (2) How to calculate a weighted average. (3) Where to find the average atomic mass.

Thinking it Through: You are told in the question that there are two isotopes of chlorine, ^{35}Cl and ^{37}Cl. You also know that the mass of chlorine on the periodic table is a weighted average (or the average atomic mass). It is given on the periodic table as 35.45 amu. Finally, you are told there are **only** two naturally occurring isotopes of chlorine, which means that the average atomic mass is comprised of a weighted average of ^{35}Cl and ^{37}Cl:

Generically: $$\left[(\text{mass isotope}_1)\frac{\%\ \text{isotope}_1}{100\%}\right]+\left[(\text{mass isotope}_2)\frac{\%\ \text{isotope}_2}{100\%}\right]=\text{average atomic mass}$$

For chlorine: $$\left[(\text{mass }^{35}\text{Cl})\frac{\%\ ^{35}\text{Cl}}{100\%}\right]+\left[(\text{mass }^{37}\text{Cl})\frac{\%\ ^{37}\text{Cl}}{100\%}\right]=\text{average atomic mass Cl}$$

$$\%^{35}\text{Cl}+\%^{37}\text{Cl}=100\%$$ (remember, because there are only 2 isotopes, the two abundances equal 100%)

Solving for $\%^{35}Cl$ in the last equation: $100\% - \%^{37}Cl = \%^{35}Cl$

And substituting: $$\left[(\text{mass }^{35}\text{Cl})\frac{100\%-\%\ ^{37}\text{Cl}}{100\%}\right]+\left[(\text{mass }^{37}\text{Cl})\frac{\%^{37}\text{Cl}}{100\%}\right]=\text{average atomic mass Cl}$$

Now you solve for $\%^{37}Cl$:

$$\%^{37}\text{Cl}=\left(\frac{\text{average atomic mass Cl}-\text{mass }^{35}\text{Cl}}{\text{mass }^{37}\text{Cl}-\text{mass }^{35}\text{Cl}}\right)\times 100\%=\left(\frac{35.45\text{ amu}-34.97\text{ amu}}{36.97\text{ amu}-34.97\text{ amu}}\right)\times 100\%=24\%$$

Finally, solve for $\%^{35}Cl$: $100\% - \%^{37}Cl = 100\% - 24\% = 76\%$ which is choice **(C).**

Choice **(A)** is not correct because the relative abundances are switched.

Choice **(B)** is not correct because the average atomic mass from the periodic table is ignored.

Choice **(D)** is not correct because the calculation uses the ratio of the masses rather than the weighted average.

Practice Questions Related to This: **PQ-14**

SQ-6. *Conceptual* The element thallium, Tl, has two stable isotopes, ^{203}Tl and ^{205}Tl. Which statement correctly describes the relationship between the relative abundances of these two isotopes?

(A) % abundance of ^{203}Tl > % abundance of ^{205}Tl

(B) % abundance of ^{203}Tl = % abundance of ^{205}Tl

(C) % abundance of ^{203}Tl < % abundance of ^{205}Tl

(D) Not enough information is provided to determine the relationship.

Knowledge Required: (1) Definition of relative abundance. (2) How to calculate a weighted average. (3) Where to find the average atomic mass.

Thinking it Through: You are told in the question that there are two isotopes of thallium, ^{203}Tl and ^{205}Tl. You also know that the mass of thallium on the periodic table is a weighted average (or the average atomic mass). It is given on the periodic table as 204.4 amu. Finally, you are told there are **only** two naturally occurring isotopes of thallium, which means that the average atomic mass is a weighted average of ^{203}Tl and ^{205}Tl. Because you are only asked for the relationship between the % abundances you do not need to do calculations. You compare the average atomic mass, 204.4 amu, to the masses of the two isotopes. The average atomic mass is closer to the 205 amu than it is to 203 amu. This means that the % abundance of ^{203}Tl isotope is less than the % abundance of ^{205}Tl isotope. The correct choice is **(C)**.

Choice **(A)** is not correct because the % abundance of ^{203}Tl must be less than the % abundance of ^{205}Tl for the average atomic mass to be 204.4 amu. Choice **(B)** is not correct because if the % abundances were equal the average atomic mass would be equal to the simple average of 203 amu and 205 amu ; 204.0 amu. Choice **(D)** is not correct because there are only two isotopes, and you can determine the relationship between the relative abundance using the average atomic mass.

Practice Questions Related to This: **PQ-15, PQ-16,** and **PQ-17**

SQ-7. Which group number contains the most diatomic molecular elements?

(A) Group 1 **(B)** Group 16 **(C)** Group 17 **(D)** Group 18

Knowledge Required: (1) Identifying the group number from the periodic table. (2) Remembering the diatomic molecular elements.

Thinking it Through: You recall that the seven diatomic molecular elements are: H_2, N_2, O_2, F_2, Cl_2, Br_2, and I_2. When you refer to the periodic table at the beginning of the study guide (p. 11) you see that four of these elements occur in Group 17. The correct choice is **(C).**

Choice **(A)** is not correct because Group 1 contains only one diatomic molecular element, H_2.

Choice **(B)** is not correct because Group 16 contains only one diatomic molecular element, O_2.

Choice **(D)** is not correct because Group 18 contains the Noble gases and these are not diatomic.

Practice Questions Related to This: **PQ-18, PQ-19, PQ-20,** and **PQ-21**

SQ-8. What is the classification of sulfur?

(A) metal **(B)** metalloid **(C)** semiconductor **(D)** nonmetal

Knowledge Required: (1) Identifying the type of element type using the periodic table.

Thinking it Through: You remember that the element symbol for sulfur is S. This element is located in the third period and Group 16. Sulfur lies above and to the right of the line that separates metals and nonmetals. Elements in this region of the periodic table are classified as nonmetals. The correct choice is **(D)**.

Choice **(A)** is not correct because the metals are located to the left of the dividing line.

Choice **(B)** is not correct because the metalloids are the elements that lie immediately above and below the dividing line. (Si, Ge, As, and Sb).

Choice **(C)** is not correct because semiconductor is not the name of a type of element. The metalloids are used in the production of semiconductors.

Practice Questions Related to This: **PQ-22, PQ-23, PQ-24,** and **PQ-25**

SQ-9. If the formula of an ionic oxide of element X is X_2O_3, what is the formula of the chloride of X?

(A) XCl_3 **(B)** XCl **(C)** X_3Cl **(D)** XCl_6

Knowledge Required: (1) Using the periodic table to predict a chemical formula. (2) Combining ions to form a neutral species.

Thinking it Through: You are told that element X forms a compound with the formula X_2O_3 and asked to predict the formula of the chloride compound containing X. You know from the periodic table that the charge oxygen commonly has in ionic compounds is –2. You also know that when ions combine to form compounds the total positive and total negative charges must cancel. This means you must determine the charge of X in the compound X_2O_3. You can do this because you know the charge of the oxide ion, –2. If there are 3 oxide ions, the total negative charge is –6. This means the total positive charge in the compound is +6. Because there are two X cations in the formula, each X cation must have a charge of +3. You also know that the chloride ion has a –1 charge. To make a neutral species when you combine X^{3+} and Cl^-, you need three Cl^- for each X^{3+}. The correct formula would be XCl_3.This is choice **(A)**.

Choice **(B)** is incorrect because the formula XCl would require X to be a +1 ion.

Choice **(C)** is incorrect because the subscript for the Cl ion goes after the symbol.

Choice **(D)** is incorrect because the formula XCl_6 would require X to have a +6 charge.

Practice Questions Related to This: **PQ-26, PQ-27, PQ-28, PQ-29,** and **PQ-30**

Practice Questions (PQ)

PQ-1. Which term best describes the relation of hydrogen (^{1}H) to deuterium (^{2}H)?

(A) allotropes **(B)** isomers **(C)** isotopes **(D)** polymers

PQ-2. Which pair represents isotopes?

(A) $^{54}_{24}Cr$ and $^{54}_{26}Fe$
(B) $^{235}_{92}U$ and $^{238}_{92}U$
(C) $^{116}_{48}Cd$ and $^{116}_{50}Sn$
(D) $^{239}_{93}Np$ and $^{239}_{94}Pu$

PQ-3. A pair of isotopes has

(A) the same number of protons and a different number of neutrons.
(B) the same number of protons and neutrons but a different number of electrons.
(C) the same number of protons and the same number of neutrons.
(D) the same number of neutrons and the same number of electrons.

PQ-4. How many neutrons are in $^{14}_{6}C$?

(A) 6 **(B)** 7 **(C)** 8 **(D)** 20

Conceptual **PQ-5.** In all neutral atoms, there are equal numbers of

(A) protons and neutrons.
(B) positrons and electrons.
(C) neutrons and electrons.
(D) electrons and protons.

PQ-6. Which element is represented by $^{56}_{24}X$?

(A) iron **(B)** germanium **(C)** barium **(D)** chromium

Conceptual **PQ-7.** Which statement is true?

(A) The nucleus of an atom contains neutrons and electrons.
(B) The atomic number of an element is the number of protons in one atom.
(C) The mass number of an atom is the number of protons in the nucleus plus the number of electrons outside.
(D) The number of electrons outside the nucleus is the same as the number of neutrons in the nucleus.

PQ-8. Which atom has the most neutrons?

(A) neon-20
(B) phosphorous-32
(C) chlorine-35
(D) sulfur-35

PQ-9. How many neutrons are in $^{37}_{17}Cl^{-}$?

(A) 17 **(B)** 20 **(C)** 21 **(D)** 37

PQ-10. Which pair of particles has the same number of electrons?

(A) F^{-}, Mg^{2+} **(B)** Ne, Ar **(C)** Br^{-}, Se **(D)** Al^{3+}, P^{3-}

PQ-11. Which ion has twenty-six electrons?

(A) Cr^{2+} **(B)** Fe^{2+} **(C)** Ni^{2+} **(D)** Cu^{2+}

Conceptual

PQ-12. A sodium ion differs from a sodium atom in that the sodium ion

(A) has fewer electrons.

(B) is an isotope of sodium.

(C) exists only in solution.

(D) has a negative charge on its nucleus.

PQ-13. Which ion has the same number of electrons as an argon atom?

(A) Mg^{2+} **(B)** Cl^{-} **(C)** Ne **(D)** Na^{+}

PQ-14. Bromine has two naturally occurring isotopes. The most abundant isotope (50.69%) is $^{79}_{35}Br$ of mass 78.9183 amu. What is the other isotope?

(A) $^{79}_{36}Br$ **(B)** $^{81}_{36}Br$ **(C)** $^{80}_{35}Br$ **(D)** $^{81}_{35}Br$

PQ-15. Two isotopes of hypothetical element X exist with abundances of 30.00% ^{100}X and 70.00% ^{101}X. what is the approximate atomic mass of X (in atomic units, amu)?

(A) 100.3 **(B)** 100.5 **(C)** 100.7 **(D)** 101.0

Conceptual

PQ-16. An enriched sample of carbon contains 20.0% ^{12}C and 80.0% ^{13}C. Which figure shows this sample?

Key: ● = ^{12}C
○ = ^{13}C

(A) (B) (C) (D)

PQ-17. Lithium has two naturally occurring isotopes, ^{6}Li and ^{7}Li, with masses of 6.015 amu and 7.016 amu, respectively. What is the relative abundance of each isotope?

(A) ^{6}Li = 7.49 % and ^{7}Li = 92.51 %

(B) ^{6}Li = 25.31 % and ^{7}Li = 74.69 %

(C) ^{6}Li = 46.16 % and ^{7}Li = 53.84 %

(D) ^{6}Li = 92.51 % and ^{7}Li = 7.49 %

PQ-18. Which group contains nonmetals, metalloids, and metals?

(A) 1 **(B)** 13 **(C)** 15 **(D)** 18

PQ-19. What is a main group element?

(A) Pu **(B)** Mg **(C)** Sc **(D)** Fe

PQ-20. The halogens are found in group

(A) 1. **(B)** 15. **(C)** 17. **(D)** 18.

PQ-21. Which compound contains a Group 14 metal?

(A) CCl_4 (B) $SnCl_2$ (C) GeF_4 (D) $ZnCl_2$

PQ-22. Which statement below is correct?

(A) Chlorine is a transition metal.

(B) Oxygen is a metal.

(C) Sodium is a metalloid.

(D) Phosphorous is a nonmetal.

PQ-23. Which compound below contains a transition metal?

(A) NaF (B) $AlCl_3$ (C) VO_2 (D) BF_3

PQ-24. Which element is an alkali metal?

(A) K (B) Mg (C) V (D) Fe

PQ-25. Which element is an actinide?

(A) Rb (B) U (C) Mo (D) Ga

PQ-26. Magnesium forms an ionic compound with an element with the formula MgX. Which ion could be X?

(A) Br^- (B) P^{3-} (C) C^{4-} (D) S^{2-}

PQ-27. Alkaline earth metals form

(A) +1 ions. (B) +2 ions. (C) +3 ions. (D) -1 ions.

PQ-28. Which compound contains a +1 ion?

(A) MgO (B) KCl (C) NO (D) AlP

PQ-29. What is the formula for the compound formed when an aluminum ion, Al^{3+}, combines with an oxide ion, O^{2-}?

(A) AlO (B) Al_2O_3 (C) Al_3O_2 (D) AlO_2

PQ-30. If gallium, atomic number 31, combines with selenium, atomic 34, what is the most likely formula based on your knowledge of the periodic nature of the elements?

(A) GaSe (B) $GaSe_2$ (C) Ga_2Se (D) Ga_2Se_3

Answers to Study Questions

1. B
2. A
3. C
4. A
5. C
6. C
7. C
8. D
9. A

Answers to Practice Questions

1. C
2. B
3. A
4. C
5. D
6. D
7. B
8. D
9. B
10. A
11. C
12. A
13. B
14. D
15. C
16. C
17. A
18. C
19. B
20. C
21. B
22. D
23. C
24. A
25. B
26. D
27. B
28. B
29. B
30. D

Chapter 2 – Electronic Structure

Chapter Summary:

This chapter will focus on electrons and their arrangement in the atom. The relationship between the arrangement of the electrons and energy is also covered. Finally, the relationship between the electronic structure of the atom and the atom's location on the periodic table is reviewed.

Specific topics covered in this chapter are:

- Light and the electromagnetic spectrum
- Quantum numbers and atomic orbitals
- Orbital diagrams and electron configurations
- Paramagnetic and diamagnetic species
- Energies of electron transitions in atoms
- Periodic properties related to the electronic structure

Previous material that is relevant to your understanding of questions in this chapter include:

- Significant figures (***Toolbox***)
- Scientific notation (***Toolbox***)
- Metals and nonmetals ***(Chapter 1)***

Common representations used in questions related to this material:

Name	Example	Used in questions related to
Atomic symbols	H	Electron configurations
Orbital box diagram	□ □□□ □□□□□ s p d	Electron configurations
Bohr model	$n = 1$, $n = 2$, $n = 3$	The Bohr model and energy levels in the hydrogen atom

Where to find this in your textbook:

The material in this chapter typically aligns to "Electronic Structure" or "Quantum" in your textbook. The name of your chapter may vary.

Practice exam:

There are practice exam questions aligned to the material in this chapter. Because there are a limited number of questions on the practice exam, a review of the breadth of the material in this chapter is advised in preparation for your exam.

How this fits into the big picture:

The material in this chapter aligns to the Big Idea of Atoms (1) as listed on page 12 of this study guide.

Study Questions (SQ)

SQ-1. What is the wavelength of light (in nm) produced by the electronic transition between levels 4 and 2 of a hydrogen atom?

(A) 182 nm **(B)** 364 nm **(C)** 486 nm **(D)** 1450 nm

Knowledge Required: (1) Ability to use the Rydberg formula. (2) Understanding of the units of energy. (3) Calculation of wavelength from an energy. (4) Relationship between the ΔE of the transition and the energy of the photon emitted.

Thinking it Through: You know that the energy of the electron energy levels in hydrogen depends on the value of the n principal quantum number. **Note: It is very likely that you will be provided this equation, and others, and any needed constants.**

$$E_n = \frac{-R_H}{n^2}$$

You also know you can calculate the energy of transition from $n = 4 \rightarrow n = 2$ using the Rydberg relationship:

$$\Delta E = R_H\left(\frac{1}{n_i^2} - \frac{1}{n_f^2}\right) \text{ where } R_H = 2.18\times10^{-18}\text{ J}, n_i = 4 \text{ and } n_f = 2$$

$$\Delta E = R_H\left(\frac{1}{4^2} - \frac{1}{2^2}\right) = 2.18\times10^{-18}\text{ J}\left(\frac{1}{16} - \frac{1}{4}\right) = -4.09\times10^{-19}\text{ J}$$

The energy of the transition is negative because the process produces a photon and the atom releases energy. The energy of the photon is the absolute value of the energy of the transition.

$$E_{\text{photon}} = |\Delta E_{\text{transition}}| = 4.09\times10^{-19}\text{ J}$$

You recall that if you know the energy of the photon, you can find the wavelength, λ, using: $E_{\text{photon}} = \frac{hc}{\lambda}$, with $h = 6.626\times10^{-34}$ J·s and $c = 2.998\times10^{-8}$ m·s^{-1}.

Solving for the wavelength you get: $\lambda = \frac{hc}{E_{\text{photon}}} = \frac{(6.626\times10^{-34}\text{ J})(2.998\times10^{8}\text{ m}\cdot\text{s}^{-1})}{4.09\times10^{-19}\text{ J}} = 4.86\times10^{-7}\text{ m}$

You can convert the wavelength to nm: $4.86\times10^{-7}\text{m}\left(\frac{10^9\text{ nm}}{1\text{ m}}\right) = 486\text{ nm}$

The correct choice is **(C)**.

Choice **(A)** is not correct because it is the wavelength calculated when the energy of the $n = 2$ level, not the difference in energy of the levels, is used to calculate the wavelength.

Choice **(B)** is not correct because it is the wavelength calculated if when the n values are not squared.

Choice **(D)** is not correct because it is the wavelength calculated when the energy of the $n = 4$ level, not the difference in energy of the levels, is used to calculate the wavelength.

Practice Questions Related to This: **PQ-1, PQ-2,** and **PQ-3**

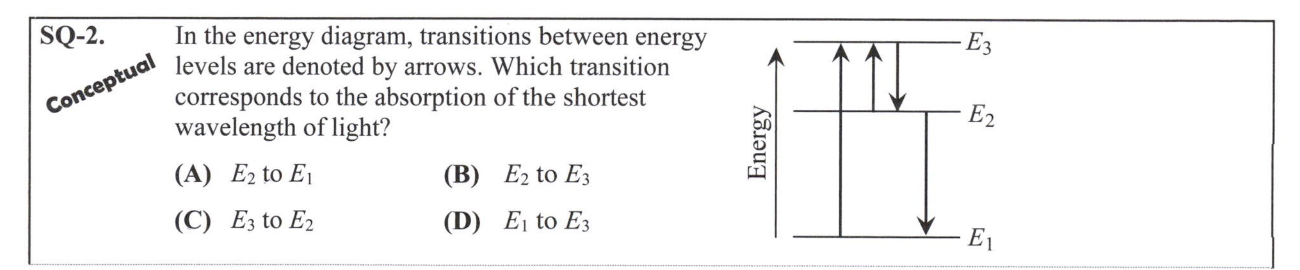

SQ-2. *Conceptual* In the energy diagram, transitions between energy levels are denoted by arrows. Which transition corresponds to the absorption of the shortest wavelength of light?

(A) E_2 to E_1 **(B)** E_2 to E_3

(C) E_3 to E_2 **(D)** E_1 to E_3

Knowledge Required: (1) Difference in absorption and emission. (2) How to interpret an energy level diagram. (3) How the energy of a photon relates to the difference in electron energy levels. (4) Relationship between energy and wavelength.

Thinking it Through: You read in the question that the process occurring is absorption. You know that this means energy is being absorbed by the atom. This tells you that you want a transition where the electron is being promoted to a higher energy level. This is represented on the diagram by an arrow pointing upwards. The farther apart the energy levels, the larger the energy of the photon involved in the transition. So, the longer the arrow the more energetic the photon. Remember that photon energy and wavelength are inversely related.

$$E_{\text{photon}} = \frac{hc}{\lambda}$$

The question is asking you to find the transition with the shortest wavelength, which means the highest energy (longest arrow). The question also wants the transition to be an absorption (upward pointing arrow). The correct choice is **(D)**.

Choice **(A)** is not correct because the transition indicated would represent the **emission** of a photon. Choice **(B)** is not correct because the transition indicated would represent the absorption of a photon, but the energy of this photon would be less than the energy (longer wavelength) of the photon absorbed in transition D. Choice **(C)** is not correct because the transition indicated would represent the **emission** of the photon with the longest wavelength.

***Practice Questions Related to This*:** **PQ-4** and **PQ-5**

SQ-3. Which set is **NOT** an allowed set of quantum numbers?

	n	ℓ	m_ℓ	m_s
(A)	3	2	–1	+ ½
(B)	4	0	–1	+ ½
(C)	3	1	–1	– ½
(D)	4	0	0	– ½

Knowledge Required: (1) Rules for quantum numbers.

Thinking it Through: You are being asked to decide which set of quantum numbers is incorrect. You remember the rules for the allowed values of the quantum numbers:

- Allowed values of n are any integer from 1 to infinity
- Allowed values of ℓ depend on the value of the n quantum number. Allowed values of ℓ are 0 up to the value of $n - 1$.
- Allowed values of m_ℓ depend on the value of the ℓ quantum number. Allowed values of m_ℓ are $-\ell$ to 0 to $+\ell$.
- Allowed values of m_s are either + ½ or – ½.

You next proceed to check each combination.

Choice **(A)** With $n = 3$, the value of 2 for ℓ is allowed. With $\ell = 2$, the value of $m_\ell = -1$ is allowed. The given value of m_s is allowed.

Choice **(B)** With $n = 4$, the value of 0 for ℓ is allowed. With $\ell = 0$, the value of $m_\ell = -1$ is **NOT** allowed. The given value of m_s is allowed.

Choice **(C)** With $n = 3$, the value of 1 for ℓ is allowed. With $\ell = 1$, the value of $m_\ell = -1$ is allowed. The given value of m_s is allowed.

Choice **(D)** With $n = 4$, the value of 0 for ℓ is allowed. With $\ell = 0$, the value of $m_\ell = 0$ is allowed. The given value of m_s is allowed.

The choice with the incorrect set of quantum numbers is choice **(B)**.

Practice Questions Related to This: **PQ-7**

SQ-4. Which type of electron is described by the quantum numbers, $n = 3$, $\ell = 2$, $m_\ell = 1$, and $m_s = +½$?

(A) 2p **(B)** 3p **(C)** 3d **(D)** 3f

Knowledge Required: (1) Relationship between principal quantum number and energy level designation. (2) Letter abbreviations for values of the ℓ quantum number.

Thinking it Through: You are given a set of quantum numbers and asked to determine what type of electron would have this set of quantum numbers. You recall that the two parts of an electron description are the energy level (n quantum number) and the orbital type (ℓ quantum number). The first part of the description is the value of n. For example, 2p describes an electron in the second energy level. The letters describing the orbital type are s, p, d, and f. The relationship between the letters and the values of the ℓ quantum number are:

$\ell = 0$ is an s orbital; $\ell = 1$ is a p orbital, $\ell = 2$ is a d orbital, and $\ell = 3$ is a f orbital.

You return to the given set of quantum numbers and decide the type of electron being described: $n = 3$ and $\ell = 2$ describes a 3d electron. The correct choice is **(C)**.

Choice **(A)** is not correct because 2p describes an electron with $n = 2$, $\ell = 1$.

Choice **(B)** is not correct because 3p describes an electron with $n = 3$, $\ell = 1$.

Choice **(D)** is not correct because 3f describes an electron with $n = 3$, $\ell = 3$ and this is not a possible set of quantum numbers.

Practice Questions Related to This: **PQ-7**

SQ-5. What is the electron configuration of the valence electrons for a ground-state Ge atom?

(A) $4p^2$ **(B)** $4s^2 3d^{10} 4p^2$

(C) $3d^{10} 4p^2$ **(D)** $4s^2 4p^2$

Knowledge Required: (1) How to write electron configurations using the periodic table. (2) Definition of valence electrons.

Thinking it Through: You are being asked for the valence electron configuration of Ge. You also make use of the definition of valence electrons: electrons in the outermost energy level. The atomic number of Ge is 32, which you know means the neutral atom has 32 electrons. Ge is also in the fourth period (row) and group (column) 14 of the periodic table. Because Ge is in the fourth period, the highest energy level containing electrons is $n = 4$ and these are the valence electrons for Ge. Group 14 is the second column in the p-block of the periodic table. You can then write the electron configuration for the valence electrons: $4s^2 4p^2$. You can check this result using the core notation: [Ar] $4s^2 3d^{10} 4p^2$. The [Ar] core has 18 electrons, the filled 3d holds 10 more, and the valence shell has 4. You have accounted for the all 32 electrons. The correct choice is **(D)**.

Choice **(A)** is not correct because it does not include the 4s electrons as valence electrons.

Choice **(B)** is not correct because electrons in the filled d block are not considered valence electrons.

Choice **(C)** is not correct because it does not include the 4s electrons and does include the filled d orbitals.

Practice Questions Related to This: **PQ-8, PQ-9,** and **PQ-10**

SQ-6. What is the ground state electron configuration for the Zr^{2+} ion?

(A) [Kr] $5s^2$ **(B)** [Kr] $4d^2$

(C) [Kr] $5s^2 4d^2$ **(D)** [Kr] $5s^2 4d^4$

Knowledge Required: (1) How to write electron configurations using the periodic table. (2) Rules for removing electrons when forming cations.

Thinking it Through: The question is asking you to write the ground-state electron configuration for a cation. You remember that ground-state means the lowest energy state, and this is written using the Aufbau principle. The Zr^{2+} ion will have two less electrons than the Zr atom. You decide that the best approach is to use the periodic table to write the ground state electron configuration for the 40 electrons in Zr. Using Zr's location on the periodic table, period 5, you quickly decide that the first 36 electrons in Zr are isoelectronic with Kr. The remaining four electrons are in the 5s and 4d orbitals. You write the ground state electron configuration as: [Kr] $5s^2 4d^2$

You know you need to remove two electrons to form the Zr^{2+} ion. The rule for removing electrons is that the electrons in the highest energy level are removed first, they have the lower ionization energy. After you remove the two 5s electrons the ground state electron configuration for Zr^{2+} is [Kr] $4d^2$. Choice **(B)** is the correct answer.

Choice **(A)** is incorrect because the highest energy electrons were not removed.
Choice **(C)** is incorrect because the electron configuration given is for the Zr atom.
Choice **(D)** is incorrect because the electron configuration reflects the addition of two extra electrons.

Practice Questions Related to This: **PQ-11, PQ-12,** and **PQ-13**

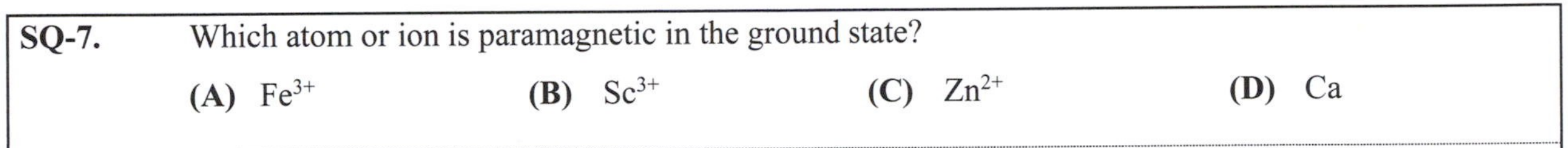

SQ-7. Which atom or ion is paramagnetic in the ground state?

(A) Fe^{3+} **(B)** Sc^{3+} **(C)** Zn^{2+} **(D)** Ca

Knowledge Required: (1) Drawing and interpreting orbital diagrams. (2) Definition of paramagnetic.

Thinking it Through: You are being asked to decide which species is paramagnetic. You recall that for a species to be paramagnetic it must have unpaired electrons. You remember that an orbital diagram is a useful model that helps show if a species has unpaired electrons. To create an orbital diagram for each species you write the electron configurations of each.

Fe^{3+} [Ar] $3d^5$ Sc^{3+} [Ar] Zn^{2+} [Ar] $3d^{10}$ Ca [Ar] $4s^2$

To draw the orbital diagram, you recall that there are five d orbitals, three p orbitals, and one s orbital.

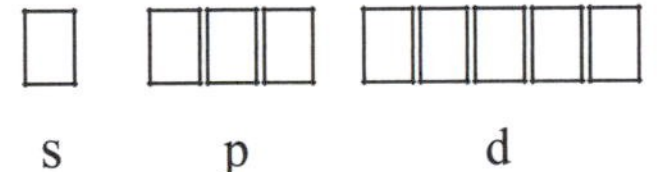

Hund's rule also tells you that each orbital will be half-filled before a second electron is added to an orbital. You draw the valence orbital diagrams for the species in the question.

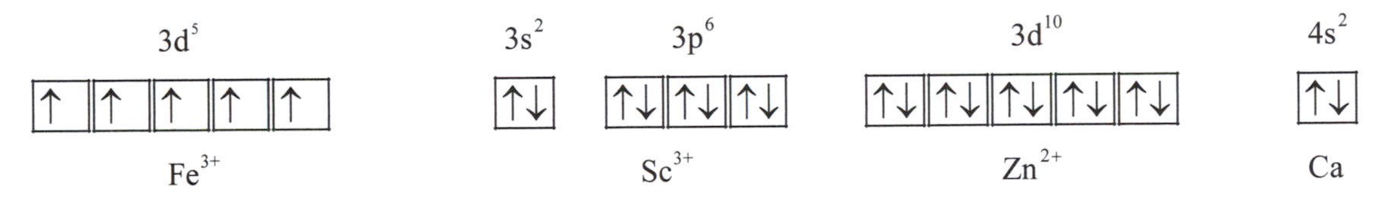

The only species with unpaired electrons is Fe^{3+}. It will be the only one listed that is paramagnetic in its ground state.

The correct choice is **(A)**.

Choices **(B)**, **(C)**, and **(D)** are not correct because they do not have unpaired electrons.

Practice Questions Related to This: **PQ-14, PQ-15,** and **PQ-16**

SQ-8. In which atom does a 2p electron experience the greatest effective nuclear charge, Z_{eff}?

(A) B **(B)** C **(C)** F **(D)** N

Knowledge Required: (1) Definition of Z_{eff}. (2) Periodic trend for Z_{eff}.

Thinking it Through: You read that the problem is asking about Z_{eff}. You recall that Z_{eff} refers to the effective nuclear charge the valence electrons experience. The valence electrons do not experience the entire nuclear charge, of the protons in the nucleus, because the inner electrons "block" (shield or screen) some of this positive charge. This can be thought of in an equation:

$$Z_{eff} = Z_{nuclear} - \text{shielding}$$

You remember that as you move across a period, the shielding ability of the inner electrons, in this case the $1s^2$ electrons, does not change. The real nuclear charge, $Z_{nuclear}$, is increasing as you move from left to right because of the increase in the number of protons. The result of this is that the periodic trend is the Z_{eff} increases as you move from left to right across a period.

You decide that F will have the largest Z_{eff}. The correct choice is **(C)**.

Choices **(A)**, **(B)** and **(D)** are not correct because they appear to the left of F in the second period.

Practice Questions Related to This: **PQ-17**

SQ-9. Which atom has the largest atomic radius?

(A) K **(B)** I **(C)** Rb **(D)** Sn

Knowledge Required: (1) Periodic trend for atomic radius (2) How Z_{eff} changes as you move from left to right in a period.

Thinking it Through: You realize that the question is asking about a periodic trend. You remember that as you go down a group (column) the atoms get larger because more energy levels are needed to house the electrons. Also, when going from left to right across a period, the value of the effective nuclear charge, Z_{eff}, is increasing. This means the electrons in the valence energy level experience a stronger attraction for the nucleus, and the size of the atom decreases as Z_{eff} increases. You start by finding the elements' positions on the periodic table:

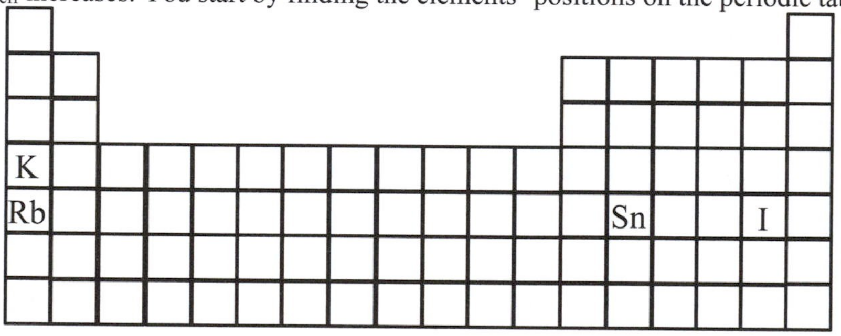

K
Rb Sn I

K and Rb are in group 1A with Rb in period 5 and K in period 4. Using the periodic trend, you know that Rb has a larger atomic radius than K. The elements Rb, Sn, and I are all in period 5. Using the periodic trend again, you know that the valence electrons in I experiences a larger Z_{eff} than the valence electrons in Sn. Similarly, the valence electrons in Sn experiences a larger Z_{eff} than those in Rb. You then select your answer because the atom with the smallest Z_{eff} will have the largest atomic radius - Rb. The correct choice is choice **(C)**.

Choice **(A)** is not correct because K is smaller than Rb. Choice **(B)** is not correct because I experiences the largest Z_{eff} of the elements in period 5, so it has the smallest atomic radius in this period. Choice **(D)** is not correct because the Z_{eff} for Sn is larger than that of I, but smaller than that of Rb. The result is that it has an atomic radius in between that of Rb and I.

Practice Questions Related to This: **PQ-18, PQ-19,** and **PQ-20**

SQ-10. Which pair correctly shows the size difference between the species?

(A) $Li^+ > Be^{2+}$ **(B)** $Li^+ > Na^+$ **(C)** $Li^+ > Li$ **(D)** $S > S^{2-}$

Knowledge Required: (1) The periodic trends in atomic radii. (2) The periodic trends in ionic radii. (3) Relationship of the size of an ion to its atom.

Thinking it Through: You recognize that this question is asking you to compare sizes of atoms and ions. You recall the periodic trends. The trend for neutral atoms is that they get smaller moving across a period (due to the increasing Z_{eff}) and larger moving down a group (due to the increasing number of electron shells). You realize that you need to consider the effect of ion formation on size. When an atom loses electrons, the nuclear charge does not change, so the same number of protons are "pulling" on the remaining electrons. This results in a decrease in the size of the electron cloud, and the positive ion is smaller than the atom ($X^+ < X$). If you remove a second electron to form a 2+ ion, the effect is even greater. For negative ions, the opposite is true. The addition of an extra electron causes the energy level to expand. The result is that negative ions are larger than their neutral atoms ($X^- > X$). This can be summarized together:

$$X^- > X > X^+$$

Using this information, you examine the options one by one:

Choice **(A)** Both are in the same period, so Li is larger than Be. However, the formation of the ions and the differences in charge must be taken in to account. The larger charge on Be^{2+} results in more contraction of the electron cloud than in Li^+. The result is that Li^+ is larger than Be^{2+}.

Choice **(B)** Again we are comparing two ions of different elements. Na and Li are in the same group and Na is larger than Li. Li^+ has only 1 energy level and Na^+ has two energy levels, so Li^+ is smaller than Na^+.

Choice **(C)** Here you are comparing a neutral atom and it cation. The neutral will be larger than its cation, Li is larger than Li^+.

Choice **(D)** To pick the larger of the two you must compare the size of an anion and its neutral atom. Negative ions of an element are larger than the atom of the element; so S^{2-} is larger than S.

Choice **(A)** is the correct answer because it is the only one with the larger species first.

Practice Questions Related to This: **PQ-21, PQ-22, PQ-23,** and **PQ-24**

SQ-11. Arrange the elements Li, Ne, Na, and Ar in increasing order of first ionization energy.

(A) Na < Li < Ar < Ne **(B)** Li < Na < Ar < Ne

(C) Na < Li < Ne < Ar **(D)** Ar < Ne < Na < Li

Knowledge Required: (1) The definition of ionization energy. (2) The relationship of ionization energy and atomic size. (3) The periodic trends in atomic radii.

Thinking it Through: You read that the question is asking about a trend in ionization energy. Ionization energy is the energy needed to remove an electron from a gaseous atom or ion. You reason that it is "harder" to remove an electron from a small atom, so the smaller the atom the larger the ionization energy. The question is asking you for the elements in order of increasing ionization energy so the first element in your list will be the element that has the lowest ionization energy (largest radius), followed by the next largest ionization energy (next largest radius), etc. Using this logic, you decide that Na has the largest atomic radius, and the lowest ionization energy, because it is below Li and to the left of Ar in the same period. Similarly, you decide that Ne has the smallest atomic radius, hence largest ionization energy, because it is above Ar and to the right of Li.
You arrange the elements in the order increasing ionization energy Na < Li < Ar < Ne.

Choice **(A)** is the correct answer.

Choice **(B)** is not correct because it has reversed the relative ionization energies of Na and Li.

Choice **(C)** is not correct because it has reversed the relative ionization energies of Ne and Ar.

Choice **(D)** is not correct because it has misapplied the trend in atomic radii across a period.

Practice Questions Related to This: **PQ-25, PQ-26, PQ-27, PQ-28, PQ-29,** and **PQ-30**

Practice Questions (PQ)

PQ-1. What is the energy of a photon with a wavelength (λ) of 656 nm?

(A) 3.03×10^{-19} J **(B)** 3.03×10^{-28} J **(C)** 4.35×10^{-31} J **(D)** 4.35×10^{-40} J

PQ-2. What is the wavelength of a photon with an energy of 6.51×10^{-19}J?

(A) 3.05×10^{-9} nm **(B)** 305 nm **(C)** 1.00×10^{15} nm **(D)** 4.52×10^{26} nm

PQ-3. When excited, a sodium atom emits a photon of frequency $5.090 \times 10^{14}\ s^{-1}$. What is the energy associated with this emission?

(A) 3.90×10^{-50} J **(B)** 3.90×10^{-48} J **(C)** 3.37×10^{-28} J **(D)** 3.37×10^{-19} J

Conceptual **PQ-4.** Which electronic transition in a hydrogen atom is associated with the largest *emission* of energy?

(A) $n = 2$ to $n = 1$ **(B)** $n = 2$ to $n = 3$ **(C)** $n = 2$ to $n = 4$ **(D)** $n = 3$ to $n = 2$

Conceptual **PQ-5.** Which emission line in the hydrogen spectrum occurs at the highest frequency?

(A) $n = 3$ to $n = 1$ **(B)** $n = 4$ to $n = 2$ **(C)** $n = 7$ to $n = 5$ **(D)** $n = 10$ to $n = 8$

Conceptual **PQ-6.** The picture at the right is an early model for the atom. This model differs from the current model because in the current model

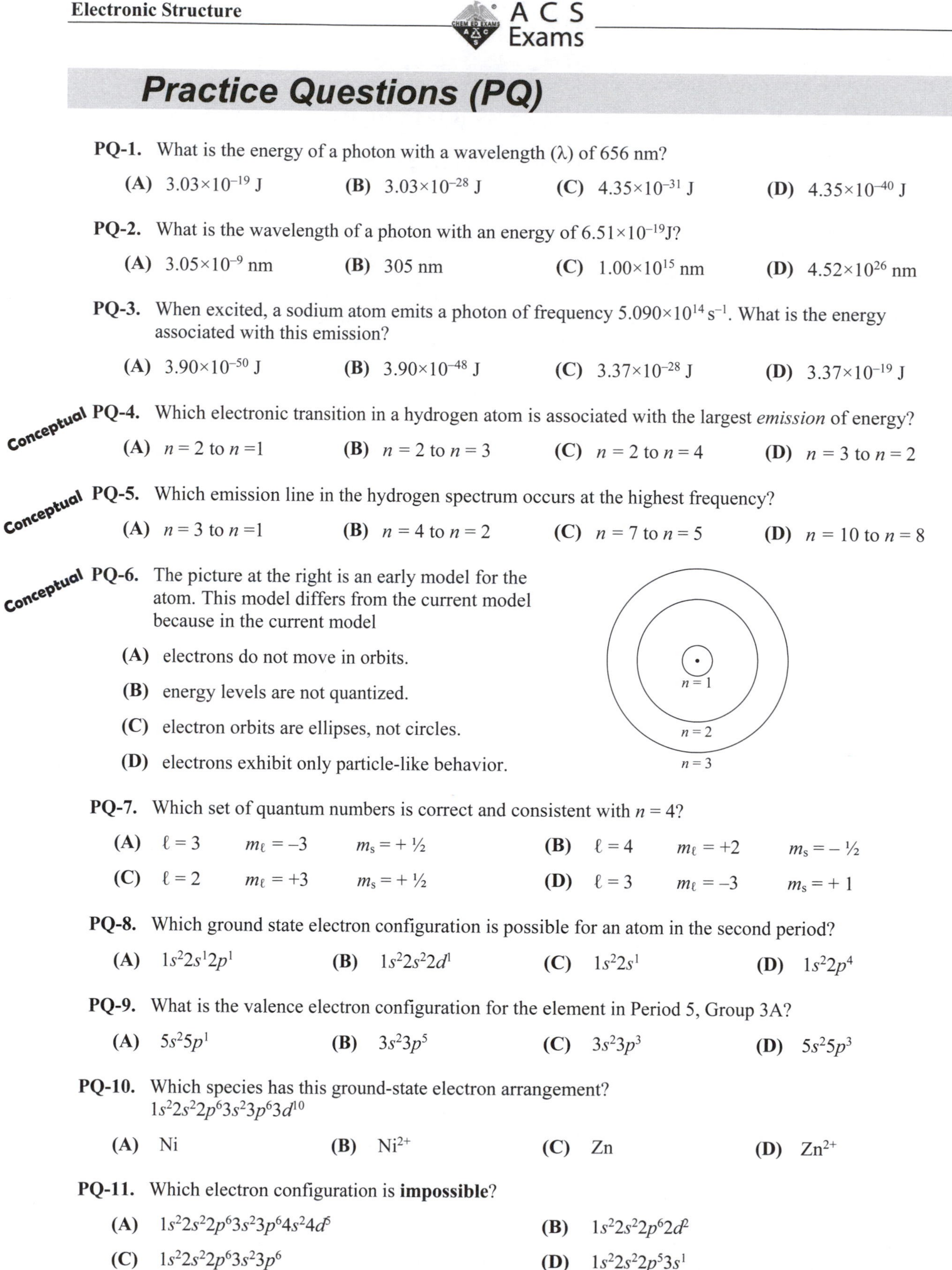

(A) electrons do not move in orbits.

(B) energy levels are not quantized.

(C) electron orbits are ellipses, not circles.

(D) electrons exhibit only particle-like behavior.

PQ-7. Which set of quantum numbers is correct and consistent with $n = 4$?

(A) $\ell = 3$ $m_\ell = -3$ $m_s = +\frac{1}{2}$ **(B)** $\ell = 4$ $m_\ell = +2$ $m_s = -\frac{1}{2}$

(C) $\ell = 2$ $m_\ell = +3$ $m_s = +\frac{1}{2}$ **(D)** $\ell = 3$ $m_\ell = -3$ $m_s = +1$

PQ-8. Which ground state electron configuration is possible for an atom in the second period?

(A) $1s^2 2s^1 2p^1$ **(B)** $1s^2 2s^2 2d^1$ **(C)** $1s^2 2s^1$ **(D)** $1s^2 2p^4$

PQ-9. What is the valence electron configuration for the element in Period 5, Group 3A?

(A) $5s^2 5p^1$ **(B)** $3s^2 3p^5$ **(C)** $3s^2 3p^3$ **(D)** $5s^2 5p^3$

PQ-10. Which species has this ground-state electron arrangement?
$1s^2 2s^2 2p^6 3s^2 3p^6 3d^{10}$

(A) Ni **(B)** Ni^{2+} **(C)** Zn **(D)** Zn^{2+}

PQ-11. Which electron configuration is **impossible**?

(A) $1s^2 2s^2 2p^6 3s^2 3p^6 4s^2 4d^5$ **(B)** $1s^2 2s^2 2p^6 2d^2$

(C) $1s^2 2s^2 2p^6 3s^2 3p^6$ **(D)** $1s^2 2s^2 2p^5 3s^1$

PQ-12. What is the ground-state electron configuration of the manganese atom, Mn?

(A) $1s^22s^22p^63s^23p^64s^24d^5$ **(B)** $1s^22s^22p^63s^23p^63d^7$

(C) $1s^22s^22p^63s^23p^64s^24p^5$ **(D)** $1s^22s^22p^63s^23p^63d^54s^2$

PQ-13. What is the maximum number of electrons that can occupy an orbital labeled d_{xy}?

(A) 1 **(B)** 2 **(C)** 3 **(D)** 4

PQ-14. An atom of Fe has two 4*s* electrons and six 3*d* electrons. How many unpaired electrons would there be in a Fe^{2+} ion?

(A) one **(B)** two **(C)** three **(D)** four

PQ-15. Which of these species (is/are) paramagnetic? Ti^{4+} Fe^{2+} Zn

(A) Fe^{2+} only **(B)** Zn only **(C)** Ti^{4+} and Fe^{2+} only **(D)** Fe^{2+} and Zn only

Conceptual **PQ-16.** Which of these orbital diagram pairs are equivalent for the ground state *p* valence electrons of carbon?

(A) ↑↓ __ __ / ↑↑ __ __

(B) ↑ ↓ __ / ↑↑ __ __

(C) ↓ ↓ __ / ↑ ↑ __

(D) ↑ ↑ __ / ↑↓ __ __

PQ-17. The size of metal atoms

(A) generally increases progressively from top to bottom in a group in the periodic table.

(B) generally increases progressively from left to right in a period.

(C) are smaller than those of the corresponding ions.

(D) do not change upon losing electrons.

PQ-18. Which of these elements has the ***smallest*** atomic radius?

(A) fluorine **(B)** chlorine **(C)** bromine **(D)** iodine

PQ-19. The radii of the ions in this series decreases because

Ion	Ionic Radii
Na^+	0.095 nm
Mg^{2+}	0.065 nm
Al^{3+}	0.050 nm

(A) the elements are in the same period.

(B) the effective nuclear charge is increasing.

(C) the atomic radius of Na decreases from Na to Al.

(D) the first ionization energies increase from Na to Al.

Conceptual **PQ-20.** Which of these atoms will be the ***smallest***?

(A) Si ($Z = 14$) **(B)** P ($Z = 15$) **(C)** Ge ($Z = 32$) **(D)** As ($Z = 33$)

PQ-21. The species F^-, Mg^{2+}, and Na^+ all have the same number of electrons. Which is the predicted order when they are arranged in order of decreasing size (largest first).

(A) $F^- > Mg^{2+} > Na^+$ **(B)** $Na^+ > Mg^{2+} > F^-$ **(C)** $Mg^{2+} > F^- > Na^+$ **(D)** $F^- > Na^+ > Mg^{2+}$

PQ-22. Which ion is the ***smallest***?

(A) Al^{3+} **(B)** Na^+ **(C)** F^- **(D)** O^{2-}

Conceptual **PQ-23.** What happens when a bromine atom becomes a bromide ion?

(A) A positive ion is formed.

(B) The bromine nucleus acquires a negative charge.

(C) The atomic number of bromine is decreased by one.

(D) The bromide ion is then larger than the bromine atom.

PQ-24. Which ion has the largest radius?

(A) Cl^- **(B)** F^- **(C)** K^+ **(D)** Cu^{2+}

PQ-25. Which element has the highest electronegativity?

(A) cesium **(B)** iodine **(C)** oxygen **(D)** lithium

PQ-26. Which element is more electronegative than arsenic and less electronegative then sulfur?

(A) chlorine **(B)** phosphorous **(C)** tin **(D)** oxygen

PQ-27. Which equation corresponds to the electron affinity of chlorine?

(A) $Cl^-(g) \rightarrow Cl(g) + e^-$ **(B)** $Cl_2(g) + e^- \rightarrow Cl_2^-(g)$

(C) $Cl(g) \rightarrow Cl^+(g) + e^-$ **(D)** $Cl(g) + e^- \rightarrow Cl^-(g)$

PQ-28. Which pair of elements is listed in order of decreasing first ionization energy?

(A) Na, Mg **(B)** Mg, Al **(C)** Al, Si **(D)** Si, P

PQ-29. When the species F^-, Na^+, and Ne are arranged in order of increasing energy for the removal of an electron, what is the correct order?

(A) $F^- < Na^+ < Ne$ **(B)** $Na^+ < Ne < F^-$ **(C)** $F^- < Ne < Na^+$ **(D)** $Ne < F^- < Na^+$

Conceptual **PQ-30.** The first three ionization energies of an element X are 590, 1145, and 4912 $kJ \cdot mol^{-1}$. What is the most likely formula for the stable ion of X?

(A) X^+ **(B)** X^{2+} **(C)** X^{3+} **(D)** X^-

Answers to Study Questions

1. C
2. D
3. B
4. C
5. D
6. B
7. A
8. C
9. C
10. A
11. A

Answers to Practice Questions

1. A
2. B
3. D
4. A
5. A
6. A
7. A
8. C
9. A
10. D
11. B
12. D
13. B
14. D
15. A
16. C
17. A
18. A
19. B
20. B
21. D
22. A
23. D
24. A
25. C
26. B
27. D
28. B
29. C
30. B

Chapter 3 – Formula Calculations and the Mole

Chapter Summary:

This chapter will focus on using molecular formulas to determine the abundance and mass of collections of molecules or atoms within a molecule, including empirical formulas, atomic ratios, and percent composition.

Specific topics covered in this chapter are:

- Empirical and molecular formulas
- Mole and Avogadro's number
- Molar mass
- Percent composition (mass percent)

Previous material that is relevant to your understanding of questions in this chapter include:

- Scientific notation (***Toolbox***)
- Significant figures (***Toolbox***)
- Unit conversions (***Toolbox***)
- Atomic mass (***Chapter 1***)
- Periodic Table (***Chapter 1***)

Common representations used in questions related to this material:

Name	Example	Used in questions related to
Compound units	$g{\cdot}mol^{-1}$	molar mass

Where to find this in your textbook:

The material in this chapter typically aligns to "Composition of Substances and Solutions" and "Stoichiometry of Chemical Reactions" in your textbook. The name of your chapter may vary.

Practice exam:

There are practice exam questions aligned to the material in this chapter. Because there are a limited number of questions on the practice exam, a review of the breadth of the material in this chapter is advised in preparation for your exam.

How this fits into the big picture:

The material in this chapter aligns to the Big Idea of Atoms (1) as listed on page 12 of this study guide.

Study Questions (SQ)

SQ-1. Avogadro's number equals the number of

Conceptual

(A) atoms in one mole of O_2.

(B) atoms in one mole of atoms.

(C) marbles in one mole of marbles.

(D) atoms in one mole of atoms and marbles in one mole of marbles.

Knowledge Required: (1) Definition of a mole.

Thinking it Through: You are asked in this question to evaluate the choices for which statement completes the sentence, "Avogadro's number equals the number of." You know that Avogadro's number is the number of items in one mole; items can refer to atoms, choice **(B)**, or marbles, choice **(C)**. Choice **(A)** is incorrect because 1

molecule of O_2 would have two moles of atoms of O. The mole is defined as 6.022×10^{23} (i.e., Avogadro's number) of anything. The correct answer is the combination of the two correct choices, **(B)** and **(C)**, which is choice **(D).**

Practice Questions Related to This: **PQ-1, PQ-2, PQ-3**

SQ-2. Which sample contains the least number of atoms?

Conceptual

(A) 1.00 g C **(B)** 1.00 g O_2 **(C)** 1.00 g H_2 **(D)** 1.00 g Fe

Knowledge Required: (1) Where to find the average atomic mass. (2) How to determine molar mass. (3) Converting from mass to moles. (4) How to use Avogadro's number.

Thinking it Through: You are asked to determine which mass of a given molecular formula has the ***least*** number of atoms.

Because the mass of the samples in the choices are the same, you can focus on the number of atoms or moles of atoms in each sample based on using (1) the molar mass (higher molar mass, lower number of atoms) and (2) the formula of the element (1 atom per formula unit for C or Fe vs. two atoms per molecule for O_2 or H_2). Combining these, you find that that molar mass of iron is the highest compared to the other substances even with only one atom per formula unit, and you predict 1.00 g of Fe would have the smallest number of atoms.

Alternatively, you know that given a molecular formula you can determine a molar mass; given a molar mass you can determine the number of moles; and, given the number of moles you can determine the number of atoms. For choice **(A)**, you can covert the mass of carbon to the moles of carbon by dividing 1.00 gram of carbon by 12.01 grams of carbon per mole: this results in 0.0833 moles of carbon. You can convert moles of carbon to atoms of carbon by multiplying by Avogadro's number: 0.0833 mol C × 6.022×10^{23} atoms C per mol C; this results in 5.01×10^{22} atoms C:

$$(1.00\ \text{g C})\left(\frac{1\ \text{mol C}}{12.01\ \text{g C}}\right)\left(\frac{6.022\times10^{23}\ \text{atoms C}}{1\ \text{mol C}}\right) = 5.01\times10^{22}\ \text{atoms C}$$

You repeat this process for each of the choices:

$$(1.00\ \text{g O}_2)\left(\frac{1\ \text{mol O}_2}{32.00\ \text{g O}_2}\right)\left(\frac{6.022\times10^{23}\ \text{molecule O}_2}{1\ \text{mol O}_2}\right)\left(\frac{2\ \text{atoms O}}{1\ \text{molecule O}_2}\right) = 3.76\times10^{22}\ \text{atoms O}$$

$$(1.00\ \text{g H}_2)\left(\frac{1\ \text{mol H}_2}{2.016\ \text{g H}_2}\right)\left(\frac{6.022\times10^{23}\ \text{molecule H}_2}{1\ \text{mol H}_2}\right)\left(\frac{2\ \text{atoms H}}{1\ \text{molecule H}_2}\right) = 5.97\times10^{23}\ \text{atoms H}$$

$$(1.00\ \text{g Fe})\left(\frac{1\ \text{mol Fe}}{55.85\ \text{g Fe}}\right)\left(\frac{6.022\times10^{23}\ \text{atoms Fe}}{1\ \text{mol Fe}}\right) = 1.08\times10^{22}\ \text{atoms Fe}$$

Therefore, you can conclude that 1.00 g Fe has the ***least*** number of atoms, choice **(D).**

Practice Questions Related to This: **PQ-4, PQ-5, PQ-6, PQ-7**

SQ-3. What is the molar mass of $K_3Fe(C_2O_4)_3$?

(A) 122.96 g·mol^{-1} **(B)** 261.15 g·mol^{-1}

(C) 359.10 g·mol^{-1} **(D)** 437.15 g·mol^{-1}

Knowledge Required: (1) Where to find the average atomic mass. (2) How to determine the number of atoms of each element in a molecular formula. (3) Definition of molar mass.

Thinking it Through: The question asks you to determine the molar mass; therefore, you need to know the moles of each element in the compound. You are given the formula of a compound in the question. From the formula, you determine that there are three potassium atoms and one iron atom in each formula unit of the compound. Next, you determine that the number of carbon and oxygen atoms is greater than their subscripts because C_2O_4 is contained within parentheses with a subscript greater than one. Therefore, the number of carbon atoms per formula unit is the product of the two subscripts, 2 × 3 = 6; and the number of oxygen atoms per formula unit is 4 × 3 = 12. You have now determined that there are three atoms of potassium, one atom of iron, six atoms of carbon, and twelve atoms of oxygen per formula unit of $K_3Fe(C_2O_4)_3$. To determine the molar mass, you multiple the number of moles of each element in the formula unit by the average atomic mass from the periodic table, then add all the masses.

$$(3 \text{ mol K})\left(\frac{39.10 \text{ g K}}{1 \text{ mol K}}\right) + (1 \text{ mol Fe})\left(\frac{55.85 \text{ g Fe}}{1 \text{ mol Fe}}\right) + (6 \text{ mol C})\left(\frac{12.01 \text{ g C}}{1 \text{ mol C}}\right) + (12 \text{ mol O})\left(\frac{16.00 \text{ g O}}{1 \text{ mol O}}\right)$$

$$= 117.30 \text{ g K} + 55.85 \text{ g Fe} + 72.06 \text{ g C} + 192.00 \text{ g O}$$

$$= 437.15 \text{ g } K_3Fe(C_2O_4)_3$$ which is choice **(D)**

Choice **(A)** is not correct because the mole ratios must be used in determining molar mass.
Choice **(B)** is not correct because the subscript of the C_2O_4 unit is not used.
Choice **(C)** is not correct because the subscript of potassium is not used.

Practice Questions Related to This: **PQ-8, PQ-9**

SQ-4. What mass of carbon is present in 1.4×10^{20} molecules of sucrose ($C_{12}H_{22}O_{11}$)? (Molar mass of $C_{12}H_{22}O_{11}$ = 342 g·mol^{-1})

(A) 2.0×10^{22} g **(B)** 1.7×10^{21} g **(C)** 3.3×10^{-2} g **(D)** 2.8×10^{-3} g

Knowledge Required: (1) How to use Avogadro's number. (2) How to determine mole ratios from molecular formulas. (3) Where to find the average atomic mass. (4). How to convert from moles to grams.

Thinking it Through: You are asked to determine the mass of carbon in a given number of sucrose molecules. You know that Avogadro's number is used to convert number of molecules to number of moles.

$$(1.4 \times 10^{20} \text{ molecules sucrose})\left(\frac{1 \text{ mol sucrose}}{6.022 \times 10^{23} \text{ molecules sucrose}}\right) = 2.3 \times 10^{-4} \text{ mol sucrose}$$

From the molecular formula, you can determine that the ratio of moles of carbon to moles of sucrose is twelve to one. Using this ratio, you can determine the number of moles of carbon:

$$(2.3 \times 10^{-4} \text{ mol sucrose})\left(\frac{12 \text{ mol C}}{1 \text{ mol sucrose}}\right) = 2.8 \times 10^{-3} \text{ mol C}$$

The average atomic mass of carbon from the periodic table provides the last needed information to convert the number of moles of carbon to the mass of carbon:

$$(2.8 \times 10^{-3} \text{ mol C})\left(\frac{12.01 \text{ g C}}{1 \text{ mol C}}\right) = 3.3 \times 10^{-2} \text{ g C}$$ which is choice **(C)**

Choice **(B)** is not correct, because the number of molecules must be converted to the number of moles of sucrose before the ratio of moles of carbon to moles of sucrose and the molar mass of carbon can be used to determine the mass of carbon.

Choice **(A)** is not correct, because the number of molecules must be converted to the number of moles of sucrose and then the number of moles of carbon before the molar mass of carbon can be used to determine the mass of carbon.

Choice **(D)** is not correct because 2.8×10^{-3} is the number of moles of carbon in the given amount of sucrose and not the mass of carbon.

Practice Questions Related to This: **PQ-10, PQ-11, PQ-12, PQ-13**

SQ-5. What mass of carbon is present in 0.500 mol of sucrose ($C_{12}H_{22}O_{11}$)?

Molar mass / g·mol⁻¹	
$C_{12}H_{22}O_{11}$	342

(A) 6.01 g **(B)** 72.0 g **(C)** 144. g **(D)** 288. g

Knowledge Required: (1) How to determine mole ratios from molecular formulas. (2) Where to find the average atomic mass. (3) How to convert from moles to mass (in grams).

Thinking it Through: You are asked to determine the mass of an element given a molecular formula and molar mass. You begin by determining the number of moles of carbon in 0.500 mol of the compound using mole ratios from the molecular formula where you see that for each molecule of sucrose, there are 12 atoms of carbon ($C_{\underline{12}}H_{22}O_{11}$) or for each mole of sucrose molecules, there are 12 moles of carbon atoms:

$$(0.500 \text{ mol sucrose})\left(\frac{12 \text{ mol C}}{1 \text{ mol sucrose}}\right) = 6.00 \text{ mol C}$$

Next, the number of moles of carbon is multiplied by the molar mass of carbon to determine the mass of carbon:

$$(6.00 \text{ mol C})\left(\frac{12.01 \text{ g C}}{1 \text{ mol C}}\right) = 72.0 \text{ g C}$$

The mass of carbon in 0.500 moles of sucrose is 72.0 g, choice **(B)**.

Choice **(A)** is not correct because the 0.500 refers to the moles of sucrose and not the moles of carbon.

Choice **(C)** is not correct because there are 0.500 moles of sucrose and not 1.00 mole of sucrose; therefore, there are half the moles of carbon and half the mass of carbon.

Choice **(D)** is not correct because the number of moles of carbon is not obtained by dividing the atomic ratio by the number of moles of the compound.

Practice Questions Related to This: **PQ-14, PQ-15, PQ-16, PQ-17**

SQ-6. What is the mass percent of oxygen in K_2CO_3? (Molar mass of K_2CO_3 = 138.2 g·mol⁻¹)

(A) 11.58% **(B)** 23.15% **(C)** 34.73% **(D)** 69.46%

Knowledge Required: (1) Atomic ratios from molecular formulas. (2) Where to find the average atomic mass.

Thinking it Through: You are asked to determine the mass percent of a particular element in a given molecular formula; you should note that the molar mass is provided and therefore you do not need to calculate it independently. You begin by reminding yourself of the definition of percent composition by mass:

$$\text{percent composition by mass} = \frac{\text{mass of one element}}{\text{total mass of the sample}} \times 100\%$$

The next thing you determine is which sample size you would like to use (for the "total mass of the sample"). You could use 100 g (and then need to determine the mass of oxygen in 100 g) or you could use 1 mole of the compound, which is given in the molar mass). You choose to use the molar mass as the sample size (138.2 g), so you now need to determine the mass of oxygen in 1 mole of K_2CO_3:

In 1 mole of K_2CO_3, there are 3 moles of O: $3 \text{ mol O}\left(\frac{16.00 \text{ g O}}{1 \text{ mol O}}\right) = 48.00 \text{ g O}$

You now determine the percent composition by mass of oxygen:

$$\left(\frac{48.00 \text{ g O}}{138.2 \text{ g } K_2CO_3}\right) \times 100\% = 34.73\%$$ which is choice **(C)**.

Choice **(A)** is not correct because the number of moles of oxygen were not used.

Choice **(B)** is not correct because the mass percent of oxygen in the compound is based on the number of moles of oxygen in the compound and not the number of moles of molecular oxygen (i.e., O_2) in the compound.

Choice **(D)** is not correct because, again, the mass percent is based on moles of oxygen and not molecular oxygen. To arrive at this answer, one would multiply the molar mass of molecular oxygen (32 g·mol⁻¹) by the subscript coefficient (i.e., 3) and then divide by the molar mass of the compound.

Practice Questions Related to This: **PQ-18 PQ-19, PQ-20, PQ-21, PQ-22, PQ-23, PQ-24, PQ-25, PQ-26**

SQ-7. A 4.08 g sample of a compound of nitrogen and oxygen contains 3.02 g of oxygen. What is the empirical formula?

(A) NO_2 **(B)** NO_3 **(C)** N_3O_2 **(D)** N_2O_5

Knowledge Required: (1) Where to find the average atomic mass. (2) Converting from mass to moles. (3) Determining atomic ratios.

Thinking it Through: You are asked to determine the empirical formula for a given mass of the two elements in a compound. You are told that there are only two elements in the compound. You are also told that the mass of oxygen is 3.02 g and the overall mass of the compound is 4.08 g; from this information you can determine that the mass of nitrogen is 1.06 g by subtracting the mass of oxygen from the total mass of the compound. From here, you divide the mass of each element by the average atomic mass of each:

$$(3.02\text{ g O})\left(\frac{1\text{ mol O}}{16.00\text{ g O}}\right) = 0.1888\text{ mol O}$$

$$(1.06\text{ g N})\left(\frac{1\text{ mol N}}{14.01\text{ g N}}\right) = 0.07566\text{ mol N}$$

To convert these mole values to an empirical formula, you need to determine the molar ratio. To achieve this, you choose the element with the least number of moles as the reference element and divide the number of moles of the other element by the least number of moles to determine the atomic ratio; this makes the element with the least number of moles a value of 1 mole in the mole ratio and all other mole values greater than 1 mole:

$$N_{0.07566}O_{0.1888} \rightarrow N_{\frac{0.07566}{0.07566}}O_{\frac{0.1888}{0.07566}} \rightarrow N_1O_{2.495} \rightarrow N_{1\times2}O_{2.498\times2} \rightarrow N_2O_5$$

As shown in the sequence above, this resulted in a formula of $N_1O_{2.5}$; however, these are not whole numbers. To achieve a correct empirical formula, you doubled the numbers such that the resulting formula is N_2O_5, choice **(D)**.

Choice **(A)** is not correct because when the atomic ratios are not close to whole numbers (e.g., 2.495 to 2.000 for this question), the ratio cannot be rounded down.

Choice **(B)** is not correct because empirical formulas are not based on mass ratios.

Choice **(C)** is not correct because 4.08 g was used as the mass of nitrogen.

Practice Questions Related to This: **PQ-27, PQ-28**

SQ-8. Ibuprofen, a common pain reliever, contains 75.7% carbon, 15.5% oxygen, and 8.80% hydrogen by mass. What is the empirical formula?

(A) C_6H_9O **(B)** C_9HO_2 **(C)** $C_{13}H_{18}O_2$ **(D)** $C_{76}H_9O_{16}$

Knowledge Required: (1) Where to find the average atomic mass. (2) Converting from mass to moles. (3) Determining atomic ratios.

Thinking it Through: You are asked to determine the empirical formula from mass percent values for three elements in Ibuprofen (i.e., carbon, oxygen, and hydrogen). First consider the mass of carbon, oxygen, and hydrogen for an arbitrary amount of the compound; in this case, you select 100 g to use the percent composition values directly from the problem.

If you had 100 g of Ibuprofen, you would have 75.5 g C, 15.5 g O, and 8.80 g H. Next, you determine the number of moles of each of the elements by dividing by the molar mass:

$$(75.7\text{ g C})\left(\frac{1\text{ mol C}}{12.01\text{ g C}}\right) = 6.30\text{ mol C}$$

$$(15.5\text{ g O})\left(\frac{1\text{ mol O}}{16.00\text{ g O}}\right) = 0.969\text{ mol O}$$

$$(8.80\text{ g C})\left(\frac{1\text{ mol C}}{1.008\text{ g C}}\right) = 8.73\text{ mol H}$$

To convert these mole values to an empirical formula, you want whole numbers for mole ratios. To achieve this, you choose the element with the least number of moles as the reference element and divide the number of moles of the remaining elements by the least number of moles to determine mole ratios.

$$\frac{6.30\text{ mol C}}{0.969\text{ mol O}} = \frac{6.50\text{ mol C}}{1.00\text{ mol O}} \qquad \frac{8.73\text{ mol H}}{0.969\text{ mol O}} = \frac{9.00\text{ mol H}}{1.00\text{ mol O}}$$

Alternatively, you can go directly to the formula using the number of moles you calculated:

$$C_{6.30}H_{8.73}O_{0.969} \rightarrow C_{\frac{6.30}{0.969}}H_{\frac{8.73}{0.969}}O_{\frac{0.969}{0.969}} \rightarrow C_{6.5}H_9O_1 \rightarrow C_{6.5\times2}H_{9\times2}O_{1\times2} \rightarrow C_{13}H_{18}O_2$$

As shown in the sequence above, this resulted in a formula $C_{6.5}H_9O$; however, these are not whole numbers. To achieve a correct empirical formula, you double the values such that the resulting formula is $C_{13}H_{18}O_2$, choice **(C)**.

Choice **(A)** is not correct because the resulting atomic ratios cannot be rounded.
Choice **(B)** is not correct because the mole ratios are not reduced mass percent values.
Choice **(D)** is not correct because the mole ratios are not mass percent values.

Practice Questions Related to This: **PQ-28, PQ-29, PQ-30**

Practice Questions (PQ)

Conceptual **PQ-1.** Which sample has the largest mass?

(A) 1 mole of marshmallows
(B) 1 mole of Pb (lead) atoms
(C) 1 mole of CO_2 (carbon dioxide) molecules
(D) All of these have the same mass.

Conceptual **PQ-2.** A 10 gram sample of which substance contains the greatest number of hydrogen atoms?

(A) 10 grams of CH_4 **(B)** 10 grams of HCl **(C)** 10 grams of H_2 **(D)** 10 grams of PH_3

PQ-3. The number of atoms in 9.0 g of aluminum is the same as the number of atoms in

(A) 8.1 g of Mg. **(B)** 9.0 g of Mg. **(C)** 12.1 of Mg. **(D)** 18.0 g Mg.

PQ-4. A single molecule of a certain compound has a mass of 3.4×10^{-22} g. Which value comes closest to the mass of a mole of this compound?

(A) 50 g·mol^{-1} **(B)** 100 g·mol^{-1} **(C)** 150 g·mol^{-1} **(D)** 200 g·mol^{-1}

PQ-5. 25.0 g of carbon dioxide contains _____ oxygen atoms. (Molar mass of CO_2 = 44.0 g·mol^{-1})

(A) 1.00×10^{24} **(B)** 6.80×10^{23} **(C)** 6.02×10^{23} **(D)** 3.40×10^{23}

PQ-6. How many atoms are in 1.50 g of Al?

(A) 0.0556 **(B)** 18.0 **(C)** 3.35×10^{22} **(D)** 2.44×10^{25}

PQ-7. A cup of coffee contains 95.0 mg of caffeine, $C_8H_{10}N_4O_2$. How many caffeine molecules are in one cup? (Molar mass of $C_8H_{10}N_4O_2$ = 194.2 g·mol^{-1})

(A) 3.06×10^{-23} molecules
(B) 2.95×10^{20} molecules
(C) 1.11×10^{25} molecules
(D) 1.23×10^{27} molecules

Conceptual **PQ-8.** A sample of a compound of xenon and fluorine contains molecules of a single type; XeF_n, where n is a whole number. If 9.03×10^{20} of these XeF_n molecules have a mass of 0.311 g, what is the value of n?

(A) 2 **(B)** 3 **(C)** 4 **(D)** 6

PQ-9. A 3.41×10^{-6} g sample is known to contain 4.67×10^{16} molecules. What is this compound?

(A) CH_4 **(B)** CO_2 **(C)** H_2O **(D)** NH_3

PQ-10. How many sodium ions are in 25.0 g of Na_2CO_3? (Molar mass of Na_2CO_3 = 105.99 g·mol^{-1})

(A) 1.13×10^{20} ions **(B)** 7.10×10^{20} ions **(C)** 1.42×10^{23} ions **(D)** 2.84×10^{23} ions

PQ-11. A typical silicon chip, such as those in electronic calculators, has a mass of 2.3×10^{-4} g. Assuming the chip is pure silicon, how many silicon atoms are in the chip?

(A) 4.9×10^{18} **(B)** 1.4×10^{20} **(C)** 3.9×10^{21} **(D)** 2.6×10^{27}

PQ-12. A single tablet of regular strength Tylenol contains 325. mg of acetaminophen ($C_8H_9NO_2$). How many moles of acetaminophen are in a single tablet of Tylenol? (Molar mass of $C_8H_9NO_2$ = 151.2 g·mol^{-1})

(A) 2.15×10^{-3} **(B)** 0.465 **(C)** 2.15 **(D)** 465.

PQ-13. How many moles of hydrogen atoms are in six moles of $Ca(OH)_2$?

(A) 2 **(B)** 6 **(C)** 8 **(D)** 12

PQ-14. What is the mass of 1.75 moles of Zn?

(A) 0.0268 g **(B)** 37.4 g **(C)** 65.4 g **(D)** 114 g

PQ-15. 3TC is a small molecule, antiretroviral medication ($C_8H_{11}N_3O_3S$). What mass (in g) of nitrogen is in 7.43×10^{-4} moles of 3TC? (Molar mass of $C_8H_{11}N_3O_3S$ = 229.26 g·mol^{-1})

(A) 3.47×10^{-3} g **(B)** 3.12×10^{-2} g **(C)** 1.70×10^{-1} g **(D)** 5.11×10^{-1} g

PQ-16. Guaifenesin ($C_{10}H_{14}O_4$) is a common over-the-counter expectorant. What mass (in g) of oxygen is in 200. mg dose of Guaifenesin? (Molar mass of $C_{10}H_{14}O_4$ = 198.2 g·mol^{-1})

(A) 0.00403 g **(B)** 0.0161 g **(C)** 0.0646 g **(D)** 3.20 g

PQ-17. A sample of ethanol (C_2H_6O) contains 3.024 g of hydrogen. How many moles of carbon are in the sample? (Molar mass of C_2H_6O = 46.07 g·mol^{-1})

(A) 8.333×10^{-2} **(B)** 0.1667 **(C)** 0.2500 **(D)** 0.9980

PQ-18. What mass (g) of oxygen does 25.5 g of aluminum carbonate, $Al_2(CO_3)_3$, contain? (Molar mass of $Al_2(CO_3)_3$ = 233.99 g·mol^{-1})

(A) 1.59 g **(B)** 1.74 g **(C)** 5.23 g **(D)** 15.7 g

Conceptual **PQ19.** In the compound $(NH_4)_2S_2O_3$, which element is present in the largest percent by mass?

(A) H **(B)** N **(C)** O **(D)** S

Conceptual **PQ-20.** Which compound contains the largest percent by mass of oxygen?

(A) CO_2 **(B)** NO_2 **(C)** SO_2 **(D)** SiO_2

Conceptual **PQ-21.** Which compound has the highest percent composition by mass of oxygen?

(A) $CaCO_3$ **(B)** $CaSO_3$ **(C)** Li_2CO_3 **(D)** Li_2SO_3

Conceptual **PQ-22.** A 10.000 g sample of water contains 11.19% H by mass. What would be the %H in a 20.000 g sample of water?

(A) 5.560% **(B)** 11.19% **(C)** 22.38% **(D)** 88.81%

PQ-23. What is the mass percent of oxygen in $Fe_2(SO_4)_3$?

(A) 15.40% **(B)** 18.76% **(C)** 30.80% **(D)** 48.01%

PQ-24. What is the percent by mass of oxygen in $Ca(NO_3)_2$?

(A) 29.3% **(B)** 47.1% **(C)** 58.5% **(D)** 94.1%

Conceptual **PQ-25.** Which compound contains the greatest percentage of nitrogen?

(A) KNO_3 **(B)** KNO_2 **(C)** $NaNO_3$ **(D)** $NaNO_2$

PQ-26. What is the mass percent of carbon in isooctane (C_8H_{18}), a common component of gasoline?

(A) 7.743% **(B)** 15.88% **(C)** 84.12% **(D)** 92.26%

PQ27. What is the empirical formula of a compound of carbon, hydrogen, and oxygen that contains 51.56% carbon and 14.09% hydrogen by mass?

(A) C_2H_7O **(B)** $C_4H_{13}O_2$ **(C)** $C_7H_2O_5$ **(D)** $C_8H_{13}O_2$

PQ-28. Tetrahydrocannabinol (THC), the active agent in marijuana, contains 80.16% carbon, 9.63% hydrogen, and 10.17% oxygen by mass. What is the empirical formula of THC?

(A) $C_7H_{10}O$ **(B)** C_8HO **(C)** $C_{21}H_{30}O_2$ **(D)** $C_{30}H_3O_5$

PQ-29. What is the empirical formula for a compound that is comprised of 54.0% sodium, 8.50% boron, and 37.5% oxygen by mass?

(A) $Na_2B_2O_3$ **(B)** Na_3BO_3 **(C)** $Na_3B_2O_2$ **(D)** Na_4BO_4

PQ-30. A molecular compound is found to consist of 30.4% nitrogen and 69.6% oxygen. If the molecule contains 2 atoms of nitrogen, what is the molar mass of the molecule?

(A) 46 $g{\cdot}mol^{-1}$ **(B)** 92 $g{\cdot}mol^{-1}$ **(C)** 154 $g{\cdot}mol^{-1}$ **(D)** 168 $g{\cdot}mol^{-1}$

Answers to Study Questions

1. D
2. D
3. D
4. C
5. B
6. C
7. D
8. C

Answers to Practice Questions

1. A
2. C
3. A
4. D
5. B
6. C
7. B
8. C
9. B
10. D
11. A
12. A
13. D
14. D
15. B
16. C
17. D
18. D
19. D
20. A
21. C
22. B
23. D
24. C
25. D
26. C
27. B
28. C
29. B
30. B

A C S
Exams

Chapter 4 – Stoichiometry

Chapter Summary:

This chapter will focus on balancing chemical equations and using mole ratios to determine the amount of products formed or the amount of starting materials necessary for a given chemical reaction.

Specific topics covered in this chapter are:

- Balancing chemical reactions
- Converting between moles of starting material(s) and moles of product(s)
- Limiting reactant, theoretical yield, experimental yield, percent yield

Previous material that is relevant to your understanding of questions in this chapter include:

- Conversions (***Toolbox***)
- Formula mass (***Chapter 2***)
- Molar mass (***Chapter 2***)
- The mole (***Chapter 2***)

Common representations used in questions related to this material:

Name	Example	Used in questions related to
Particulate representations	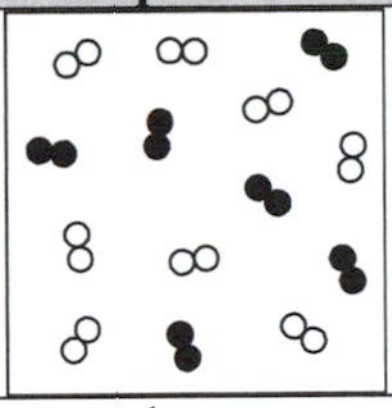	Balancing reactions and limiting reactant
Compound units	$g \cdot mol^{-1}$	molar mass

Where to find this in your textbook:

The material in this chapter typically aligns to "Stoichiometry of Chemical Reactions" in your textbook. The name of your chapter(s) may vary.

Practice exam:

There are practice exam questions aligned to the material in this chapter. Because there are a limited number of questions on the practice exam, a review of the breadth of the material in this chapter is advised in preparation for your exam.

How this fits into the big picture:

The material in this chapter aligns to the Big Idea of Atoms (1) and Reactions (5) as listed on page 12 of this study guide.

Study Questions (SQ)

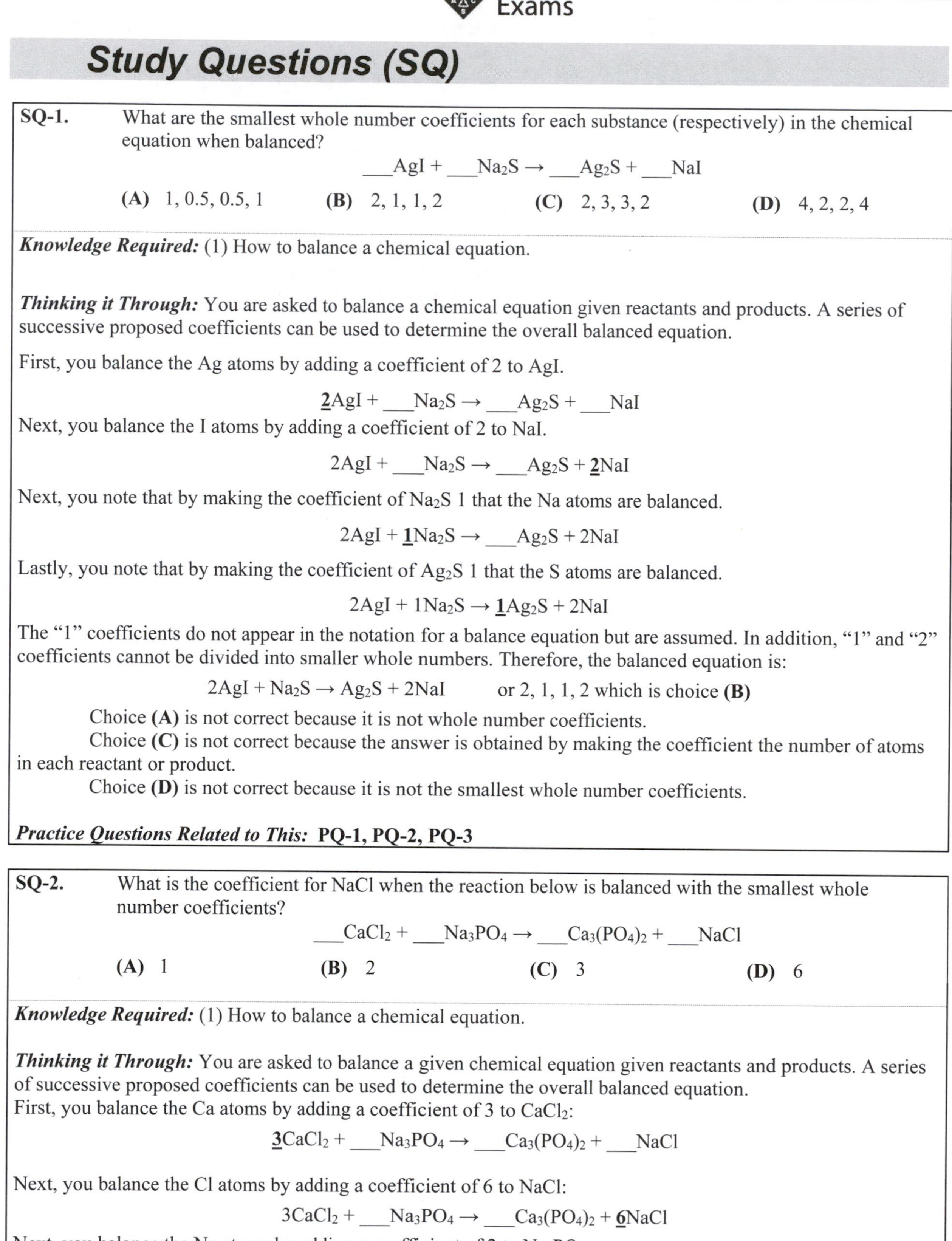

SQ-1. What are the smallest whole number coefficients for each substance (respectively) in the chemical equation when balanced?

$$\_\_AgI + \_\_Na_2S \rightarrow \_\_Ag_2S + \_\_NaI$$

(A) 1, 0.5, 0.5, 1 **(B)** 2, 1, 1, 2 **(C)** 2, 3, 3, 2 **(D)** 4, 2, 2, 4

Knowledge Required: (1) How to balance a chemical equation.

Thinking it Through: You are asked to balance a chemical equation given reactants and products. A series of successive proposed coefficients can be used to determine the overall balanced equation.

First, you balance the Ag atoms by adding a coefficient of 2 to AgI.

$$\underline{2}AgI + \_\_Na_2S \rightarrow \_\_Ag_2S + \_\_NaI$$

Next, you balance the I atoms by adding a coefficient of 2 to NaI.

$$2AgI + \_\_Na_2S \rightarrow \_\_Ag_2S + \underline{2}NaI$$

Next, you note that by making the coefficient of Na_2S 1 that the Na atoms are balanced.

$$2AgI + \underline{1}Na_2S \rightarrow \_\_Ag_2S + 2NaI$$

Lastly, you note that by making the coefficient of Ag_2S 1 that the S atoms are balanced.

$$2AgI + 1Na_2S \rightarrow \underline{1}Ag_2S + 2NaI$$

The "1" coefficients do not appear in the notation for a balance equation but are assumed. In addition, "1" and "2" coefficients cannot be divided into smaller whole numbers. Therefore, the balanced equation is:

$$2AgI + Na_2S \rightarrow Ag_2S + 2NaI$$ or 2, 1, 1, 2 which is choice **(B)**

Choice **(A)** is not correct because it is not whole number coefficients.

Choice **(C)** is not correct because the answer is obtained by making the coefficient the number of atoms in each reactant or product.

Choice **(D)** is not correct because it is not the smallest whole number coefficients.

Practice Questions Related to This: **PQ-1, PQ-2, PQ-3**

SQ-2. What is the coefficient for NaCl when the reaction below is balanced with the smallest whole number coefficients?

$$\_\_CaCl_2 + \_\_Na_3PO_4 \rightarrow \_\_Ca_3(PO_4)_2 + \_\_NaCl$$

(A) 1 **(B)** 2 **(C)** 3 **(D)** 6

Knowledge Required: (1) How to balance a chemical equation.

Thinking it Through: You are asked to balance a given chemical equation given reactants and products. A series of successive proposed coefficients can be used to determine the overall balanced equation.

First, you balance the Ca atoms by adding a coefficient of 3 to $CaCl_2$:

$$\underline{3}CaCl_2 + \_\_Na_3PO_4 \rightarrow \_\_Ca_3(PO_4)_2 + \_\_NaCl$$

Next, you balance the Cl atoms by adding a coefficient of 6 to NaCl:

$$3CaCl_2 + \_\_Na_3PO_4 \rightarrow \_\_Ca_3(PO_4)_2 + \underline{6}NaCl$$

Next, you balance the Na atoms by adding a coefficient of 2 to Na_3PO_4:

$$3CaCl_2 + \underline{2}Na_3PO_4 \rightarrow \_\_Ca_3(PO_4)_2 + 6NaCl$$

To balance the number of PO_4^{3-} ions, $Ca_3(PO_4)_2$ will have a coefficient of 1:

$$3CaCl_2 + 2Na_3PO_4 \rightarrow \underline{1}Ca_3(PO_4)_2 + 6NaCl$$

The "1" coefficients do not appear in the notation for a balance equation but are assumed. In addition, "1," "3," and "6" coefficients cannot be divided into smaller whole numbers. The balanced equation is:

$$3CaCl_2 + 2Na_3PO_4 \rightarrow Ca_3(PO_4)_2 + 6NaCl$$

The question asked for the coefficient on NaCl, which is "6," Choice **(D)**

Choice **(A)** is not correct because it is the coefficient of $Ca_3(PO_4)_2$.
Choice **(B)** is not correct because it is the coefficient of Na_3PO_4.
Choice **(C)** is not correct because it is the coefficient of $CaCl_3$.

Practice Questions Related to This: **PQ-4, PQ-5, PQ-6**

SQ-3. Determine the number of moles of carbon dioxide (CO_2) formed when 2.50 mol ethanol (C_2H_6O) undergoes combustion in excess oxygen.

$$C_2H_6O(l) + 3O_2(g) \rightarrow 2CO_2(g) + 3H_2O(g)$$

(A) 1.25 mol **(B)** 1.67 mol
(C) 2.50 mol **(D)** 5.00 mol

Knowledge Required: (1) How to use coefficients of a balanced chemical reaction to determine mole ratios.

Thinking it Through: The question asks you to determine the number of moles of product given a specific number of moles of a reactant (ethanol). The question prompt states that one of the two reactants is in excess (oxygen); this is important because it identifies that ethanol is then the limiting reactant. You begin by noting that 2.50 moles of ethanol is combusted in the presence of oxygen; in the balanced equation, ethanol has a coefficient of 1. You also note that the coefficient of carbon dioxide is 2. Therefore, the ratio of moles of ethanol consumed in the reaction to the moles of carbon dioxide produced is 1 to 2; you use this ratio to calculate the number of moles of CO_2 given 2.50 mol ethanol.

$$2.50 \text{ mol ethanol}\left(\frac{2 \text{ mol } CO_2}{1 \text{ mol ethanol}}\right) = 5.00 \text{ mol } CO_2 \quad \text{Choice } \mathbf{(D)}$$

Choice **(A)** is not correct because the answer is obtained by inverting the mole ratio (1:2 instead of 2:1).
Choice **(B)** is not correct because the answer is obtained by using the mole ratio of carbon dioxide to oxygen (2:3).
Choice **(C)** is not correct because the mole ratio of ethanol to carbon dioxide is not 1:1.

Practice Questions Related to This: **PQ-7, PQ-8, PQ-9, PQ-10**

SQ-4. What is the mass (in g) of ammonia (NH_3) necessary to produce 0.340 mol NO in the presence of excess oxygen?

$$4NH_3 + 5O_2 \rightarrow 4NO + 6H_2O$$

Molar Mass / $g \cdot mol^{-1}$	
H_2O	18.02
NH_3	17.03
NO	30.01
O_2	32.00

(A) 1.45 g **(B)** 5.21 g **(C)** 5.79 g **(D)** 10.2 g

Knowledge Required: (1) How to use coefficients of a balanced chemical reaction to determine mole ratios. (2) How to convert from moles to mass.

Thinking it Through: You are asked to determine the mass of a starting material needed to produce a given amount (in moles) of a product. You are told that the other starting material is in excess and therefore does not limit the amount of product produced. The two species of interest are ammonia (a starting material) and NO (a product); from the balanced equation,the mole ratio of ammonia to NO is 4:4.

$$0.340 \text{ mol NO}\left(\frac{4 \text{ mol } NH_3}{4 \text{ mol NO}}\right)\left(\frac{17.03 \text{ g } NH_3}{1 \text{ mol } NH_3}\right) = 5.79 \text{ g } NH_3 \quad \text{Choice } \mathbf{(C)}$$

Choice **(A)** is not correct because the answer is obtained using only the coefficient of NO and assuming the coefficient of NH_3 is 1 (so uses the ratio of 4:1).

Choice **(B)** is not correct because the answer is obtained by using a 9:10 ratio that is the ratio of the sum of the starting material coefficients to the sum of the product coefficients.

Choice **(D)** is not correct because the answer is obtained using the molar mass of NO instead of NH_3.

Practice Questions Related to This: **PQ-11, PQ-12, PQ-13, PQ-14, PQ-15, PQ-16, PQ-17, PQ18, PQ-19**

SQ-5. What is the theoretical yield of NO that can be produced from the reaction of 3.50 mol NH_3 and 4.01 mol O_2?

$$4NH_3 + 5O_2 \rightarrow 4NO + 6H_2O$$

(A) 3.21 mol **(B)** 3.50 mol **(C)** 4.46 mol **(D)** 5.01 mol

Knowledge Required: (1) How to use coefficients of a balanced chemical reaction to determine mole ratios. (2) How to determine limiting reactant and theoretical yield.

Thinking it Through: You are asked to determine the theoretical yield given amounts of both reactants. This is a limiting reactant question. For each reactant, you will calculate the potential number of moles of product; you will assume that the other reactant is in excess. The lowest calculated amount of product will be the theoretical yield.

$$3.50 \text{ mol } NH_3\left(\frac{4 \text{ mol NO}}{4 \text{ mol } NH_3}\right) = 3.50 \text{ mol NO}$$

$$4.01 \text{ mol } O_2\left(\frac{4 \text{ mol NO}}{5 \text{ mol } O_2}\right) = 3.21 \text{ mol NO}$$

Considering these two results, the 4.01 mol O_2 will lead to a smaller amount of NO; therefore, this amount of NO is the theoretical yield given the amounts of reactants, and choice **(A)** is correct.

Choice **(B)** is not correct because it is the amount of NO produced based on using NH_3 as the limiting reactant; however, NH_3 is not the limiting reactant.

Choice **(C)** is not correct because the answer is obtained by using a 9:10 ratio that is the ratio of the sum of the reactant coefficients to the sum of the product coefficients.

Choice **(D)** is not correct because the answer is obtained with the inverted mole ratio of O_2 to NO.

Practice Questions Related to This: **PQ-20, PQ-21, PQ-22, PQ-23, PQ-24, PQ-25, PQ-26**

SQ-6. What is the mass (in g) of TiO_2 formed when 0.476 g $TiCl_4$ are reacted with 0.0501 g H_2O?

$$TiCl_4 + 2H_2O \rightarrow TiO_2 + 4HCl$$

Molar Mass / g·mol^{-1}	
H_2O	18.02
HCl	36.46
$TiCl_4$	189.7
TiO_2	79.88

(A) 0.111 g **(B)** 0.200 g **(C)** 0.222 g **(D)** 1.13 g

Knowledge Required: (1) How to use coefficients of a balanced chemical reaction to determine mole ratios. (2) How to convert from moles to mass. (3) How to determine limiting reactant and theoretical yield.

Thinking it Through: You are asked to determine the mass of a product formed (theoretical yield) given the mass of each of the reactants. You will need to therefore determine the limiting reactant in order to determine the theoretical yield. For each reactant, you will calculate the number of moles, potential number of moles of product, and potential mass of product; you will assume that the other reactant is in excess for each. The lowest mass of product calculated will be the mass of the product formed.

$$4.76\text{ g TiCl}_4\left(\frac{1\text{ mol TiCl}_4}{189.7\text{ g TiCl}_4}\right)\left(\frac{1\text{mol TiO}_2}{1\text{ mol TiCl}_4}\right)\left(\frac{79.88\text{ g TiO}_2}{1\text{ mol TiO}_2}\right) = 0.200\text{ g TiO}_2$$

$$0.0501\text{ g H}_2\text{O}\left(\frac{1\text{ mol H}_2\text{O}}{18.02\text{ g H}_2\text{O}}\right)\left(\frac{1\text{mol TiO}_2}{2\text{ mol H}_2\text{O}}\right)\left(\frac{79.88\text{ g TiO}_2}{1\text{ mol TiO}_2}\right) = 0.111\text{ g TiO}_2$$

Considering these two results, the 0.0501 g of H_2O led to a smaller mass of TiO_2; therefore, this amount of TiO_2 is the maximum amount of mass that could be formed given the amounts of reactant and is the theoretical yield. Water is the limiting reagent. Choice **(A)**.

Choice **(B)** is not correct because it is the mass of TiO_2 produced based on using $TiCl_4$ as the limiting reactant; however, $TiCl_4$ is not the limiting reactant.

Choice **(C)** is not correct because the mole ratio of TiO_2 to H_2O was incorrectly determined or assumed to be 1:1.

Choice **(D)** is not correct because the answer is obtained based on using $TiCl_4$ as the limiting reactant and inverting the molar masses of both $TiCl_4$ and TiO_2.

Practice Questions Related to This: **PQ-27, PQ-28, PQ-29**

SQ-7. *Conceptual*

An initial reaction mixture of N_2 and H_2 is shown to the right. Which representation shows what would be present after the reaction is complete?

$$N_2(g) + 3H_2(g) \rightarrow 2NH_3(g)$$

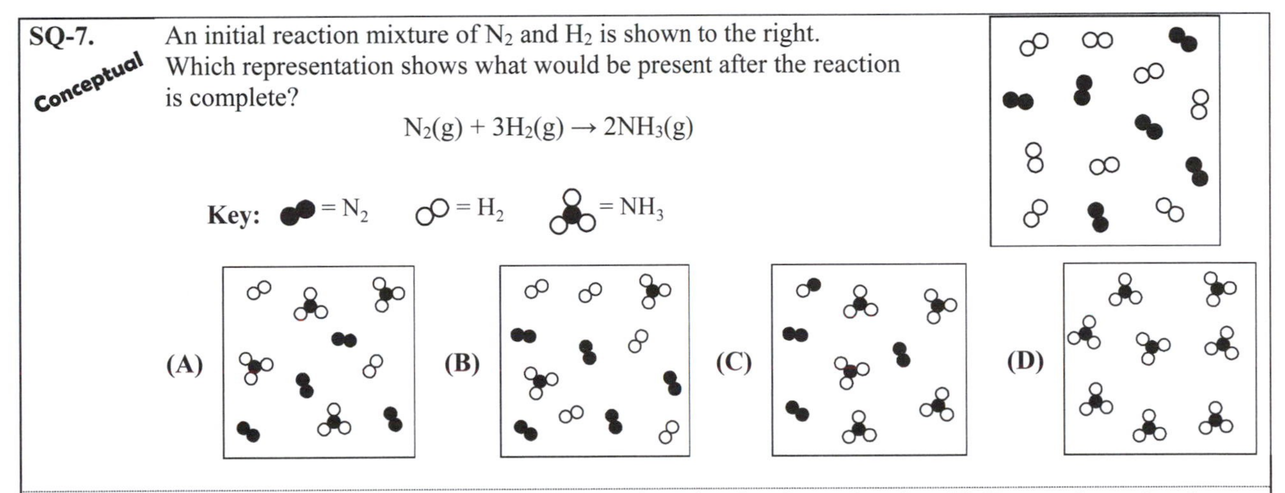

Knowledge Required: (1) The relationship between number of molecules and moles. (2) How to determine the limiting reagent using number of molecules.

Thinking it Through: You are asked to choose which particulate nature of matter representation shows what would be present after a given chemical reaction. You will use the number of molecules and atoms shown in the representation to determine the limiting reagent for the reaction. Given the notation about hydrogen and nitrogen shown in the representation, you can determine that there are six molecules of nitrogen and eight molecules of hydrogen. Based on the balanced equation, hydrogen will be the limiting reagent. Six molecules of hydrogen will react with two molecules of nitrogen to form four molecules of NH_3. This will leave two unreacted hydrogen molecules and four unreacted nitrogen molecules. From inspecting the answer options, the representation that best depicts the outcome of this reaction is Choice **(A)**. You might also see a question like this where there is one more molecule of H_2 which would then react and make two more molecules of ammonia.

Choice **(B)** is not correct because the representation does not show the reaction at completion.

Choice **(C)** is not correct because the representation shows a lone NH molecule that is not given in the chemical reaction.

Choice **(D)** is not correct because the representation does not account for unreacted starting materials and has too many product molecules given the amount of starting materials.

Practice Questions Related to This: **PQ-30**

Practice Questions (PQ)

Conceptual **PQ-1.** The limiting reagent for a given reaction can be recognized because it is the reagent that

(A) has the smallest coefficient in the balanced equation for the reaction.

(B) has the smallest mass in the reaction mixture.

(C) is present in the smallest molar quantity.

(D) would be used up first.

PQ-2. What are the coefficients for each substance (respectively) in the chemical reaction when balanced?

$$\_\_\_C_7H_{14}O_2 + \_\_\_ O_2 \rightarrow \_\_\_CO_2 + \_\_\_H_2O$$

(A) 2, 21, 14, 14 **(B)** 2, 19, 14, 14 **(C)** 2, 19, 7, 7 **(D)** 1, 10, 7, 7

PQ-3. What are the smallest whole number coefficients for each substance (respectively) in the chemical reaction when balanced?

$$\_\_\_NaBr + \_\_\_ Cl_2 \rightarrow \_\_\_NaCl + \_\_\_Br_2$$

(A) 1, 1, 2, 2 **(B)** 1, 2, 2, 2 **(C)** 2, 1, 2, 1 **(D)** 2, 4, 2, 4

PQ-4. When this equation is balanced with the smallest set of whole numbers, what is the coefficient for N_2?

$$\_\_\_N_2H_4(g) + \_\_\_N_2O_4(g) \rightarrow \_\_\_N_2(g) + \_\_\_H_2O(g)$$

(A) 1 **(B)** 2 **(C)** 3 **(D)** 4

PQ-5. What is the coefficient for ZnO when the reaction is balanced with smallest whole number coefficients?

$$\_\_\_ZnS + \_\_\_O_2 \rightarrow \_\_\_ZnO + \_\_\_SO_2$$

(A) 1 **(B)** 2 **(C)** 3 **(D)** 4

PQ-6. What are the smallest whole number coefficients for each substance (respectively) in the chemical reaction when balanced?

$$\_\_\_CH_4 + \_\_\_ O_2 \rightarrow \_\_\_CO_2 + \_\_\_H_2O$$

(A) 1, 2, 1, 2 **(B)** 1, 3, 1, 2 **(C)** 2, 4, 2, 4 **(D)** 4, 2, 2, 2

PQ-7. How many moles of iron are necessary to react completely with 1.75 mol of oxygen gas?

$$3O_2(g) + 4Fe(s) \rightarrow 2FeO_3(s)$$

(A) 1.31 mol **(B)** 1.75 mol **(C)** 2.33 mol **(D)** 5.25 mol

PQ-8. What mass of NO_2 is formed by the complete reaction of 2 mol O_3 with excess NO?

$$O_3 + NO \rightarrow O_2 + NO_2$$

Molar mass / g·mol^{-1}	
O_3	48.0
NO_2	46.0

(A) 23.0 g **(B)** 46.0 g **(C)** 92.0 g **(D)** 96.0 g

PQ-9. How many moles of Ba_3N_2 are necessary to react completely with 1.5 mol of water?

$$Ba_3N_2 + 6H_2O \rightarrow 3Ba(OH)_2 + 2NH_3$$

(A) 0.25 mol **(B)** 1 mol **(C)** 6 mol **(D)** 9 mol

PQ-10. How many moles of water are necessary to react completely with 8.0 mol PCl_5?

$$PCl_5 + 4H_2O \rightarrow H_3PO_4 + 5HCl$$

(A) 2.0 mol **(B)** 8.0 mol **(C)** 16 mol **(D)** 32 mol

PQ-11. What mass of SbF_3 (179 g·mol^{-1}) is needed to produce 1.00 g of Freon–12, CCl_2F_2 (molar mass = 121 g·mol^{-1})?

$$3CCl_4 + 2SbF_3 \rightarrow 3CCl_2F_2 + 2SbCl_3$$

(A) 0.667 g **(B)** 0.986 g **(C)** 1.48 g **(D)** 2.22 g

PQ-12. What is the theoretical yield of aluminum chloride that could be obtained from 6.00 mol of barium chloride and excess aluminum sulfate?

$$Al_2(SO_4)_3 + 3BaCl_2 \rightarrow 3BaSO_4 + 2AlCl_3$$

Molar mass / g·mol^{-1}	
$AlCl_3$	133.3
$BaCl_2$	208.23
$Al_2(SO_4)_3$	342.15

(A) 1250 g **(B)** 801 g **(C)** 534 g **(D)** 134 g

PQ-13. A self-contained breathing apparatus uses potassium superoxide, KO_2, to convert the carbon dioxide and water in exhaled air into oxygen.

Molar mass / g·mol^{-1}	
CO_2	44.01

$$4KO_2(s) + 4CO_2(g) + 4H_2O(g) \rightarrow 4KHCO_3(s) + 3O_2(g)$$

How many molecules of oxygen gas will be produced from 0.0468 g of carbon dioxide exhaled in a typical breath in excess KO_2 and water?

(A) 4.8×10^{20} molecules **(B)** 6.4×10^{20} molecules

(C) 8.5×10^{20} molecules **(D)** 1.9×10^{21} molecules

PQ-14. If 365.5 g $H_2O(l)$ is combined with excess KO_2, what mass of $O_2(g)$ is expected to be produced?

$$2KO_2(s) + 2H_2O(l) \rightarrow 2KOH(s) + O_2(g) + H_2O_2(l)$$

Molar mass / g·mol^{-1}	
H_2O	18.02
O_2	32.00

(A) 11.42 g **(B)** 162.4 g **(C)** 324.7 g **(D)** 659.4 g

PQ-15. How many water molecules are produced if 0.34 mol of propane, C_3H_8, combusts in excess oxygen?

$$C_3H_8(g) + 5O_2(g) \rightarrow 3CO_2(g) + 4H_2O(g)$$

(A) 5.1×10^{22} **(B)** 2.1×10^{23} **(C)** 8.2×10^{23} **(D)** 2.4×10^{24}

PQ-16. What is the maximum mass (in g) of NO_2 formed when 7.5 g of N_2O_5 are consumed?

$$2N_2O_5(g) \rightarrow 4NO_2(g) + O_2(g)$$

Molar mass / g·mol^{-1}	
N_2O_5	108.0
NO_2	46.0

(A) 1.6 g **(B)** 3.2 g **(C)** 6.4 g **(D)** 15 g

PQ-17. What mass of Sn is formed by the complete reaction of 1.75 mol H_2 with excess SnO_2? $SnO_2 + 2H_2 \rightarrow Sn + 2H_2O$

Molar mass / g·mol^{-1}	
Sn	118.7

(A) 104 g **(B)** 527 g **(C)** 151 g **(D)** 301 g

PQ-18. What mass of FeS is necessary to form 1.08 g Fe_2O_3 with excess O_2?

$$4FeS + 7O_2 \rightarrow 2Fe_2O_3 + 4SO_2$$

Molar mass / g·mol^{-1}	
FeS	87.92
Fe_2O_3	159.7

(A) 0.0136 g **(B)** 0.298 g **(C)** 1.19 g **(D)** 2.16 g

PQ-19. A 2.32 g sample of $Na_2SO_4 \cdot nH_2O$ yields 1.42 g Na_2SO_4 upon heating.

$$Na_2SO_4 \cdot nH_2O(s) \rightarrow Na_2SO_4(s) + nH_2O(g)$$

What is the value of n?

Molar mass / g·mol^{-1}	
Na_2SO_4	142

(A) 2 **(B)** 3 **(C)** 5 **(D)** 10

PQ-20. A mixture of 9 mol F_2 and 4 mol S are allowed to react.

$$3F_2 + S \rightarrow SF_6$$

How many moles of F_2 remain after 3 mol of S have reacted?

(A) 3 **(B)** 2 **(C)** 1 **(D)** 0

PQ-21. What is the maximum amount of ammonia that can be produced when 0.35 mol N_2 reacts with 0.90 mol H_2?

$$N_2(g) + 3H_2(g) \rightarrow 2NH_3(g)$$

(A) 0.18 mol NH_3 **(B)** 0.60 mol NH_3 **(C)** 0.70 mol NH_3 **(D)** 1.4 mol NH_3

Conceptual **PQ-22.** Equimolar amounts of nitrogen, hydrogen, and argon are placed in a reaction chamber. The amount of which substance(s) will determine the amount of ammonia produced in the reaction below if the reaction goes to completion?

$$N_2(g) + 3H_2(g) \rightarrow 2NH_3(g)$$

(A) argon **(B)** hydrogen **(C)** nitrogen **(D)** nitrogen and hydrogen

Conceptual **PQ-23.** How many total molecules will be present when the reaction of 5,000 molecules of CS_2 with 15,000 molecules of O_2 goes to completion?

$$CS_2(l) + 3\ O_2(g) \rightarrow CO_2(g) + 2SO_2(g)$$

(A) 5,000 **(B)** 10,000 **(C)** 15,000 **(D)** 20,000

PQ-24. What amount of the excess reagent remains when 0.30 mol NH_3 reacts with 0.40 mol O_2 to produce NO and H_2O?

$$4NH_3 + 5O_2 \rightarrow 4NO + 6H_2O$$

(A) 0.10 mol NH_3 **(B)** 0.10 mol O_2 **(C)** 0.025 mol NH_3 **(D)** 0.025 mol O_2

PQ-25. What amount of Al_2O_3 is produced from the reaction of 3.0 mol Al with 2.0 mol Fe_2O_3?

$$2Al + Fe_2O_3 \rightarrow Al_2O_3 + 2Fe$$

(A) 1.5 mol **(B)** 2.0 mol **(C)** 3.0 mol **(D)** 6.0 mol

PQ-26. What amount of the excess reagent remains when 0.500 mol Li reacts with 0.350 mol N_2?

$$6Li(s) + N_2(g) \rightarrow 2Li_3N(s)$$

(A) 0.0833 mol N_2 **(B)** 0.267 mol N_2 **(C)** 0.150 mol Li **(D)** 0.417 mol Li

PQ-27. The combustion of C_3H_8O with O_2 is represented by this equation.

$$2C_3H_8O + 9O_2 \rightarrow 6CO_2 + 8H_2O$$

When 3.00 g C_3H_8O and 7.38 g O_2 are combined, what is the excess reagent and how many moles of that reagent remain?

Molar mass / g·mol^{-1}	
C_3H_8O	60.1
O_2	32.0

(A) 0.006 mol O_2 **(B)** 0.24 mol O_2

(C) 0.024 mol C_3H_8O **(D)** 0.18 mol C_3H_8O

PQ-28. Consider this reaction for the production of lead.

$$2PbO(s) + PbS(s) \longrightarrow 3Pb(s) + SO_2(g)$$

What is the theoretical yield of lead that can be obtained by the reaction of 57.33 g PbO and 33.80 g of PbS?

Molar mass / g·mol⁻¹	
Pb	207.2
PbO	223.2
PbS	239.3

(A) 43.48 g **(B)** 72.75 g **(C)** 79.83 g **(D)** 87.80 g

PQ-29. What amount of excess reagent remains when 4.0 g zinc react with 2.0 g phosphorus?

$$3Zn + 2P \rightarrow Zn_3P_2$$

Molar mass / g·mol⁻¹	
P	30.97
Zn	63.39
Zn_3P_2	258.1

(A) 0.70 g P **(B)** 1.3 g P **(C)** 0.22 g Zn **(D)** 4.2 g Zn

Conceptual **PQ-30.** An initial reaction mixture of O_2 and F_2 is shown to the right. Which representation shows what would be present after the reaction is complete?

$$O_2 + 2F_2 \rightarrow 2OF_2$$

○ = oxygen ● = fluorine

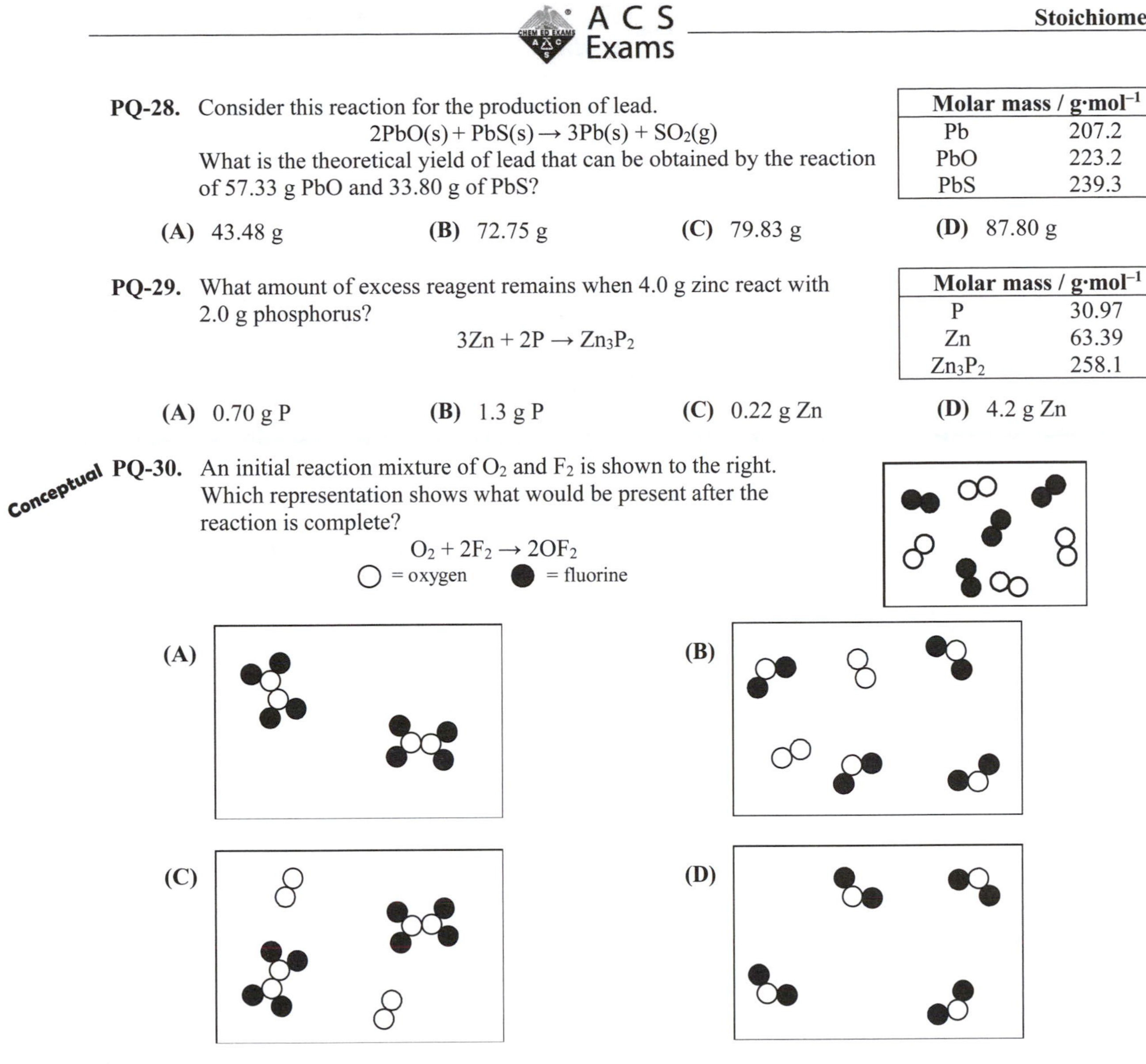

(A) **(B)**

(C) **(D)**

Answers to Study Questions

1. B
2. D
3. D
4. C
5. A
6. A
7. A

Answers to Practice Questions

1. D
2. B
3. C
4. C
5. B
6. A
7. C
8. C
9. A
10. D
11. B
12. C
13. A
14. C
15. C
16. C
17. A
18. C
19. C
20. D
21. B
22. B
23. C
24. D
25. A
26. B
27. A
28. C
29. A
30. B

Chapter 5 – Solutions and Aqueous Reactions, Part 1

Chapter Summary:

This chapter will focus on substances in solutions including qualitative and quantitative properties of solutions and reactions of substances in solution.

Specific topics covered in this chapter are:

- Molar concentration including:
  - Preparation of solutions
  - Dilutions
  - Concentration of ions in solutions
- Solution stoichiometry including:
  - Gravimetric analysis
  - Titrations
  - Limiting reactants, theoretical yield, experimental yield and percent yield
- Reactions
  - Precipitation reactions
  - Acid/base reactions
  - Oxidation and reduction reactions
- Reaction representations
  - Molecular, total and net ionic equations

Previous material that is relevant to your understanding of questions in this chapter include:

- The mole and formula calculations (***Chapter 3***)
- Stoichiometry and balancing equations (***Chapter 4***)

Common representations used in questions related to this material:

Name	Example	Used in questions related to
Particulate representations *(NOTE: Water is often omitted from these representations for clarity.)*	Cl^- Ca^{2+} NO_3^- Cl^- NO_3^- NO_3^- Ca^{2+} Ca^{2+} NO_3^- Ag^+ Cl^- Ag^+ Cl^- Ag^+ Cl^- Ag^+ Cl^-	Reactions, solution properties

Where to find this in your textbook:

The material in this chapter typically aligns to "Aqueous Reactions" in your textbook. The name of your chapter may vary.

Practice exam:

There are practice exam questions aligned to the material in this chapter. Because there are a limited number of questions on the practice exam, a review of the breadth of the material in this chapter is advised in preparation for your exam.

How this fits into the big picture:

The material in this chapter aligns to the Big Idea of Intermolecular Interactions (4) and Reactions (5) as listed on page 12 of this study guide.

Study Questions (SQ)

SQ-1. Which compound is a weak electrolyte when dissolved in water?

(A) HNO_3 **(B)** KOH **(C)** NH_3 **(D)** $NaNO_3$

Knowledge Required: (1) Definition of an electrolyte. (2) Classification of strong vs. weak electrolyte.

Thinking it Through: The definition of an electrolyte is a substance or solution that will conduct electricity. For this question, the solutes are all dissolved in water. You also know the reason the solutions conduct electricity is that the substance (or solute) produces ions when dissolved in water, creating an electrolytic solution. A strong electrolyte will dissociate or ionize completely in water while a weak electrolyte will only partially ionize in water.

Considering the responses therefore, you are looking for a substance that would be classified as a weak acid or base as these substances would partially ionize in water. Ammonia or NH_3 is a weak base, and therefore is a weak electrolyte. The correct choice is **(C)**.

Nitric acid, HNO_3, is a strong acid and therefore a strong electrolyte, so choice **(A)** is incorrect.

Potassium hydroxide, KOH, is a strong base and also a strong electrolyte, so choice **(B)** is incorrect. Potassium hydroxide is also ionic as is sodium nitrate, $NaNO_3$ or choice **(D)**. Ionic compounds when in solution are present as ions, so are strong electrolytes making choice **(D)** incorrect as well.

Practice Questions Related to This: **PQ-1**

SQ-2. What mass (in g) of magnesium nitrate, $Mg(NO_3)_2$, are required to produce 250.0 mL of a 0.0750 M solution?

(A) 0.0188 g **(B)** 0.0445 g **(C)** 1.61 g **(D)** 2.79 g

Knowledge Required: (1) Calculation of molar mass. (2) Definition of molar concentration.

Thinking it Through: The first thing you recognize when reading this question is that you are given a volume (250.0 mL) and molar concentration (0.0750 M). You know that "M" or molar concentration is:

$$\text{molar concentration }(\text{M}) = \frac{\text{mol solute}}{\text{vol (L) solution}} \qquad 0.0750\text{ M} = \frac{0.0750\text{ mol Mg(NO}_3)_2}{1\text{ L solution}}$$

Then, given the volume, this allows you to determine the number of moles of solute ($Mg(NO_3)_2$):

$$\left(\frac{0.0750\text{ mol Mg(NO}_3)_2}{1\text{ L solution}}\right)(0.2500\text{ L solution}) = 0.0188\text{ mol Mg(NO}_3)_2$$

Use the molar mass of magnesium nitrate to determine the mass:

$$(0.0188\text{ mol Mg(NO}_3)_2)\left(\frac{148.33\text{ g Mg(NO}_3)_2}{\text{mol Mg(NO}_3)_2}\right) = 2.79\text{ g Mg(NO}_3)_2$$

This corresponds to choice **(D)**.

Choice **(A)** is not correct because the molar mass was not used (this is the number of moles).

Choice **(B)** is not correct because 250.0 mL was not converted to liters and the number of moles was incorrectly calculated.

Choice **(C)** is not correct because the wrong molar mass was used (for $MgNO_3$, not $Mg(NO_3)_2$).

Practice Questions Related to This: **PQ-2, PQ-3, PQ-4,** and **PQ-5**

SQ-3. What mass (in g) of sodium ions, Na^+, are there in 25 mL of 0.75 M Na_2SO_3?

Molar mass / $g \cdot mol^{-1}$	
Na^+	22.99
Na_2SO_3	126.1

(A) 0.019 g **(B)** 0.87 g **(C)** 2.4 g **(D)** 4.8 g

Knowledge Required: (1) Interpreting a chemical formula. (2) Definition of molar concentration.

Thinking it Through: Similar to **SQ-2** previously, you will use the molar concentration and volume of the solution given in the problem to determine number of moles. Here, you are going to pay special attention to the substance: $\left(\frac{0.75 \text{ mol Na}_2\text{SO}_3}{1 \text{ L solution}}\right)(0.025 \text{ L solution}) = 0.019 \text{ mol Na}_2\text{SO}_3$

From here, if you used the molar mass of Na_2SO_3, you would determine the mass of sodium sulfite needed to make this solution, which is not what the question is asking. Rather, you are going to use the chemical formula and the ratio of sodium ions (Na^+) to the formula unit (Na_2SO_3) where the "2" subscript tells you there are 2 Na^+ to every 1 Na_2SO_3 or 2 mol Na^+ to 1 mol Na_2SO_3. Using this, you determine the number of moles of Na^+ and then the mass of Na^+: $(0.019 \text{ mol Na}_2\text{SO}_3)\left(\frac{2 \text{ mol Na}^+}{1 \text{ mol Na}_2\text{SO}_3}\right)\left(\frac{22.99 \text{ g Na}^+}{\text{mol Na}^+}\right) = 0.87 \text{ g Na}^+$ which is choice **(B)**.

Choice **(A)** corresponds to the number of moles of Na_2SO_3. Choice **(C)** is the mass of sodium sulfite, Na_2SO_3. Choice **(D)** uses the correct mole ratio but the wrong molar mass.

Practice Questions Related to This: **PQ-6** and **PQ-7**

SQ-4. As water is being added to a concentrated NaOH solution

Conceptual

(A) the moles of solute are decreasing, and the molarity is decreasing.

(B) the moles of solute are decreasing, and the volume is increasing.

(C) the volume of the solution is increasing, and the molarity is decreasing.

(D) the volume of the solution is increasing and the molarity is increasing.

Knowledge Required: (1) Definition of molarity. (2) Process of dilution.

Thinking it Through: Molarity or molar concentration is defined as:

$$\text{molar concentration (M)} = \frac{\text{mol solute}}{\text{vol (L) solution}}$$

When adding water or solvent to a solution, the number of moles of solute (n) will remain constant. So, if considering two solutions (labeled here as 1 and 2):

$$\text{Solution (1) concentration, M}_1 = \frac{\text{mol solute, } n}{\text{volume, V}_1} \qquad \text{Solution (2) concentration, M}_2 = \frac{n}{\text{V}_2}$$

Solving for moles of solute (n) and combining the equations, you arrive at the dilution equation:

$$\textit{Solution}(1):\ n = \text{M}_1\text{V}_1 \qquad \textit{Solution}(2):\ n = \text{M}_2\text{V}_2 \qquad \text{M}_1\text{V}_1 = \text{M}_2\text{V}_2$$

Now you see that as the volume increases (adding solvent or water), the concentration will decrease. This corresponds to choice **(C)**.

Choices **(A)** or **(B)** incorrectly have the number of moles changing (it remains constant). Choice **(D)** has the molar concentration increasing rather than decreasing.

Practice Questions Related to This: **PQ-8, PQ-9,** and **PQ-10**

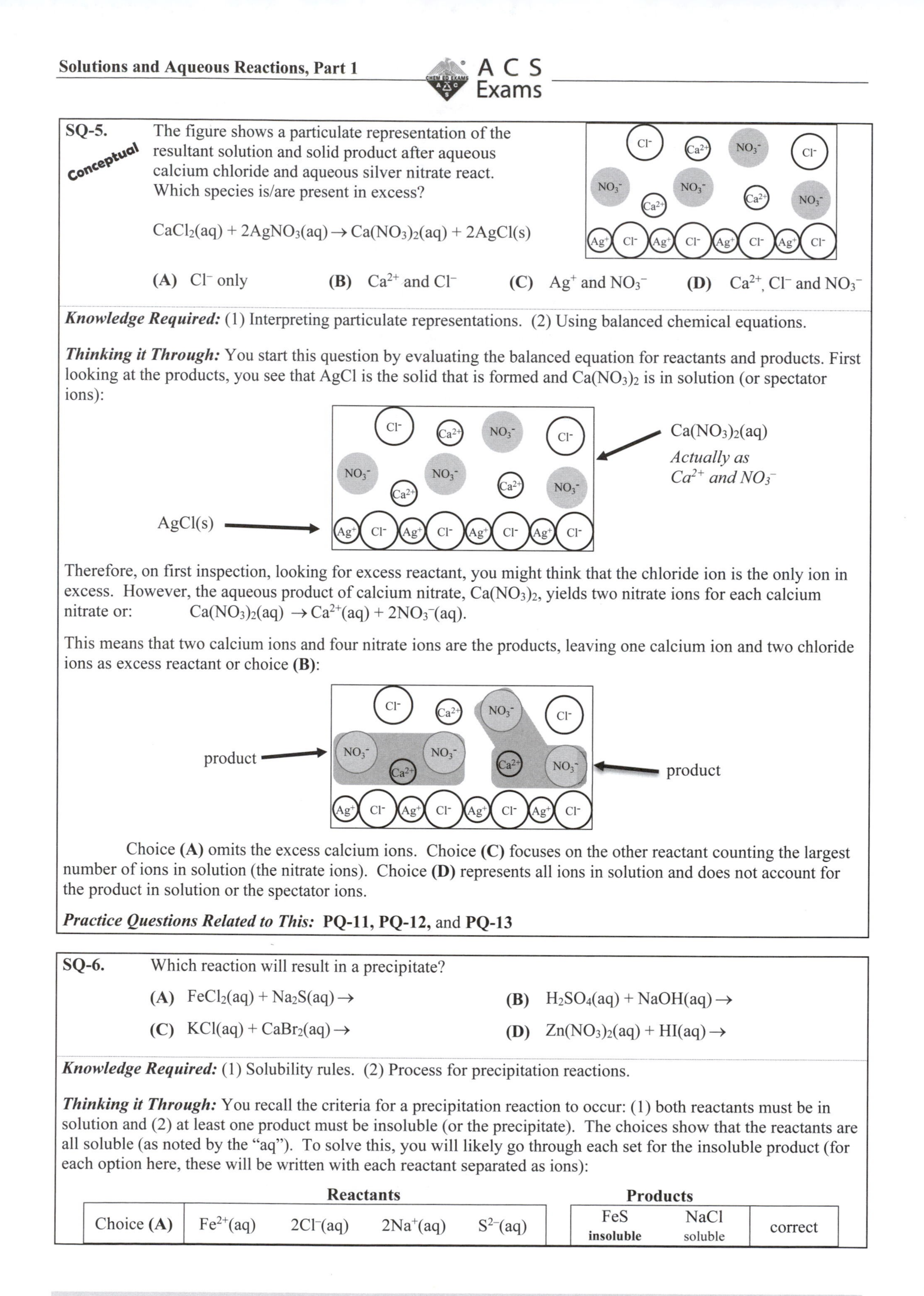

SQ-5. *Conceptual* The figure shows a particulate representation of the resultant solution and solid product after aqueous calcium chloride and aqueous silver nitrate react. Which species is/are present in excess?

$CaCl_2(aq) + 2AgNO_3(aq) \rightarrow Ca(NO_3)_2(aq) + 2AgCl(s)$

(A) Cl^- only **(B)** Ca^{2+} and Cl^- **(C)** Ag^+ and NO_3^- **(D)** Ca^{2+}, Cl^- and NO_3^-

Knowledge Required: (1) Interpreting particulate representations. (2) Using balanced chemical equations.

Thinking it Through: You start this question by evaluating the balanced equation for reactants and products. First looking at the products, you see that AgCl is the solid that is formed and $Ca(NO_3)_2$ is in solution (or spectator ions):

Therefore, on first inspection, looking for excess reactant, you might think that the chloride ion is the only ion in excess. However, the aqueous product of calcium nitrate, $Ca(NO_3)_2$, yields two nitrate ions for each calcium nitrate or: $Ca(NO_3)_2(aq) \rightarrow Ca^{2+}(aq) + 2NO_3^-(aq)$.

This means that two calcium ions and four nitrate ions are the products, leaving one calcium ion and two chloride ions as excess reactant or choice **(B)**:

Choice **(A)** omits the excess calcium ions. Choice **(C)** focuses on the other reactant counting the largest number of ions in solution (the nitrate ions). Choice **(D)** represents all ions in solution and does not account for the product in solution or the spectator ions.

Practice Questions Related to This: **PQ-11, PQ-12,** and **PQ-13**

SQ-6. Which reaction will result in a precipitate?

(A) $FeCl_2(aq) + Na_2S(aq) \rightarrow$ **(B)** $H_2SO_4(aq) + NaOH(aq) \rightarrow$

(C) $KCl(aq) + CaBr_2(aq) \rightarrow$ **(D)** $Zn(NO_3)_2(aq) + HI(aq) \rightarrow$

Knowledge Required: (1) Solubility rules. (2) Process for precipitation reactions.

Thinking it Through: You recall the criteria for a precipitation reaction to occur: (1) both reactants must be in solution and (2) at least one product must be insoluble (or the precipitate). The choices show that the reactants are all soluble (as noted by the "aq"). To solve this, you will likely go through each set for the insoluble product (for each option here, these will be written with each reactant separated as ions):

	Reactants				Products		
Choice **(A)**	$Fe^{2+}(aq)$	$2Cl^-(aq)$	$2Na^+(aq)$	$S^{2-}(aq)$	FeS **insoluble**	NaCl soluble	correct

Reactants					Products		
Choice **(B)**	$H^+(aq)$	$HSO_4^-(aq)$	$Na^+(aq)$	$OH^-(aq)$	H_2O liquid	Na_2SO_4 soluble	incorrect
Choice **(C)**	$K^+(aq)$	$Cl^-(aq)$	$Ca^{2+}(aq)$	$2Br^-(aq)$	KBr soluble	$CaCl_2$ soluble	incorrect
Choice **(D)**	$Zn^{2+}(aq)$	$NO_3^-(aq)$	$H^+(aq)$	$I^-(aq)$	HNO_3 soluble	ZnI_2 soluble	incorrect

Practice Questions Related to This: **PQ-14, PQ-15,** and **PQ-16**

SQ-7. What is the net ionic equation for the reaction between $Na_2S(aq)$ and $Pb(NO_3)_2(aq)$?

(A) $S^{2-}(aq) + Pb^{2+}(aq) \rightarrow PbS(s)$

(B) $Na_2S(aq) + Pb(NO_3)_2(aq) \rightarrow 2NaNO_3(aq) + PbS(s)$

(C) $S^{2-}(aq) + Pb^{2+}(aq) + 2NO_3^-(aq) \rightarrow 2NO_3^-(aq) + PbS(s)$

(D) $2Na^+(aq) + S^{2-}(aq) + Pb^{2+}(aq) + 2NO_3^-(aq) \rightarrow 2Na^+(aq) + 2NO_3^-(aq) + PbS(s)$

Knowledge Required: (1) Precipitation reactions. (2) Definition of net ionic equations.

Thinking it Through: You start this problem by first determining the molecular equation. To do this, you recognize that the reactants could participate in a precipitation or double-displacement reaction and then determine the products based on solubility rules:

$$\underset{\text{soluble}}{Na_2S(aq)} + \underset{\text{soluble}}{Pb(NO_3)_2(aq)} \rightarrow \underset{\text{soluble}}{2NaNO_3(aq)} + \underset{\text{insoluble}}{PbS(s)}$$

The next step will be to determine the total ionic equation. You do this by separating the soluble compounds into the dissociated ions (but keep the insoluble compound or the precipitate together):

$$2Na^+(aq) + S^{2-}(aq) + Pb^{2+}(aq) + 2NO_3^-(aq) \rightarrow 2Na^+(aq) + 2NO_3^-(aq) + PbS(s)$$

Finally, you cancel all ions that appear the same as both reactants and products:

$$\cancel{2Na^+}(aq) + S^{2-}(aq) + Pb^{2+}(aq) + \cancel{2NO_3^-}(aq) \rightarrow \cancel{2Na^+}(aq) + \cancel{2NO_3^-}(aq) + PbS(s)$$

You have canceled the spectator ions (Na^+ and NO_3^-), leaving the net ionic equation of:

$S^{2-}(aq) + Pb^{2+}(aq) \rightarrow PbS(s)$ which is choice **(A)**.

Choice **(B)** is the molecular equation and choice **(D)** is the total ionic equation. Choice **(C)** cancels only one spectator ion (Na^+) but not the nitrate ion.

Practice Questions Related to This: **PQ-17, PQ-18,** and **PQ-19**

SQ-8. *Conceptual* Which best illustrates how the weak acid, HF, exists in aqueous solution (water molecules not shown)?

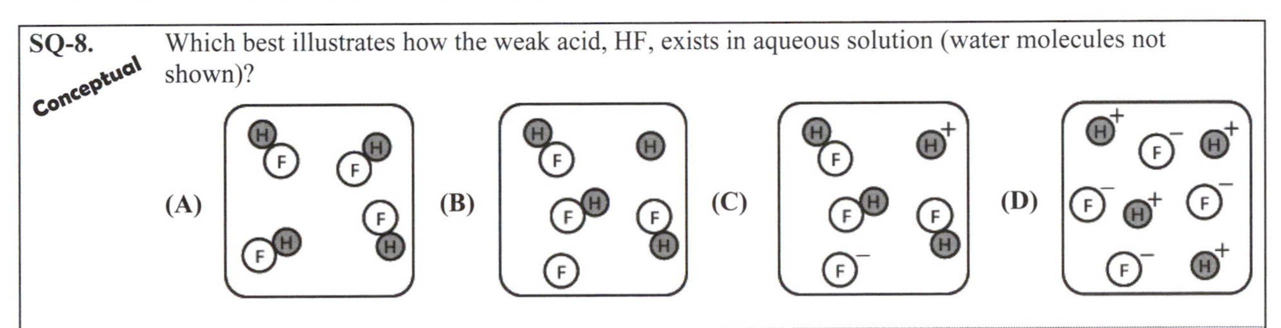

Knowledge Required: (1) Definition of a weak acid. (2) Interpreting particulate representations.

Thinking it Through: Because you are told HF is a weak acid, you can use either the Arrhenius definition or the Bronsted-Lowry definition. For the Arrhenius definition, an acid is a substance that will increase the hydrogen ion concentration in water with a weak acid partially ionizing to do this. For the Bronsted-Lowry definition, an acid is

a substance that donates a proton, again with a weak acid partially ionizing to do this. Choices **(B)** and **(C)** both appear to show the partial ionization, however, on closer inspection you see that choice **(C)** has the hydrogen ion produced while choice **(B)** is showing hydrogen atoms. Therefore, choice **(C)** is correct.

Choice **(D)** is also showing the production of hydrogen ions but is implying that HF is a strong acid rather than a weak acid, so this is incorrect. Choice **(A)** is showing no ionization and therefore, no production of the hydrogen ion, so this is also incorrect.

Practice Questions Related to This: **PQ-20** and **PQ-21**

SQ-9. What is the oxidation number of Mo in MoO_4^{2-}?

(A) +2 **(B)** +4 **(C)** +6 **(D)** +8

Knowledge Required: (1) Rules for assigning oxidation numbers.

Thinking it Through: To assign oxidation numbers, the sum of the individual oxidation numbers of each atom must equal the total charge on the species, which is –2 here (the charge on the ion). You need to know that oxygen is almost always assigned an oxidation number of –2. That means with four oxygen atoms, there will be a net oxidation value of –8 for oxygen and this leaves +6 for the oxidation number on Mo:

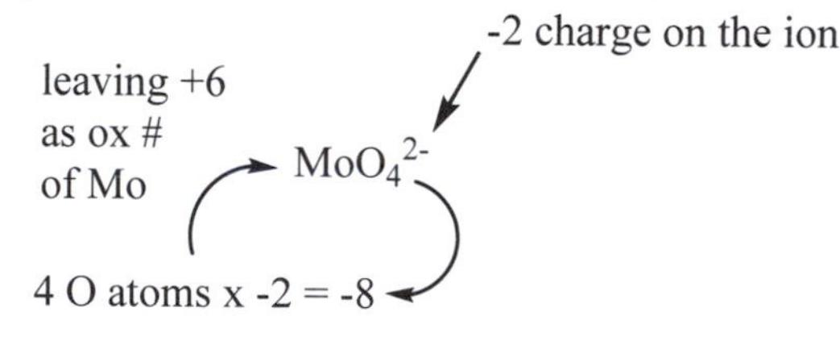

You might also solve this using a mathematical expression: [4 O atoms × –2 each] + [1 Mo atom × ox #] = –2. Solving for the ox # of Mo, you also get +6 which is choice **(C)**.

Choice **(A)** balances the charge on the ion to get +2. Choice **(B)** balances the number of oxygen atoms. Choice **(D)** omits using charge on the ion and only balances the oxidation number from the oxygen atoms.

Practice Questions Related to This: **PQ-22, PQ-23,** and **PQ-24**

SQ-10. Which combination is correct for the reaction represented by this equation?

$$Cu(s) + Zn^{2+}(aq) \rightarrow Zn(s) + Cu^{2+}(aq)$$

	oxidized	reduced
(A)	Cu	Zn^{2+}
(B)	Cu^{2+}	Zn
(C)	Zn	Cu^{2+}
(D)	Zn^{2+}	Cu

Knowledge Required: (1) Definition of oxidation number. (2) Definition of oxidation and reduction.

Thinking it Through: You start this question by assigning oxidation numbers to all species because you know that for something to be oxidized, it will increase in oxidation number as it loses electrons and for something to be reduced, it will decrease in oxidation number as it gains electrons:

Cu(s)	+	Zn^{2+}(aq)	→	Zn(s)	+	Cu^{2+}(aq)
0		+2		0		+2

This now allows you to split these into half reactions based on (1) keeping the same elements together and (2) using the definitions above to assign these as oxidation or reduction:

oxidation	$Cu(s)$ (0)	→	$Cu^{2+}(aq) + 2e^-$ (+2)	*losing electrons; increasing in oxidation number*
reduction	$Zn^{2+}(aq) + 2e^-$ (+2)	→	$Zn(s)$ (0)	*gaining electrons; decreasing in oxidation number*

Therefore, the reactant of Cu(s) is oxidized and Zn^{2+} is reduced or choice **(A)**.

Choice **(B)** reverses this assignment. Choice **(C)** and **(D)** use products rather than reactants.

Practice Questions Related to This: **PQ-25, PQ-26,** and **PQ-27**

SQ-11. For this reaction, what are the oxidizing and reducing agents?

$$2AuCl_3(aq) + 3H_2(g) \rightarrow 2Au(s) + 6HCl(aq)$$

	oxidizing agent	reducing agent
(A)	$AuCl_3$	H_2
(B)	H_2	$AuCl_3$
(C)	Au	HCl
(D)	HCl	Au

Knowledge Required: (1) Definitions of oxidizing and reducing agents.

Thinking it Through: Similar to the previous study question, you can approach this the same way and then extend this into assigning oxidizing or reducing agent. Start this question by assigning oxidation numbers to all species. This question will be slightly more challenging as there is a spectator ion present (Cl^-) which we will remove below for the next few steps:

$$\underset{+3}{Au^{3+}(aq)} + \underset{0}{H_2(g)} \rightarrow \underset{0}{Au(s)} + \underset{+1}{H^+(aq)}$$

Notice that we have also removed the stoichiometric coefficients.

This again allows you to split these into half reactions based on (1) keeping the same elements together and (2) using the definitions above to assign these as oxidation or reduction:

oxidation	$H_2(g)$ (0)	→	$2H^+(aq) + 2e^-$ (+1)	*Note you mass balanced first and then added the e^-*
reduction	$Au^{3+}(aq) + 3e^-$ (+3)	→	$Au(s)$ (0)	

Oxidizing agents cause oxidation by undergoing reduction. Au^{3+} was reduced, therefore the oxidizing agent is $AuCl_3$. Reducing agents cause reduction by undergoing oxidation. H_2 was oxidized, therefore the reducing agent is H_2 or choice **(A)**.

Choice **(B)** reverses this assignment. Choice **(C)** and **(D)** use products rather than reactants.

Practice Questions Related to This: **PQ-28, PQ-29,** and **PQ-30**

Practice Questions (PQ)

PQ-1. What is the strongest electrolyte in dilute aqueous solution?

(A) $HClO_4$ **(B)** HCN **(C)** HF **(D)** HNO_2

Conceptual **PQ-2.** Which diagram represents the most concentrated solution?
Note: ● represents a solute particle and ⚬●⚬ represents a water molecule

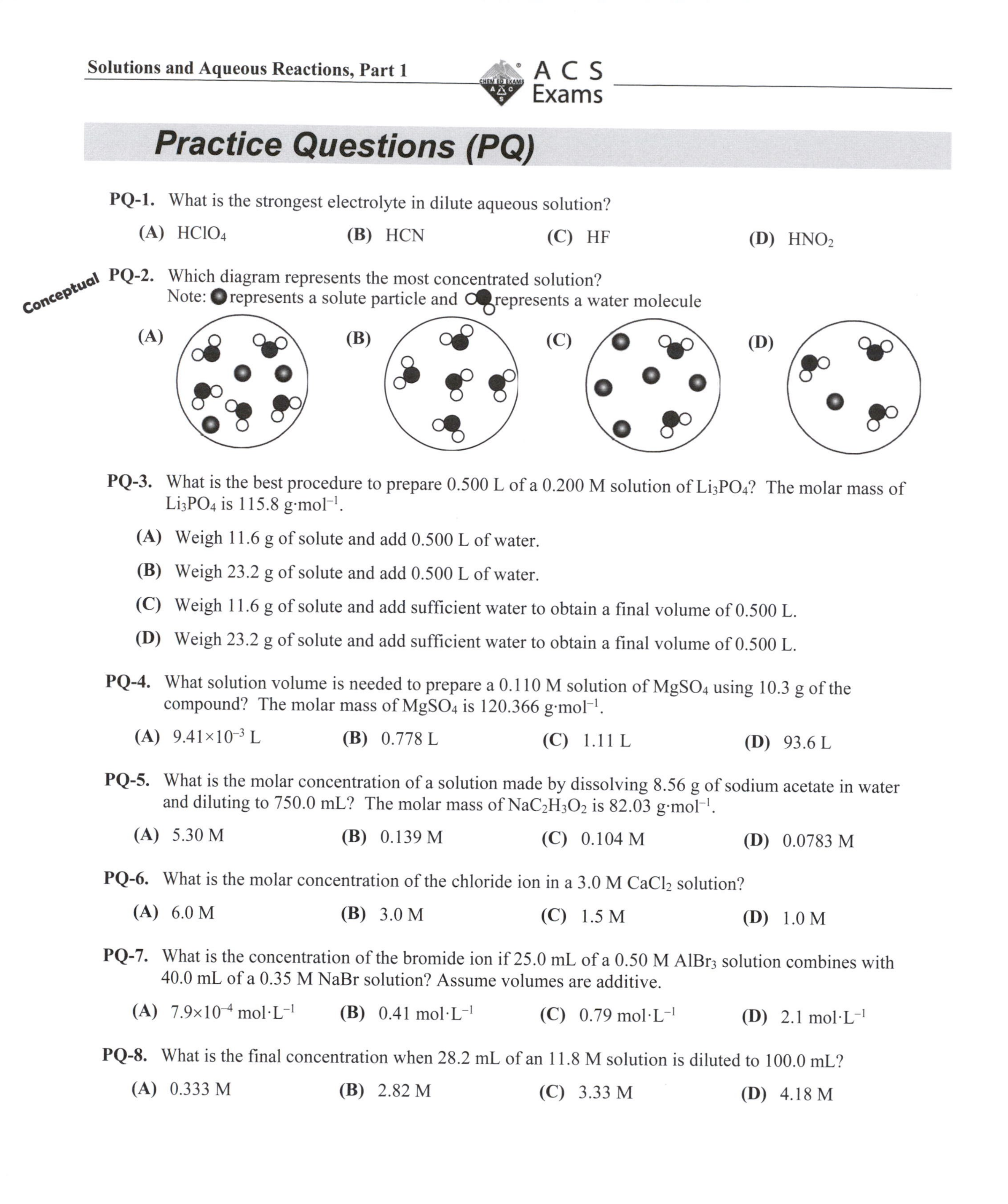

PQ-3. What is the best procedure to prepare 0.500 L of a 0.200 M solution of Li_3PO_4? The molar mass of Li_3PO_4 is 115.8 $g{\cdot}mol^{-1}$.

(A) Weigh 11.6 g of solute and add 0.500 L of water.

(B) Weigh 23.2 g of solute and add 0.500 L of water.

(C) Weigh 11.6 g of solute and add sufficient water to obtain a final volume of 0.500 L.

(D) Weigh 23.2 g of solute and add sufficient water to obtain a final volume of 0.500 L.

PQ-4. What solution volume is needed to prepare a 0.110 M solution of $MgSO_4$ using 10.3 g of the compound? The molar mass of $MgSO_4$ is 120.366 $g{\cdot}mol^{-1}$.

(A) $9.41{\times}10^{-3}$ L **(B)** 0.778 L **(C)** 1.11 L **(D)** 93.6 L

PQ-5. What is the molar concentration of a solution made by dissolving 8.56 g of sodium acetate in water and diluting to 750.0 mL? The molar mass of $NaC_2H_3O_2$ is 82.03 $g{\cdot}mol^{-1}$.

(A) 5.30 M **(B)** 0.139 M **(C)** 0.104 M **(D)** 0.0783 M

PQ-6. What is the molar concentration of the chloride ion in a 3.0 M $CaCl_2$ solution?

(A) 6.0 M **(B)** 3.0 M **(C)** 1.5 M **(D)** 1.0 M

PQ-7. What is the concentration of the bromide ion if 25.0 mL of a 0.50 M $AlBr_3$ solution combines with 40.0 mL of a 0.35 M NaBr solution? Assume volumes are additive.

(A) $7.9{\times}10^{-4}$ $mol{\cdot}L^{-1}$ **(B)** 0.41 $mol{\cdot}L^{-1}$ **(C)** 0.79 $mol{\cdot}L^{-1}$ **(D)** 2.1 $mol{\cdot}L^{-1}$

PQ-8. What is the final concentration when 28.2 mL of an 11.8 M solution is diluted to 100.0 mL?

(A) 0.333 M **(B)** 2.82 M **(C)** 3.33 M **(D)** 4.18 M

PQ-9. A solution is made by dissolving 60 g of NaOH (molar mass = 40 $g \cdot mol^{-1}$) in enough distilled water to make 300 mL of a stock solution. What volumes of this solution and distilled water, when mixed, will result in a solution that is approximately 1 M NaOH?

	mL stock solution	mL distilled water
(A)	20	80
(B)	20	100
(C)	60	30
(D)	60	90

PQ-10. What is the molar concentration of a solution when 7.00 mL of a 3.25 M aqueous solution is diluted to 25.00 mL?

(A) 0.280 M **(B)** 0.910 M **(C)** 1.10 M **(D)** 2.15 M

PQ-11. What volume of 0.200 M $K_2C_2O_4$ is required to react completely with 30.0 mL of 0.100 M $Fe(NO_3)_3$?

$$2Fe(NO_3)_3 + 3K_2C_2O_4 \rightarrow Fe_2(C_2O_4)_3 + 6KNO_3$$

(A) 10.0 mL **(B)** 15.0 mL **(C)** 22.5 mL **(D)** 30.0 mL

PQ-12. What volume (in mL) of a 0.0557 M $Sr(OH)_2$ solution is needed to neutralize 36.4 mL of a 0.0750 M HNO_3 solution?

(A) 24.5 mL **(B)** 36.4 mL **(C)** 49.0 mL **(D)** 98.0 mL

PQ-13. What volume of 0.100 M HCl is needed to neutralize 25.0 mL of 0.100 M $Ba(OH)_2$?

$$2HCl + Ba(OH)_2 \rightarrow BaCl_2 + 2H_2O$$

(A) 0.250 mL **(B)** 12.5 mL **(C)** 25.0 mL **(D)** 50.0 mL

PQ-14. When 1.0 M aqueous solutions of KI, $NaC_2H_3O_2$, and $Pb(NO_3)_2$ are mixed together, what precipitate forms?

(A) $KC_2H_3O_2$ **(B)** $NaNO_3$ **(C)** $Pb(C_2H_3O_2)_2$ **(D)** PbI_2

Conceptual **PQ-15.** An aqueous solution contains Mg^{2+}, Ag^+, and Fe^{3+} cations. Aqueous solutions of K_2S, NaOH, and HCl are available. In what order should the reagents be added to precipitate each cation one at a time?

Ion	General Solubility Rule
Cl^-, Br^-, I^-	All are soluble except Ag^+, Pb^{2+}, and Hg_2^{2+}
OH^-	All hydroxides are insoluble except those of Na^+, K^+; hydroxides of Ba^{2+} and Ca^{2+} are slightly soluble
S^{2-}	All sulfides are insoluble except those of Na^+, K^+, NH_4^+ and Mg^{2+}, Ca^{2+}, Sr^{2+}, Ba^{2+}

	1st	2nd	3rd
(A)	HCl	K_2S	NaOH
(B)	HCl	NaOH	K_2S
(C)	K_2S	HCl	NaOH
(D)	NaOH	K_2S	HCl

Please note, you may not be provided with a solubility table on an ACS exam.

PQ-16. Which combination will result is a precipitate forming?

(A) NaOH(aq) and HCl(aq) **(B)** NaOH(aq) and $FeCl_3$(aq)

(C) $NaNO_3$(aq) and $FeCl_3$(aq) **(D)** Zn(s) and HCl(aq)

PQ-17. What is/are the spectator ion(s) in this reaction?

$$HC_2H_3O_2(aq) + NaOH(aq) \rightarrow NaC_2H_3O_2(aq) + H_2O(l)$$

(A) $C_2H_3O_2^-$ only **(B)** H^+ and OH^- **(C)** Na^+ only **(D)** Na^+ and $C_2H_3O_2^-$

PQ-18. Which molecular equation would give this net ionic equation?

$$H^+(aq) + OH^-(aq) \rightarrow H_2O(l)$$

(A) $H_2SO_4(aq) + 2NH_3(aq) \rightarrow (NH_4)_2SO_4(aq)$

(B) $HCN(aq) + KOH(aq) \rightarrow H_2O(l) + KCN(aq)$

(C) $3HBr(aq) + Fe(OH)_3(s) \rightarrow 3H_2O(l) + FeBr_3(aq)$

(D) $2HClO_4(aq) + Ba(OH)_2(aq) \rightarrow 2H_2O(l) + Ba(ClO_4)_2(aq)$

PQ-19. What is the balanced net ionic equation for the reaction of $CuCl_2(aq)$ and $H_2(g)$?

(A) $Cu^{2+}(aq) + H_2(g) \rightarrow Cu(s) + 2H^+(aq)$

(B) $CuCl_2(aq) + H_2(g) \rightarrow Cu(s) + 2HCl(aq)$

(C) $Cu^{2+}(aq) + Cl_2(aq) + H_2(g) \rightarrow Cu(s) + 2H^+(aq) + 2Cl^-(aq)$

(D) $Cu^{2+}(aq) + 2Cl^-(aq) + H_2(g) \rightarrow Cu(s) + 2H^+(aq) + 2Cl^-(aq)$

PQ-20. What is a weak acid in water?

(A) HBr **(B)** HCl **(C)** HF **(D)** HI

Conceptual **PQ-21.** What particles would be present in $NH_3(aq)$?

(A) water molecules, NH_4^+ ions and OH^- ions

(B) water molecules, H_3O^+ ions and NH_2^- ions

(C) water molecules and NH_3 molecules

(D) water molecules and NH_3 molecules and a small number of NH_4^+ ions and OH^- ions

PQ-22. What is the oxidation number of chlorine in $NaClO_2$?

(A) –1 **(B)** +1 **(C)** +3 **(D)** +5

PQ-23. What is the oxidation number of chromium in $Na_2Cr_2O_7$?

(A) +12 **(B)** +6 **(C)** +3 **(D)** –2

PQ-24. In which compound does vanadium have the lowest oxidation state?

(A) V_2O_5 **(B)** V_2O_3 **(C)** VO_2 **(D)** VO

PQ-25. Which is an example of an oxidation-reduction reaction?

(A) $HCl(aq) + RbOH(aq) \rightarrow H_2O(l) + RbCl(aq)$

(B) $(NH_4)_2CO_3(s) \rightarrow 2NH_3(g) + CO_2(g) + H_2O(g)$

(C) $Pb(NO_3)_2(aq) + Na_2SO_4(aq) \rightarrow PbSO_4(s) + 2NaNO_3(aq)$

(D) $Pb(s) + PbO_2(s) + 2H_2SO_4(aq) \rightarrow 2PbSO_4(s) + 2H_2O(l)$

PQ-26. In which reaction is carbon reduced?

(A) $CO(g) + O_2(g) \rightarrow CO_2(g)$

(B) $CO_2(g) + H_2O(l) \rightarrow H_2CO_3(aq)$

(C) $CH_4(g) + 2O_2(g) \rightarrow CO_2(g) + 2H_2O(l)$

(D) $C_2H_2(g) + H_2(g) \rightarrow C_2H_4(g)$

PQ-27. Which is an example of an oxidation-reduction reaction?

(A) $CaCO_3(s) \rightarrow CaO(s) + CO_2(g)$

(B) $2H_2O_2(aq) \rightarrow O_2(g) + 2H_2O(l)$

(C) $NaHCO_3(s) + HCl(aq) \rightarrow NaCl(aq) + CO_2(g) + H_2O(l)$

(D) $2Pb^{2+}(aq) + Cr_2O_7^{2-}(aq) + H_2O(l) \rightarrow 2PbCrO_4(s) + 2H^+(aq)$

PQ-28. What are the oxidizing agent and the reducing agent in this reaction?

$$Cu(NO_3)_2(aq) + 2V(NO_3)_2(aq) \rightarrow Cu(s) + 2V(NO_3)_3(aq)$$

	oxidizing agent	reducing agent
(A)	$V(NO_3)_2$	$Cu(NO_3)_2$
(B)	$Cu(NO_3)_2$	$V(NO_3)_2$
(C)	Cu	$V(NO_3)_2$
(D)	$V(NO_3)_2$	Cu

PQ-29. What is the oxidizing agent in this reaction?

$$Pt^{2+}(aq) + CH_3CH_2OH(aq) \rightarrow CH_3CHO(aq) + Pt(s) + 2H^+(aq)$$

(A) $Pt^{2+}(aq)$

(B) $Pt(s)$

(C) $H^+(aq)$

(D) $CH_3CH_2OH(aq)$

PQ-30. Which statement about this redox reaction is correct?

$$2MnO_4^-(aq) + 5H_2O_2(aq) + 6H^+(aq) \rightarrow 2Mn^{2+}(aq) + 5O_2(g) + 8H_2O(l)$$

(A) O_2 acts as the oxidant in this reaction.

(B) H_2O_2 acts as a reducing agent.

(C) H_2O_2 acts as an oxidizing agent.

(D) Only oxidation has taken place.

Answers to Study Questions

1. C
2. D
3. B
4. C
5. B
6. A
7. A
8. C
9. C
10. A
11. A

Answers to Practice Questions

1. A
2. C
3. C
4. B
5. B
6. A
7. C
8. C
9. A
10. B
11. C
12. A
13. D
14. D
15. A
16. B
17. C
18. D
19. A
20. C
21. D
22. C
23. B
24. D
25. D
26. D
27. B
28. B
29. A
30. B

Chapter 6 – Heat and Enthalpy

Chapter Summary:

This chapter will focus on energy and enthalpy. The transfer of energy during chemical and physical changes will be calculated. The usefulness of enthalpy as a state function will be used to calculate the values of enthalpy changes using Hess's law and enthalpies of formation.

Specific topics covered in this chapter are:

- Calorimetry and specific heat
- Enthalpy and heat
- First law of Thermodynamics
- Hess's law
- Exothermic and endothermic processes

Previous material that is relevant to your understanding of questions in this chapter include:

- Significant figures (***Toolbox***)
- Scientific notation (***Toolbox***)
- Balancing chemical reactions ***(Chapter 3)***
- Stoichiometry ***(Chapter 4)***

Common representations used in questions related to this material:

Name	Example	Used in questions related to
compound units	$kJ \cdot mol^{-1}$	density, enthalpy changes for reactions (ΔH)

Where to find this in your textbook:

The material in this chapter typically aligns to "Thermochemistry". The name of your chapter may vary.

Practice exam:

There are practice exam questions aligned to the material in this chapter. Because there are a limited number of questions on the practice exam, a review of the breadth of the material in this chapter is advised in preparation for your exam.

How this fits into the big picture:

The material in this chapter aligns to the Big Idea of Energy and Thermodynamics (6) as listed on page 12 of this study guide.

Study Questions (SQ)

SQ-1. What is the energy change for a system that releases 100. kJ of heat to its surroundings and has 25 kJ of work done on it?

(A) 125 kJ **(B)** 75 kJ **(C)** –75 kJ **(D)** –125 kJ

Knowledge Required: (1) First law of thermodynamics. (2) Sign conventions for heat and work.

Thinking it Through: The question asks you to determine the energy change for a system that has undergone some processes. You recall the first law of thermodynamics says that the change in internal energy (ΔE) of a system is equal to the amount of thermal energy transferred (q) plus the amount of work involved (w). In equation form:

$$\Delta E = q + w$$

The question tells you that the system releases 100. kJ of energy as heat. You know that when a system **releases** energy the sign is negative, so $q = -100.$ kJ. The question also tells you that the system has 25 kJ of work done on it. When work is done **on** a system, this is energy that is being added to the system, so $w = +25$ kJ. As a reference, when a system **absorbs** heat, the sign is positive (so $q > 0$), and when work is done **by** a system, the sign is negative (so $w < 0$).

You can now calculate the change in energy using the first law equation:

$$\Delta E = -100.\text{ kJ} + 25\text{ kJ} = -75\text{ kJ} \quad \text{Choice } \textbf{(C)}$$

Choice **(A)** is incorrect because it has the sign for q reversed. It uses $q = +100.$ kJ
Choice **(B)** is incorrect because it has the sign for both q and w reversed.
Choice **(D)** is incorrect because it has the sign for w reversed. It uses $w = -25$ kJ.

Practice Questions Related to This: **PQ-1** and **PQ-2**

SQ-2. Conceptual

One-gram samples of carbon and copper, originally at room temperature, reach equilibrium after being placed in 200 mL of boiling water. Which statement is correct?

Specific heat / $J \cdot g^{-1} \cdot K^{-1}$	
carbon	0.709
copper	0.385

(A) Copper would absorb the same amount of energy as carbon and attain the same final temperature as carbon.

(B) Copper would absorb the same amount of energy as carbon and attain a higher final temperature than carbon.

(C) Copper would absorb less energy than carbon and attain the same final temperature as carbon.

(D) Copper would absorb more energy than carbon and attain the same final temperature as carbon.

Knowledge Required: (1) Definition of specific heat. (2) Relationship between heat, specific heat, temperature change, and mass.

Thinking it Through: You are being asked to select the correct statement about the scenario. You recognize that the scenario involves heat flow for two different materials. You also recognize that the substance with the higher specific heat is more resistant to temperature change than a substance with a lower specific heat. You also read that the substances are allowed to reach equilibrium with the boiling water; this tells you that the two substances come to the same final temperature. Because the samples are starting at room temperature and are being placed into a boiling water bath, you recognize that each sample will absorb energy to reach thermal equilibrium.

When you examine the values of the specific heat of the two substances you see that copper has a lower specific heat than carbon. This tells you that it will take less energy to raise the temperature of the copper to the final equilibrium temperature than it will take to raise the temperature of carbon to the same final temperature. You can make this comparison just using the specific heat values, because each sample has the same mass. You now reach the conclusion that the copper will absorb less energy than the carbon and will attain the same final temperature as the carbon. Therefore, the correct answer is choice **(C)**.

Choice **(A)** is incorrect because the two samples will not absorb the same amount of energy because they have a different specific heat and have the same mass.

Choice **(B)** is incorrect because the two samples will not absorb the same amount of energy because they have a different specific heat and have the same mass and the final temperatures will be the same.

Choice **(D)** is incorrect because copper has the lower specific heat and since the masses are the same, it will absorb less energy than carbon to get the same final temperature.

Practice Questions Related to This: **PQ-3** and **PQ-4**

SQ-3. A 10.0 g sample of silver is heated to 100.0 °C and then added to 20.0 g of water at 23.0 °C in an insulated calorimeter. At thermal equilibrium, the temperature if the system was measured as 25.0 °C. What is the specific heat of silver?

	Specific heat / $J{\cdot}g^{-1}{\cdot}°C^{-1}$
$H_2O(l)$	4.184

(A) 0.053 $J{\cdot}g^{-1}{\cdot}C^{-1}$ **(B)** 0.22 $J{\cdot}g^{-1}{\cdot}C^{-1}$ **(C)** 4.5 $J{\cdot}g^{-1}{\cdot}C^{-1}$ **(D)** 8.4 $J{\cdot}g^{-1}{\cdot}C^{-1}$

Knowledge Required: (1) Relationship between mass, thermal energy, specific heat, and temperature change. (2) First law of thermodynamics.

Thinking it Through: You are being asked to determine the specific heat of a substance. You are given a description of the experiment. In this experiment you know that energy will move from the warmer silver to the cooler water. You also know that according to the first law of thermodynamics, all the energy lost by the silver must be absorbed by the water (assuming this is a fully insulated container).

Mathematically this is represented as: $q_{Ag} + q_{water} = 0$

The heat gained or lost can be calculated using: $q = cm\Delta T$, where c is the specific heat capacity.

Substituting this into the first relationship gives: $c_{Ag}m_{Ag}\Delta T_{Ag} + c_{water}m_{water}\Delta T_{water} = 0$

When you go back to the question you identify the quantity you are being asked for as c_{Ag}.

Solving the expression for c_{Ag} and remembering that $\Delta T = T_{final} - T_{initial}$

$$c_{Ag} = \frac{-c_{water}m_{water}\Delta T_{water}}{m_{Ag}\Delta T_{Ag}} = \frac{-(4.184\ J \cdot g^{-1} \cdot {}^{\circ}C^{-1})(20.0\ g)(25.0\ {}^{\circ}C - 23.0\ {}^{\circ}C)}{(10.0\ g)(25.0\ {}^{\circ}C - 100.0\ {}^{\circ}C)} = 0.22\ J \cdot g^{-1} \cdot {}^{\circ}C^{-1}$$

The correct answer is choice **(B)**.

Choice **(A)** is incorrect because it doesn't make use of the specific heat of water. Choice **(C)** is incorrect because it is the answer you get if you do the algebra for solving for c_{Ag} incorrectly. Choice **(D)** is incorrect because it incorrectly calculates the change in temperature (ΔT).

Practice Questions Related to This: **PQ-5, PQ-6, PQ-7,** and **PQ-8**

SQ-4. What mass of $Cl_2(g)$ is needed to release 45.2 kJ of energy as heat during the reaction of carbon and chlorine gas? $C(s) + 2Cl_2(g) \rightarrow CCl_4(l)$ $\Delta H^{\circ}_{rxn} = -135.4\ kJ \cdot mol^{-1}$

(A) 0.67 g **(B)** 11.8 g **(C)** 23.7 g **(D)** 47.3 g

Knowledge Required: (1) Stoichiometry. (2) Concept of energy per mole of reaction

Thinking it Through: You are asked to calculate the amount of a reactant needed to produce a certain amount of energy using a thermochemical equation. You recognize that the equation indicates that for every mole of reaction, –135.4 kJ of energy is released. You recall that you can use energy in a reaction in the same way you use moles of reactants and products. You can write conversion factors involving energy and moles of reactants and products. The problem is asking about the amount of Cl_2 needed to release 45.2 kJ of energy. So, you can write the conversion factors:

1 mol rxn = 2 mol Cl_2 1 mol rxn = –135.4 kJ 1 mol Cl_2 = 70.90 g Cl_2

Because the energy is being released it is – 45.2 kJ. Using this and the conversion factors above you calculate the mass of Cl_2 needed:

$$-45.2\ kJ\left(\frac{1\ mol\ rxn}{-135.4\ kJ}\right)\left(\frac{2\ mol\ Cl_2}{1\ mol\ rxn}\right)\left(\frac{70.90\ g\ Cl_2}{1\ mol\ Cl_2}\right) = 47.3\ g\ Cl_2$$

This is choice **(D)**.

Choice **(A)** is incorrect because the answer is the moles of Cl_2, not the mass. Choice **(B)** is incorrect because it used the moles reaction to moles Cl_2 conversion incorrectly. Choice **(C)** is incorrect because it does not take into account the 2 moles of Cl_2 per mole of reaction.

Practice Questions Related to This: **PQ-9** and **PQ-10**

SQ-5. What mass of benzene, $C_6H_6(l)$, must be burned in a bomb calorimeter to raise its temperature by 15 °C?

Thermodynamic Data	
$\Delta H^\circ_{combustion}(C_6H_6(l))$	$-41.9\ kJ \cdot g^{-1}$
calorimeter constant	$1.259\ kJ \cdot °C^{-1}$

(A) 0.45 g **(B)** 2.2 g **(C)** 19 g **(D)** 500 g

Knowledge Required: (1) Principles of bomb calorimetry. (2) The first law of thermodynamics.

Thinking it Through: The question says that benzene is being burned in a bomb calorimeter, and you are given the calorimeter constant. You remember that for a bomb calorimeter, a calorimeter constant is essentially a heat capacity for the device. You also recognize that the first law of thermodynamics tells you that any heat lost by this combustion reaction, must be absorbed by the calorimeter:

$$q_{calorimeter} = -q_{reaction} \text{ which also means } q_{calorimeter} + q_{reaction} = 0$$

You can calculate the heat given off by the reaction using $q_{rxn} = m\Delta H_{combustion}$ where m is the mass of benzene combusted.

The heat absorbed by the calorimeter will be $q_{calorimeter} = C_{calorimeter}\Delta T$, where $C_{calorimeter}$ is the calorimeter constant and ΔT is the change in temperature of the calorimeter.

You then write the expression: $C_{calorimeter}\Delta T + m\Delta H_{combustion} = 0$

Solving for the mass of bezene, m:

$$m = \frac{-C_{calorimeter}\Delta T}{\Delta H^\circ_{combustion}} = \frac{-(1.259\ kJ \cdot °C^{-1})(15\ °C)}{-41.9\ kJ \cdot g^{-1}} = 0.45\ g$$ This is choice **(A)**.

Choice **(B)** is incorrect because this answer comes from an algebra mistake made when solving for the mass of benzene. Choice **(C)** is incorrect because this is the amount of energy gained by the calorimeter. Choice **(D)** is incorrect because this value comes from switching the value of the calorimeter constant for the value of $\Delta H_{combustion}$.

Practice Questions Related to This: **PQ-11** and **PQ-12**

SQ-6. *Conceptual*

What is ΔH°_f for HF(g)?

$$2HF(g) \rightarrow H_2(g) + F_2(g) \qquad \Delta H^\circ_{rxn}$$

(A) $\Delta H^\circ_f = -\Delta H^\circ_{rxn}$ **(B)** $\Delta H^\circ_f = ½(-\Delta H^\circ_{rxn})$

(C) $\Delta H^\circ_f = ½(\Delta H^\circ_{rxn})$ **(D)** $\Delta H^\circ_f = \Delta H^\circ_{rxn}$

Knowledge Required: (1) Definition of the enthalpy of formation. (2) Rules for manipulating enthalpy values.

Thinking it Through: You are given a thermochemical equation and are being asked for the enthalpy of formation of HF(g). You recall that the enthalpy of formation is defined as the enthalpy of a reaction that produces 1 mole of the substance from its elements in their standard states. You can write the formation reaction for HF(g)

$$½\ H_2(g) + ½\ F_2(g) \rightarrow HF(g)$$

You notice that this reaction is related to the thermochemical equation you were given. To convert the given thermochemical equation to this formation reaction you need to flip the reaction. You also remember that when you flip a reaction, you change the sign on the ΔH° value.

$$H_2(g) + F_2(g) \rightarrow 2HF(g) \quad -\Delta H^\circ_{rxn}$$

This is a reaction that produces HF(g) from its elements in their standard state, but it is not a formation reaction since it produces 2 moles of HF. To address this, you need to multiply the reaction by ½, so the coefficient for HF(g) is 1. You also remember that when you multiply a reaction, you also multiply the ΔH° value.

$$\tfrac{1}{2}\,(H_2(g) + F_2(g) \rightarrow 2HF(g)\,) \quad -\tfrac{1}{2}\Delta H^\circ_{rxn}$$

The result is the desired formation reaction of HF(g).

$$\tfrac{1}{2}\,H_2(g) + \tfrac{1}{2}\,F_2(g) \rightarrow HF(g) \quad -\tfrac{1}{2}\Delta H^\circ_{rxn} = \Delta H^\circ_f$$

Choice **(B)** is the correct answer.

Choice **(A)** is incorrect because it is the ΔH° for the reaction that produces 2 moles of HF(g). Choice **(C)** is incorrect because the sign of ΔH° is incorrect. Choice **(D)** is incorrect because the sign is incorrect and the reaction has not been multiplied by ½.

Practice Questions Related to This: **PQ-13** and **PQ-14**

SQ-7. *Conceptual*

Given: Equation 1: $SO_2(g) \rightarrow S(s) + O_2(g)$ $\quad \Delta H^\circ_1$

Equation 2: $2SO_2(g) + O_2(g) \rightarrow 2SO_3(g)$ $\quad \Delta H^\circ_2$

Use this information to calculate the enthalpy of formation of $SO_3(g)$.

(A) $\Delta H^\circ = \Delta H^\circ_1 + \tfrac{1}{2}\,\Delta H^\circ_2$

(B) $\Delta H^\circ = -\,\Delta H^\circ_1 + \tfrac{1}{2}\,\Delta H^\circ_2$

(C) $\Delta H^\circ = \Delta H^\circ_1 + \Delta H^\circ_2$

(D) $\Delta H^\circ = -\,\Delta H^\circ_1 - \tfrac{1}{2}\,\Delta H^\circ_2$

Knowledge Required: (1) Definition of the enthalpy of formation. (2) Rules for manipulating enthalpy values.

Thinking it Through: You need to use the given information to find the enthalpy of formation for $SO_3(g)$. You begin by writing the formation reaction for $SO_3(g)$. You remember that the definition of a formation reaction is a reaction that produces 1 mole of product from its elements in their standard states.

$$S(s) + {}^{3}/_{2}O_2(g) \rightarrow SO_3(g)$$

This is the target reaction and you realize that you need to rearrange the reactions given so that they add up to this reaction. This is Hess's law.

Starting with equation 1 you realize that the S(s) needs to be a reactant so you reverse this reaction – remembering to change the sign: $SO_2(g) \rightarrow S(s) + O_2(g) \quad \Delta H^\circ_1$

becomes: $S(s) + O_2(g) \rightarrow SO_2(g) \quad \Delta H^\circ = -\,\Delta H^\circ_1$

You recognize that equation 2 has the $SO_3(g)$ as a product but the coefficient is a 2. To make this a coefficient of 1 you multiply the reaction by ½ (or divide by 2). You also remember to multiply the ΔH value by this factor as well:

$$2SO_2(g) + O_2(g) \rightarrow 2SO_3(g) \quad \Delta H^\circ_2$$

becomes $SO_2(g) + \tfrac{1}{2}O_2(g) \rightarrow SO_3(g) \quad \Delta H^\circ = \tfrac{1}{2}\,\Delta H^\circ_2$

When you add these new equations you get:

$$S(s) + O_2(g) + SO_2(g) + \tfrac{1}{2}O_2(g) \rightarrow SO_2(g) + SO_3(g)$$

$$\Delta H^\circ = -\,\Delta H^\circ_1 + \tfrac{1}{2}\,\Delta H^\circ_2$$

After canceling substances that appear on both sides and combining like terms you get:

$$S(s) + 3/2O_2(g) \rightarrow SO_3(g)$$

Which is the desired formation reaction for $SO_3(g)$. The correct answer is **(B)**.

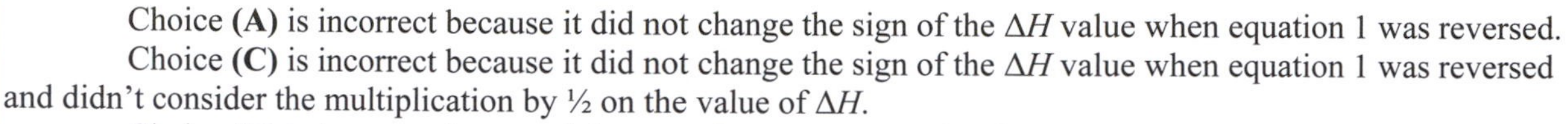

Choice **(A)** is incorrect because it did not change the sign of the ΔH value when equation 1 was reversed.

Choice **(C)** is incorrect because it did not change the sign of the ΔH value when equation 1 was reversed and didn't consider the multiplication by ½ on the value of ΔH.

Choice **(D)** is incorrect because it incorrectly reverses equation 2.

Practice Questions Related to This: **PQ-15** and **PQ-16**

SQ-8. Given this thermochemical data, what is ΔH_{rxn} for $CO(g) + ½O_2(g) \rightarrow CO_2(g)$?

$C(s) + O_2(g) \rightarrow CO_2(g)$	$\Delta H° = –394\ kJ{\cdot}mol^{-1}$
$2C(s) + O_2(g) \rightarrow 2CO(g)$	$\Delta H° = –221\ kJ{\cdot}mol^{-1}$

(A) $–615\ kJ{\cdot}mol^{-1}$ **(B)** $–284\ kJ{\cdot}mol^{-1}$ **(C)** $–173\ kJ{\cdot}mol^{-1}$ **(D)** $173\ kJ{\cdot}mol^{-1}$

Knowledge Required: (1) Hess's Law. (2) Rules for manipulating enthalpy values.

Thinking it Through: You need to use the given information to find the enthalpy of a given reaction. Note that this problem is very similar to the previous study question. The difference is that in the previous study question you were expected to be able to write the reaction of interest. In this question you have been given the desired reaction. You realize that you need to rearrange the reactions given so that they add up to this reaction. This is Hess's law.

Starting with the first given equation you recognize that the product, $CO_2(g)$, is also a product in the target equation and the coefficient is correct in the given reaction. (Note that this is the chemical equation for the combustion of CO.)You can use this equation as it is written.

$$C(s) + O_2(g) \rightarrow CO_2(g) \qquad \Delta H° = –394\ kJ{\cdot}mol^{-1}$$

In the second reaction the CO(g) is a product but it needs to be a reactant in the target reaction. It also needs a coefficient of 1 in the target reaction. Therefore, you decide to reverse this reaction and multiply it by ½ . You also remember to multiply the ΔH value by this factor as well.

$$2C(s) + O_2(g) \rightarrow 2CO(g) \qquad \Delta H° = –221\ kJ{\cdot}mol^{-1}$$

becomes

$$½\times(2CO(g) \rightarrow 2C(s) + O_2(g)) \qquad \Delta H° = ½\times(221\ kJ{\cdot}mol^{-1})$$

When you add these new equations, you get:

$$C(s) + O_2(g) + CO(g) \rightarrow CO_2(g) + C(s) + ½O_2(g) \qquad \Delta H° = –394\ kJ{\cdot}mol^{-1} + 110.5\ kJ{\cdot}mol^{-1} = –284\ kJ{\cdot}mol^{-1}$$

After canceling substances that appear on both sides and combining like terms you get:

$$CO(g) + ½O_2(g) \rightarrow CO_2(g) \qquad \Delta H° = –284\ kJ{\cdot}mol^{-1}$$

Which is the desired reaction. The correct answer is **(B)**.

Choice **(A)** is incorrect because it did not change the sign of the ΔH value when the second equation was reversed nor did it multiply the value by ½.

Choice **(C)** is incorrect because it didn't multiply the value of ΔH for equation 2 by ½ .

Choice **(D)** is incorrect because it didn't multiply the value of ΔH for equation 2 by ½ and it reversed the signs of both equations.

Practice Questions Related to This: **PQ-17, PQ-18, PQ-19, PQ-20, PQ-21,** and **PQ-22**

SQ-9. Use the enthalpies of formation, $\Delta H°_f$, in the table to determine ΔH for this reaction.

$$2C_2H_2(g) + 5O_2(g) \rightarrow 4CO_2(g) + 2H_2O(g)$$

	$\Delta H°_f$ / kJ·mol^{-1}
$C_2H_2(g)$	227
$H_2O(g)$	–242
$CO_2(g)$	–393

(A) $–2510\ kJ{\cdot}mol^{-1}$ **(B)** $–2283\ kJ{\cdot}mol^{-1}$ **(C)** $–862\ kJ{\cdot}mol^{-1}$ **(D)** $2510\ kJ{\cdot}mol^{-1}$

Knowledge Required: (1) Hess's law. (2) Rules for manipulating enthalpy values.

Thinking it Through: You recall that Hess's law allows you to calculate the ΔH of a reaction using tabulated enthalpies of formation values. This works because we can imagine any reaction as consisting of first converting the reactants back into their elements and then forming the products from these elements. This is, in effect reversing the formation reactions of the reactants and then using the formation reactions of the products. You recall that this process is represented in the relationship:

$$\Delta H^{\circ}_{rxn} = \sum m\Delta H^{\circ}_{f}(\text{products}) - \sum n\Delta H^{\circ}_{f}(\text{reactants})$$

You recall that the symbol $\sum$ means to sum and n and m are the coefficients of the reactants and products. You also notice that there is no standard enthalpy of formation given for O_2(g), this is because you know the standard enthalpy of formation of an element in its standard state is zero. The value of ΔH can be calculated as:

$$\Delta H^{\circ}_{rxn} = \left[4\Delta H^{\circ}_{f}(CO_2(g)) + 2\Delta H^{\circ}_{f}(H_2O(g))\right] - \left[2\Delta H^{\circ}_{f}(C_2H_2(g)) + 5\Delta H^{\circ}_{f}(O_2(g))\right]$$

$$\Delta H^{\circ}_{rxn} = \left[4 \text{ mol } CO_2\left(\frac{-393 \text{ kJ}}{\text{mol } CO_2}\right) + 2 \text{ mol } H_2O\left(\frac{-242 \text{ kJ}}{\text{mol } H_2O}\right)\right] - \left[2 \text{ mol } C_2H_2\left(\frac{227 \text{ kJ}}{\text{mol } C_2H_2}\right) + 5 \text{ mol } O_2\left(\frac{0 \text{ kJ}}{\text{mol } O_2}\right)\right]$$

$$\Delta H^{\circ}_{rxn} = -2510 \text{ kJ} \cdot \text{mol}^{-1}$$

Choice **(A)** is the correct answer.

Choice **(B)** is incorrect because the answer adds the reactants to the products.

Choice **(C)** is incorrect because it does not use the coefficients in the balanced equation.

Choice **(D)** is incorrect because is calculated the answer using reactants minus products.

Practice Questions Related to This: **PQ-23, PQ-24, PQ-25, PQ-26, PQ-27, PQ-28, PQ-29,** and **PQ-30**

Practice Questions (PQ)

PQ-1. If a system's internal energy increases by 250 kJ after the addition of 375 kJ of energy as heat, what was the value of the work in the process?

(A) –625 kJ **(B)** –125 kJ **(C)** 125 kJ **(D)** 625 kJ

PQ-2. If 345 kJ of energy is transferred to a system as heat while the system does 42 kJ of work, what will be the change in internal energy of the system after the processes are complete?

(A) –303 kJ **(B)** –387 kJ **(C)** 303 kJ **(D)** 387 kJ

Conceptual **PQ-3.** Two metal samples, of the same mass and initially at 25.0 °C, are heated so that each metal receives the same amount of thermal energy. Which metal will have the highest final temperature?

Specific heat / $J \cdot g^{-1} \cdot K^{-1}$	
X	0.350
Z	0.895

(A) X

(B) Z

(C) Both will have the same final temperature.

(D) It cannot be determined.

Conceptual **PQ-4.** When a cold block of aluminum at 0 °C is placed in room temperature water at 21 °C in an insulated cup, the water and the aluminum end up at a final temperature of 19 °C. Assuming there was no heat loss to the surroundings, which statement best describes the energy exchanged in this process?

(A) The block of aluminum lost the energy that the water gained.

(B) The block of aluminum gained the energy that the water lost.

(C) The block of aluminum gained less energy compared to the energy lost by the water.

(D) The block of aluminum gained more energy compared to the energy lost by the water.

PQ-5. A piece of iron is heated to 95.0 °C and then placed in an insulated vessel containing 250. g of water at 25.0 °C. when the system comes to equilibrium the temperature of the system is 35.0 °C. What is the mass (in g) of the iron? Assume no heat is lost to the surroundings.

Specific heat / $J \cdot g^{-1} \cdot K^{-1}$	
iron	0.450
water	4.184

(A) 4.48 g **(B)** 41.7 g **(C)** 387 g **(D)** 612 g

PQ-6. When a 45.0 g sample of an alloy at 100.0 °C is dropped into 100.0 g of water at 25.0 °C, the final temperature is 37.0 °C. What is the specific heat of the alloy?

Specific heat / $J \cdot g^{-1} \cdot K^{-1}$	
$H_2O(l)$	4.184

(A) 0.423 $J \cdot g^{-1} \cdot °C^{-1}$ **(B)** 1.77 $J \cdot g^{-1} \cdot °C^{-1}$ **(C)** 9.88 $J \cdot g^{-1} \cdot °C^{-1}$ **(D)** 48.8 $J \cdot g^{-1} \cdot °C^{-1}$

PQ-7. When 68.00 J of energy are added to a sample of gallium that is initially at 25.0 °C, the temperature rises to 38.0 °C. What is the volume of the sample?

Data for Gallium, Ga	
specific heat	0.372 $J \cdot g^{-1} \cdot °C^{-1}$
density	5.904 $g \cdot cm^{-3}$

(A) 2.38 cm^3 **(B)** 4.28 cm^3 **(C)** 14.1 cm^3 **(D)** 31.0 cm^3

Conceptual **PQ-8.** A student mixes 100 mL of 0.50 M NaOH with 100 mL of 0.50 M HCl in a Styrofoam® cup and observes a temperature increase of ΔT_1. When she repeats the experiment using 200 mL of each solution, she observes a temperature change of ΔT_2. If no heat is lost to the surroundings or absorbed by the Styrofoam® cup, what is the relationship between ΔT_1 and ΔT_2?

(A) $\Delta T_2 = 4\Delta T_1$ **(B)** $\Delta T_2 = 2\Delta T_1$ **(C)** $\Delta T_2 = 0.5\Delta T_1$ **(D)** $\Delta T_2 = \Delta T_1$

PQ-9. Iron can be converted to Fe_2O_3 according to the reaction below. How much heat (in kJ) is required to convert 10.5 g of iron to Fe_2O_3?

$$2Fe(s) + 3CO(g) \rightarrow Fe_2O_3(s) + 3CO_2(g) \qquad \Delta H = 26.8\ kJ \cdot mol^{-1}$$

(A) 1.76 kJ **(B)** 2.52 kJ **(C)** 5.04 kJ **(D)** 13.4 kJ

PQ-10. A 0.152-mol sample of CH_2O_2 reacted completely according to the reaction shown, and 38.7 kJ of heat was released. What is $\Delta H°$ for this reaction?

$$2CH_2O_2 + O_2 \rightarrow 2CO_2 + 2H_2O$$

(A) 509 $kJ \cdot mol^{-1}$ **(B)** 254 $kJ \cdot mol^{-1}$ **(C)** –254 $kJ \cdot mol^{-1}$ **(D)** –509 $kJ \cdot mol^{-1}$

PQ-11. A 1.00 g sample of NH_4NO_3 is decomposed in a bomb calorimeter. The temperature increases by 6.12 °C. What is the molar heat of decomposition of NH_4NO_3?

Table of Data	
NH_4NO_3	80.0 $g \cdot mol^{-1}$
calorimeter constant	1.23 $kJ \cdot °C^{-1}$

(A) –602 $kJ \cdot mol^{-1}$ **(B)** –398 $kJ \cdot mol^{-1}$ **(C)** 7.53 $kJ \cdot mol^{-1}$ **(D)** 164 $kJ \cdot mol^{-1}$

PQ-12. A 1.00 g sample of glucose, $C_6H_{12}O_6$, is burned in a bomb calorimeter, the temperature of the calorimeter rises by 9.40 °C. What is the heat capacity of the calorimeter?

Table of Data	
$C_6H_{12}O_6(s)$	180.2 $g \cdot mol^{-1}$
$\Delta H_{combustion}$	-2.83×10^3 $kJ \cdot mol^{-1}$

(A) –301 $kJ \cdot °C^{-1}$ **(B)** –1.67 $kJ \cdot °C^{-1}$ **(C)** 1.67 $kJ \cdot °C^{-1}$ **(D)** 301 $kJ \cdot °C^{-1}$

Conceptual **PQ-13.** The molar enthalpy of formation for $H_2O(l)$ is –285.8 $kJ \cdot mol^{-1}$. Which expression describes the enthalpy change for the reaction:

$$2H_2O(l) \rightarrow 2H_2(g) + O_2(g) \qquad \Delta H° = ?$$

(A) $-\frac{1}{2}(\Delta H_f^\circ)$ **(B)** $-(\Delta H_f^\circ)$ **(C)** $-2(\Delta H_f^\circ)$ **(D)** $1/(\Delta H_f^\circ)$

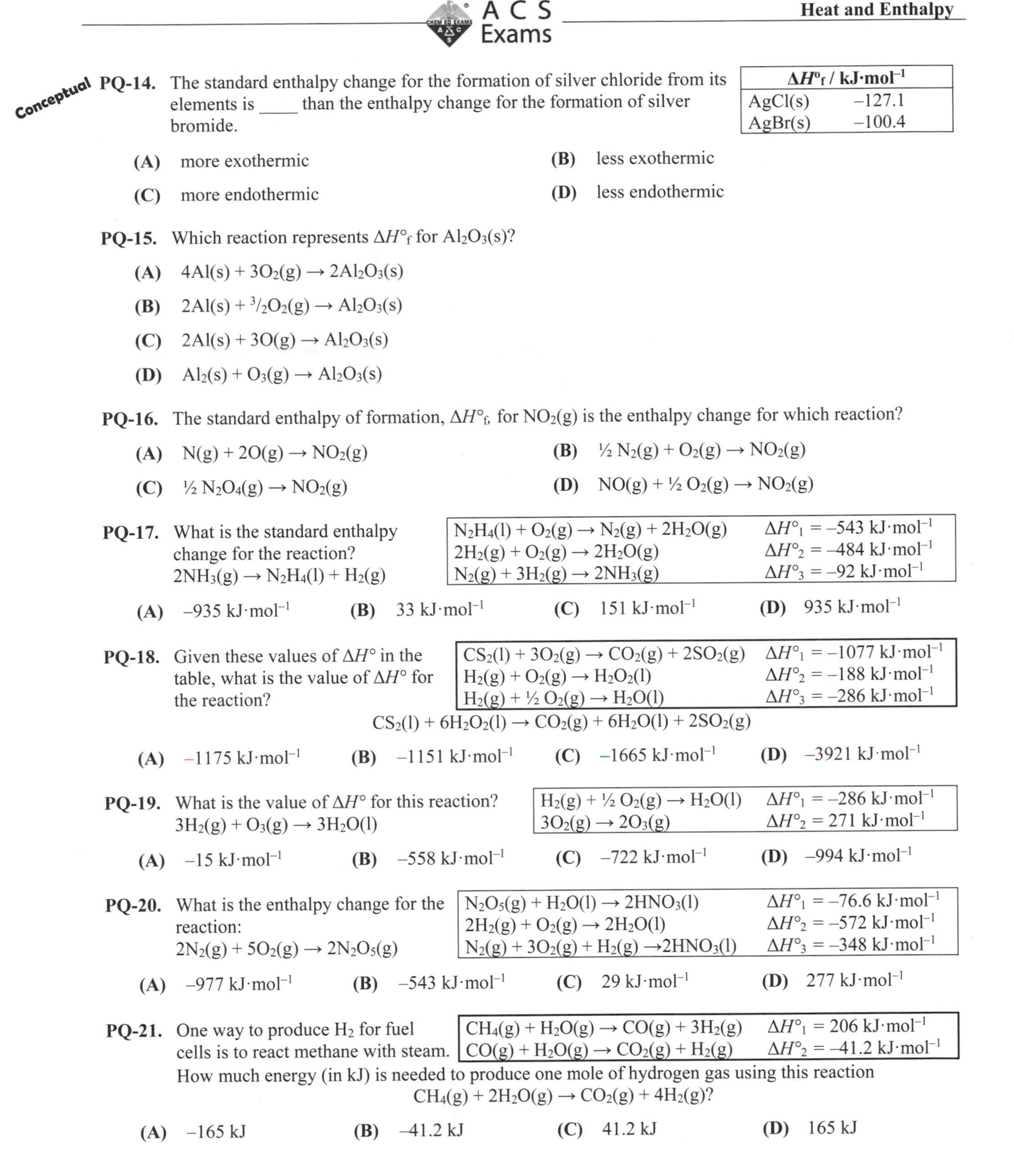

Conceptual **PQ-14.** The standard enthalpy change for the formation of silver chloride from its elements is ____ than the enthalpy change for the formation of silver bromide.

ΔH°_f / kJ·mol^{-1}	
AgCl(s)	–127.1
AgBr(s)	–100.4

(A) more exothermic
(B) less exothermic
(C) more endothermic
(D) less endothermic

PQ-15. Which reaction represents ΔH°_f for $Al_2O_3(s)$?

(A) $4Al(s) + 3O_2(g) \rightarrow 2Al_2O_3(s)$
(B) $2Al(s) + {}^3/_2O_2(g) \rightarrow Al_2O_3(s)$
(C) $2Al(s) + 3O(g) \rightarrow Al_2O_3(s)$
(D) $Al_2(s) + O_3(g) \rightarrow Al_2O_3(s)$

PQ-16. The standard enthalpy of formation, ΔH°_f, for $NO_2(g)$ is the enthalpy change for which reaction?

(A) $N(g) + 2O(g) \rightarrow NO_2(g)$
(B) $\frac{1}{2} N_2(g) + O_2(g) \rightarrow NO_2(g)$
(C) $\frac{1}{2} N_2O_4(g) \rightarrow NO_2(g)$
(D) $NO(g) + \frac{1}{2} O_2(g) \rightarrow NO_2(g)$

PQ-17. What is the standard enthalpy change for the reaction?
$2NH_3(g) \rightarrow N_2H_4(l) + H_2(g)$

Reaction	ΔH°
$N_2H_4(l) + O_2(g) \rightarrow N_2(g) + 2H_2O(g)$	$\Delta H^\circ_1 = -543$ kJ·mol^{-1}
$2H_2(g) + O_2(g) \rightarrow 2H_2O(g)$	$\Delta H^\circ_2 = -484$ kJ·mol^{-1}
$N_2(g) + 3H_2(g) \rightarrow 2NH_3(g)$	$\Delta H^\circ_3 = -92$ kJ·mol^{-1}

(A) –935 kJ·mol^{-1} **(B)** 33 kJ·mol^{-1} **(C)** 151 kJ·mol^{-1} **(D)** 935 kJ·mol^{-1}

PQ-18. Given these values of ΔH° in the table, what is the value of ΔH° for the reaction?

Reaction	ΔH°
$CS_2(l) + 3O_2(g) \rightarrow CO_2(g) + 2SO_2(g)$	$\Delta H^\circ_1 = -1077$ kJ·mol^{-1}
$H_2(g) + O_2(g) \rightarrow H_2O_2(l)$	$\Delta H^\circ_2 = -188$ kJ·mol^{-1}
$H_2(g) + \frac{1}{2} O_2(g) \rightarrow H_2O(l)$	$\Delta H^\circ_3 = -286$ kJ·mol^{-1}

$$CS_2(l) + 6H_2O_2(l) \rightarrow CO_2(g) + 6H_2O(l) + 2SO_2(g)$$

(A) –1175 kJ·mol^{-1} **(B)** –1151 kJ·mol^{-1} **(C)** –1665 kJ·mol^{-1} **(D)** –3921 kJ·mol^{-1}

PQ-19. What is the value of ΔH° for this reaction?
$3H_2(g) + O_3(g) \rightarrow 3H_2O(l)$

Reaction	ΔH°
$H_2(g) + \frac{1}{2} O_2(g) \rightarrow H_2O(l)$	$\Delta H^\circ_1 = -286$ kJ·mol^{-1}
$3O_2(g) \rightarrow 2O_3(g)$	$\Delta H^\circ_2 = 271$ kJ·mol^{-1}

(A) –15 kJ·mol^{-1} **(B)** –558 kJ·mol^{-1} **(C)** –722 kJ·mol^{-1} **(D)** –994 kJ·mol^{-1}

PQ-20. What is the enthalpy change for the reaction:
$2N_2(g) + 5O_2(g) \rightarrow 2N_2O_5(g)$

Reaction	ΔH°
$N_2O_5(g) + H_2O(l) \rightarrow 2HNO_3(l)$	$\Delta H^\circ_1 = -76.6$ kJ·mol^{-1}
$2H_2(g) + O_2(g) \rightarrow 2H_2O(l)$	$\Delta H^\circ_2 = -572$ kJ·mol^{-1}
$N_2(g) + 3O_2(g) + H_2(g) \rightarrow 2HNO_3(l)$	$\Delta H^\circ_3 = -348$ kJ·mol^{-1}

(A) –977 kJ·mol^{-1} **(B)** –543 kJ·mol^{-1} **(C)** 29 kJ·mol^{-1} **(D)** 277 kJ·mol^{-1}

PQ-21. One way to produce H_2 for fuel cells is to react methane with steam. How much energy (in kJ) is needed to produce one mole of hydrogen gas using this reaction

Reaction	ΔH°
$CH_4(g) + H_2O(g) \rightarrow CO(g) + 3H_2(g)$	$\Delta H^\circ_1 = 206$ kJ·mol^{-1}
$CO(g) + H_2O(g) \rightarrow CO_2(g) + H_2(g)$	$\Delta H^\circ_2 = -41.2$ kJ·mol^{-1}

$$CH_4(g) + 2H_2O(g) \rightarrow CO_2(g) + 4H_2(g)?$$

(A) –165 kJ **(B)** –41.2 kJ **(C)** 41.2 kJ **(D)** 165 kJ

PQ-22. What is $\Delta H°$ for the reaction

$CH_4(g) + 2O_2(g) \rightarrow CO_2(g) + 2H_2O(l)$	$\Delta H°_1 = -890.3\ kJ \cdot mol^{-1}$
$2CO(g) + O_2(g) \rightarrow 2CO_2(g)$	$\Delta H°_2 = -566.0\ kJ \cdot mol^{-1}$

$2CH_4(g) + 3O_2(g) \rightarrow 2CO(g) + 4H_2O(l)$?

(A) $-2346.6\ kJ \cdot mol^{-1}$
(B) $-1456.3\ kJ \cdot mol^{-1}$
(C) $-1214.6\ kJ \cdot mol^{-1}$
(D) $-324.3\ kJ \cdot mol^{-1}$

PQ-23. The combustion of ammonia is represented by this equation.
$4NH_3(g) + 5O_2(g) \rightarrow 4NO(g) + 6H_2O(g)$ $\Delta H°_{rxn} = -904.8\ kJ \cdot mol^{-1}$
What is the standard enthalpy of formation of $NH_3(g)$?

$\Delta H°_f$ / kJ·mol^{-1}	
NO(g)	90.3
$H_2O(g)$	–241.8

(A) $-449\ kJ \cdot mol^{-1}$ (B) $-46.1\ kJ \cdot mol^{-1}$ (C) $-184\ kJ \cdot mol^{-1}$ (D) $227\ kJ \cdot mol^{-1}$

PQ-24. Use the standard enthalpies of formation in the table to calculate $\Delta H°_{rxn}$ for this reaction.

$$2CrO_4^{2-}(aq) + 2H^+(aq) \rightarrow Cr_2O_7^{2-}(aq) + H_2O(l)$$

$\Delta H°_f$ / kJ·mol^{-1}	
$CrO_4^{2-}(aq)$	–881.2
$Cr_2O_7^{2-}(aq)$	–1490.3
$H^+(aq)$	0
$H_2O(l)$	–285.8

(A) $272.1\ kJ \cdot mol^{-1}$ (B) $13.7\ kJ \cdot mol^{-1}$ (C) $-13.7\ kJ \cdot mol^{-1}$ (D) $-272.1\ kJ \cdot mol^{-1}$

PQ-25. Calculate the enthalpy of combustion of ethylene, C_2H_4, at 25 °C and one atmosphere of pressure.

$$C_2H_4(g) + 3O_2(g) \rightarrow 2CO_2(g) + 2H_2O(l)$$

$\Delta H°_f$ / kJ·mol^{-1}	
$C_2H_4(g)$	52.3
$CO_2(g)$	–393.5
$H_2O(l)$	–285.8

(A) $-1411\ kJ \cdot mol^{-1}$ (B) $-1254\ kJ \cdot mol^{-1}$ (C) $-732\ kJ \cdot mol^{-1}$ (D) $-627\ kJ \cdot mol^{-1}$

PQ-26. Use the given heats of formation to calculate the enthalpy change for this reaction.

$$B_2O_3(g) + 3COCl_2(g) \rightarrow 2BCl_3(g) + 3CO_2(g)$$

$\Delta H°_f$ / kJ·mol^{-1}	
$B_2O_3(g)$	–1272.8
$COCl_2(g)$	–218.8
$CO_2(g)$	–393.5
$BCl_3(g)$	–403.8

(A) $649.3\ kJ \cdot mol^{-1}$ (B) $354.9\ kJ \cdot mol^{-1}$ (C) $-58.9\ kJ \cdot mol^{-1}$ (D) $-3917.3\ kJ \cdot mol^{-1}$

PQ-27. Using the given thermochemical data, what is the $\Delta H°$ for this reaction?

$$2CH_3OH(l) + O_2(g) \rightarrow HC_2H_3O_2(l) + 2H_2O(l)$$

$\Delta H°_f$ / kJ·mol^{-1}	
$CH_3OH(l)$	–238
$HC_2H_3O_2(l)$	–487
$H_2O(l)$	–286

(A) $583\ kJ \cdot mol^{-1}$ (B) $535\ kJ \cdot mol^{-1}$ (C) $-583\ kJ \cdot mol^{-1}$ (D) $-535\ kJ \cdot mol^{-1}$

PQ-28. What is the $\Delta H°$ for this reaction?

$$Hg(l) + 2Ag^+(aq) \rightarrow Hg^{2+}(aq) + 2Ag(s)$$

$\Delta H°_f$ / kJ·mol^{-1}	
$Ag^+(aq)$	105.6
$Hg^{2+}(aq)$	171.1

(A) $65.5\ kJ \cdot mol^{-1}$ (B) $40.1\ kJ \cdot mol^{-1}$ (C) $-40.1\ kJ \cdot mol^{-1}$ (D) $-65.5\ kJ \cdot mol^{-1}$

PQ-29. What is the enthalpy of formation of hydrazine, $N_2H_4(l)$?
$3N_2H_4(l) \rightarrow 4NH_3(g) + N_2(g)$ $\Delta H°_{rxn} = -336\ kJ \cdot mol^{-1}$

$\Delta H°_f$ / kJ·mol^{-1}	
$NH_3(g)$	–46.3

(A) $-521\ kJ \cdot mol^{-1}$ (B) $-112\ kJ \cdot mol^{-1}$ (C) $50.3\ kJ \cdot mol^{-1}$ (D) $290\ kJ \cdot mol^{-1}$

PQ-30. What is the standard enthalpy change for this reaction?

$$Mg(s) + 2HCl(aq) \rightarrow MgCl_2(aq) + H_2(g)$$

ΔH°_f / kJ·mol^{-1}	
HCl(aq)	–167.2
$MgCl_2$(aq)	–641.6

(A) –307.2 kJ·mol^{-1} **(B)** 307.2 kJ·mol^{-1} **(C)** –474.4 kJ·mol^{-1} **(D)** 474.4 kJ·mol^{-1}

Answers to Study Questions

1. C
2. C
3. B
4. D
5. A
6. B
7. B
8. B
9. A

Answers to Practice Questions

1. B
2. C
3. A
4. B
5. C
6. B
7. A
8. D
9. B
10. D
11. A
12. C
13. C
14. A
15. B
16. B
17. C
18. C
19. D
20. C
21. C
22. C
23. B
24. C
25. A
26. C
27. C
28. C
29. C
30. A

Chapter 7 – Structure and Bonding

Chapter Summary:

This chapter will focus on concepts related to chemical bonding. This includes both ionic and covalent bonding extending into bonding theories and applications. Although this chapter is traditionally included in first-term chemistry, the concepts in this chapter are critical to your understanding of other concepts throughout general chemistry.

Specific topics covered in this chapter are:

- Lattice energy
- Electronegativity and ionic vs. covalent bonding
- Structures
- Formal charge
- Resonance
- Bond enthalpy, bond length and bond strength
- Molecular shape and geometry
- Polarity
- Valence bond theory
- Molecular orbital theory

Previous material that is relevant to your understanding of questions in this chapter include:

- Atomic orbitals and number of valence electrons (***Chapter 2***)
- Balancing equations (***Chapter 4***)
- Enthalpy and sign conventions (***Chapter 6***)

Common representations used in questions related to this material:

Name	Example	Used in questions related to
Lewis dot structures	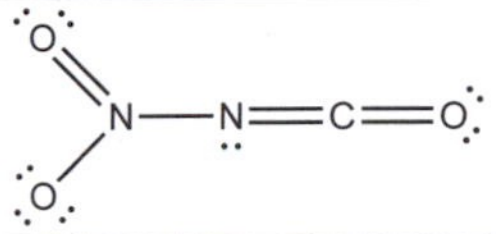 O, O, N—N=C=O	Structures, formal charge, resonance, bond enthalpy, shape, polarity, valence bond theory
Atomic orbitals	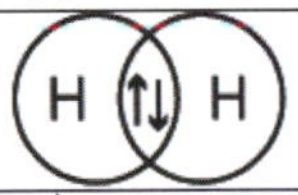 H ↑↓ H	Valence bond theory, molecular orbital theory
Molecular orbital diagrams	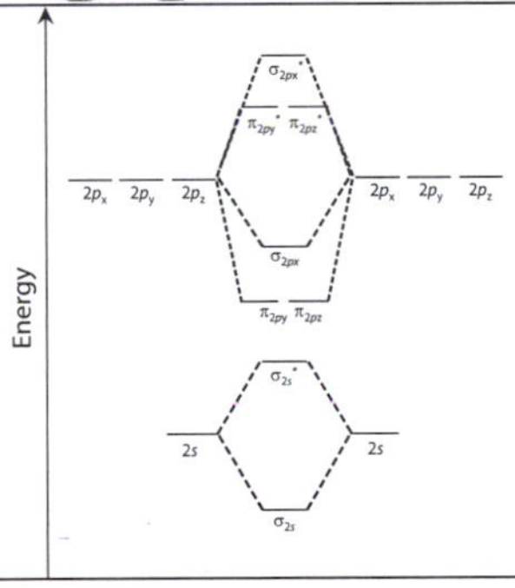 Energy; 2p	Molecular orbital theory

Where to find this in your textbook:

The material in this chapter typically aligns to one or two chapters on bonding (could be labeled as "Chemical Bonding") in your textbook. The name of your chapter may vary.

Practice exam:

There are practice exam questions aligned to the material in this chapter. Because there are a limited number of questions on the practice exam, a review of the breadth of the material in this chapter is advised in preparation for your exam.

How this fits into the big picture:

The material in this chapter aligns to the Big Idea of Bonding (2) as listed on page 12 of this study guide.

Study Questions (SQ)

SQ-1. What is the order of ScN, MgO, $CaCl_2$, and NaBr in order of increasing magnitude of lattice energy?

Conceptual

(A) $CaCl_2$ < NaBr < ScN < MgO

(B) $CaCl_2$ < NaBr < MgO < ScN

(C) NaBr < $CaCl_2$ < MgO < ScN

(D) NaBr < $CaCl_2$ < ScN < MgO

Knowledge Required: (1) Definition of lattice energy. (2) Charges and sizes of ions (based on periodic trends).

Thinking it Through: Because this question requires you to order by magnitude of lattice energy, the first thing you recall when starting to solve this is the definition of lattice energy: the amount of energy needed to completely separate 1 mole of a solid ionic compound into its gaseous ions. You also know this can be represented as a balanced equation (shown for calcium chloride): $CaCl_2(s) \rightarrow Ca^{2+}(g) + 2Cl^{-}(g)$

You then extend this into thinking about the energy associated with this process and know that you can think about this in terms of Coulombic attraction: $E \propto \frac{Q_+Q_-}{r}$ which is proportional to the energy released when the ions bond; the opposite would then be the lattice energy (separation of ions). From this proportionality, you see that the lattice energy will increase as the charges on the ions increase and decrease as the size of ions increases.

Now, you evaluate the compounds in the question based on these two criteria: (1) ionic charge and (2) ionic size. Both nitride and scandium have charges of –3 and +3 respectively. Both oxygen and magnesium have charges of –2 and +2 respectively. Calcium has a charge of +2 while chloride has a charge of –1. Finally, sodium and bromide have charges of +1 and –1 respectively. Therefore, the highest lattice energy corresponds to ScN, then MgO, then $CaCl_2$ with NaBr as the lowest. This corresponds to response **(C)**.

Choice **(A)** incorrectly orders based on molar mass. Choice **(B)** ranks $CaCl_2$ (1:2) lowest followed by ordering of charges. Choice **(D)** reverses ScN and MgO.

Practice Questions Related to This: **PQ-1** and **PQ-2**

SQ-2. The electronegativity of C is 2.5 and that of F is 4.0. What is the best description of a chemical bond between these two elements?

(A) ionic bond

(B) metallic bond

(C) nonpolar covalent bond

(D) polar covalent bond

Knowledge Required: (1) Definition of bond types. (2) Use of electronegativity to determine bond type.

Thinking it Through: Given the types of elements introduced in the question, your first response may be to say that carbon is a non-metal and fluorine is a nonmetal, so this is a covalent bond. However, multiple answer choices offer covlent as an option. As you move to considering the electronegativities provided in the question, you know that difference in electronegativity provides more of a continuum for evaluating the type of bond between atoms as opposed to the ionic vs. covalent bond categories alone.

type of bond	Ionic	Polar covalent	Nonpolar covalent
type of element (and electronegativities)	metal (low) nonmetal (high)	nonmetal (high) nonmetal (high)	one nonmetal (high)
difference in electronegativities	large (>2)	moderate to low (1.6 to 0.4)	small <0.4

Therefore, using this, the difference in electronegativities is 1.5, The difference in electronegativity is too small for an ionic bond (choice **(A)**) and is also too large for the bond to be considered nonpolar covalent (choice **(C)**). The electronegativity difference is large enough for the bond to be considered polar covalent, so choice **(D)** is correct. Finally, this is not metallic bonding (present in metals), choice **(B)**.

Practice Questions Related to This: **PQ-3** and **PQ-4**

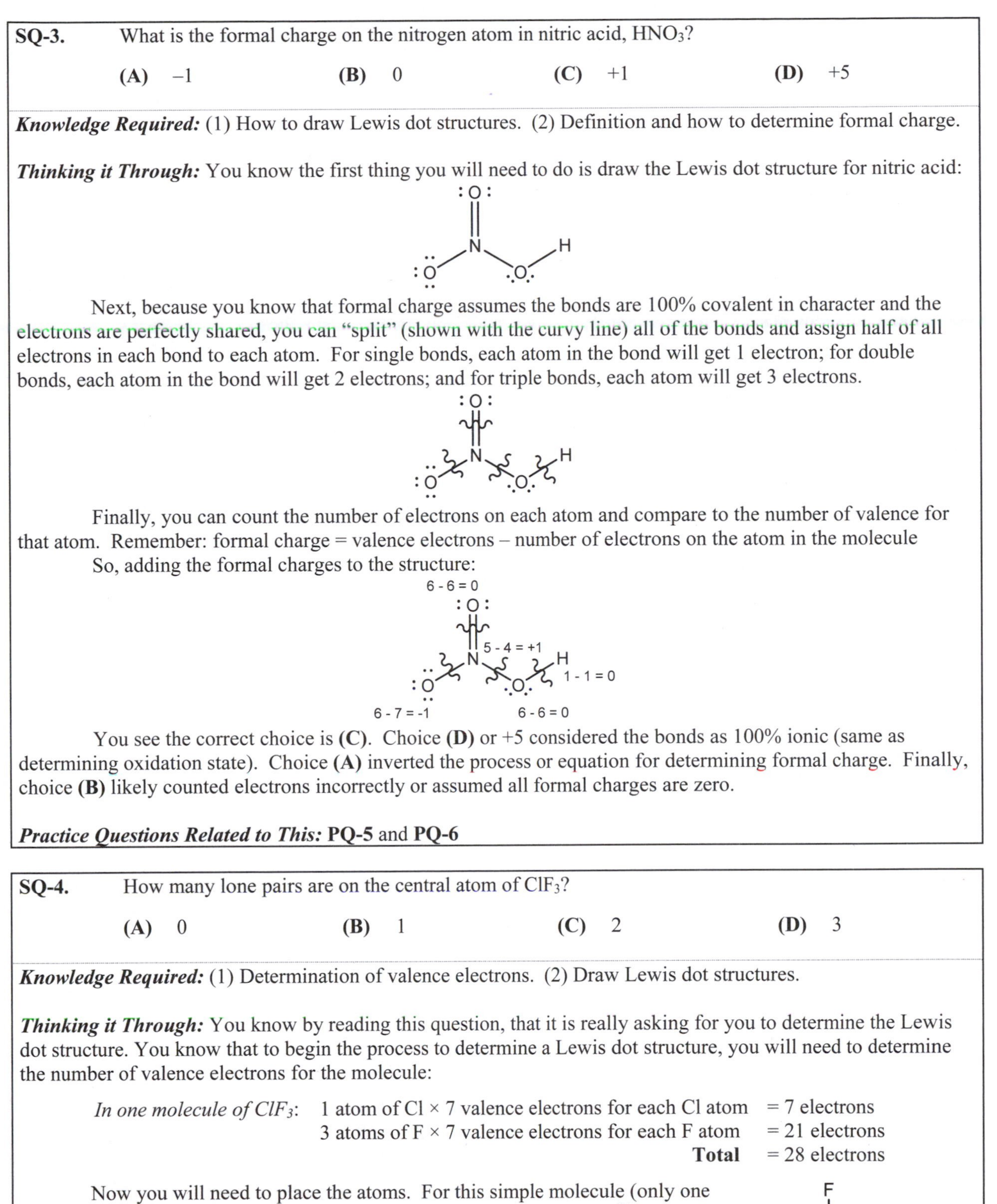

SQ-3. What is the formal charge on the nitrogen atom in nitric acid, HNO_3?

(A) −1 **(B)** 0 **(C)** +1 **(D)** +5

Knowledge Required: (1) How to draw Lewis dot structures. (2) Definition and how to determine formal charge.

Thinking it Through: You know the first thing you will need to do is draw the Lewis dot structure for nitric acid:

Next, because you know that formal charge assumes the bonds are 100% covalent in character and the electrons are perfectly shared, you can "split" (shown with the curvy line) all of the bonds and assign half of all electrons in each bond to each atom. For single bonds, each atom in the bond will get 1 electron; for double bonds, each atom in the bond will get 2 electrons; and for triple bonds, each atom will get 3 electrons.

Finally, you can count the number of electrons on each atom and compare to the number of valence for that atom. Remember: formal charge = valence electrons – number of electrons on the atom in the molecule

So, adding the formal charges to the structure:

You see the correct choice is **(C)**. Choice **(D)** or +5 considered the bonds as 100% ionic (same as determining oxidation state). Choice **(A)** inverted the process or equation for determining formal charge. Finally, choice **(B)** likely counted electrons incorrectly or assumed all formal charges are zero.

Practice Questions Related to This: **PQ-5** and **PQ-6**

SQ-4. How many lone pairs are on the central atom of ClF_3?

(A) 0 **(B)** 1 **(C)** 2 **(D)** 3

Knowledge Required: (1) Determination of valence electrons. (2) Draw Lewis dot structures.

Thinking it Through: You know by reading this question, that it is really asking for you to determine the Lewis dot structure. You know that to begin the process to determine a Lewis dot structure, you will need to determine the number of valence electrons for the molecule:

In one molecule of ClF_3:	1 atom of Cl × 7 valence electrons for each Cl atom	= 7 electrons
	3 atoms of F × 7 valence electrons for each F atom	= 21 electrons
	Total	= 28 electrons

Now you will need to place the atoms. For this simple molecule (only one central atom), the central atom is listed first (the least electronegative atom) and all other atoms are terminal. Following this, bond all of the terminal atoms to the central atom with single bonds.

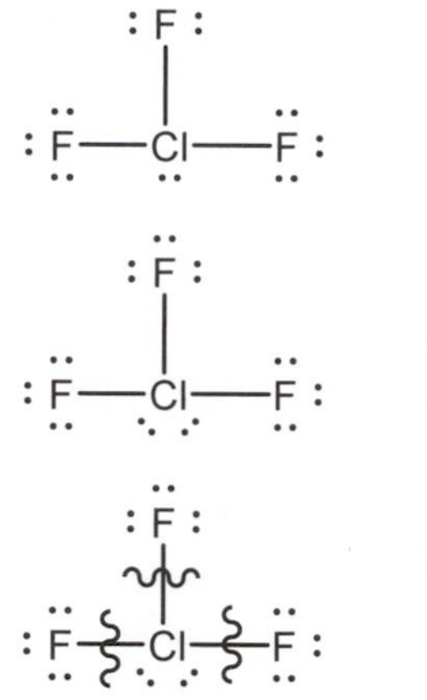

Next, complete the octets of each atom by adding lone pairs of electrons. For each fluorine atom, you will add three lone pairs and for chlorine, you will add one.

Now, you count the number of electrons showing in your structure so far. For this one, you now have 26 which is less than 28.

To increase to 28, you will need to add a lone pair of electrons. You cannot add this to fluorine (which cannot expand its octet), so the only option is to add this to chlorine. When you add the number of electrons showing in your structure (28) and compare to the number available in the molecule (28), this confirms that this is a possible structure.

Finally, as a check that the lone pair is added to the most reasonable location, you can use formal charge. Remembering that formal charge assumes the bonds are 100% covalent or perfectly shared, you can "split" the bonds to count the number of electrons for each atom from each bond.

Now, you can count the number of electrons on each atom and compare to the number of valence for that atom. Remember: formal charge = valence electrons – number of electrons on the atom in the molecule. So, with all formal charges equal to zero, this structure is reasonable and the number of lone pairs on chlorine is 2 or choice **(C)**.

Practice Questions Related to This: **PQ-7, PQ-8, PQ-9,** and **PQ-10**

SQ-5. *Conceptual* What is a resonance structure for the molecule shown?

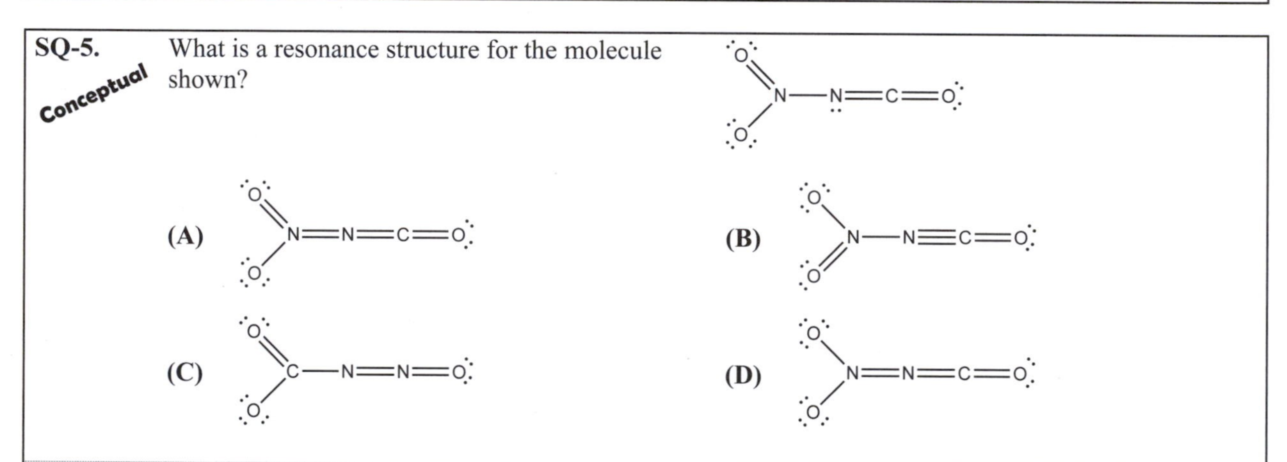

Knowledge Required: (1) Draw Lewis dot structures. (2) Definition of resonance structures. (3) Calculation of formal charges.

Thinking it Through: You can see from reading the question, that you will need to know the definition of resonance structures. Two or more structures are resonance structures when:

1. They are the same molecule – they must have the same formula (so CH_3CH_2OH and CH_3COOH cannot be resonance structures).

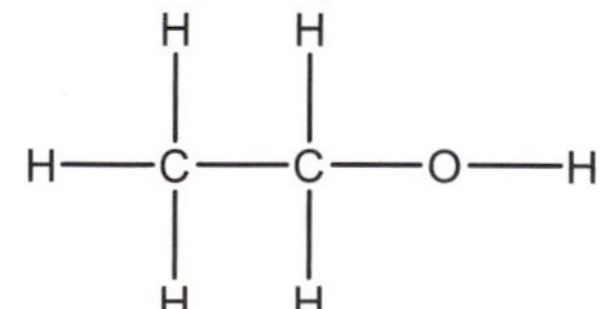

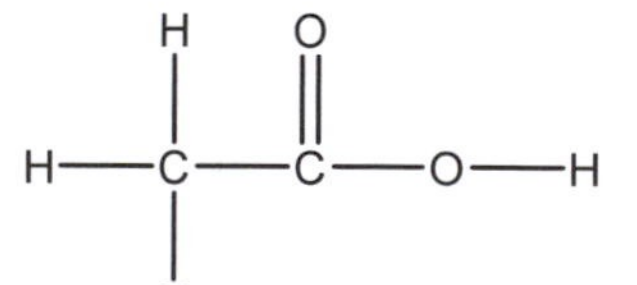

2. Only the electrons can move (not the atoms) (so CH_3CH_2OH and CH_3OCH_3 cannot be resonance structures).

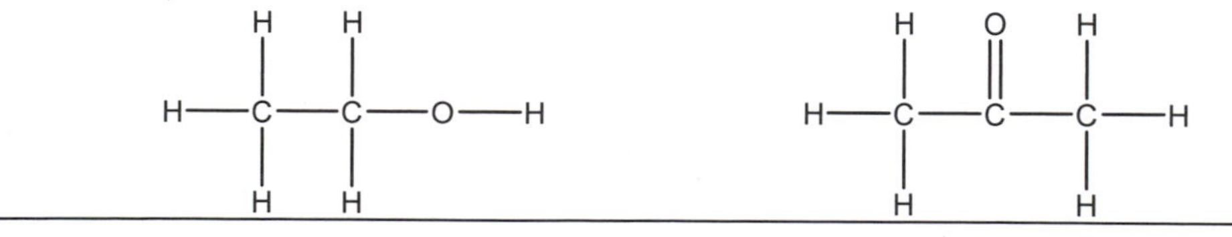

3. They must be reasonable structures. To determine this, you will use formal charges. Remembering that structures with formal charges as close to zero as possible are more reasonable and when a formal charge cannot be zero, the more electronegative atom will have a negative formal charge. So, the carbonate ion has three equivalent resonance structures.

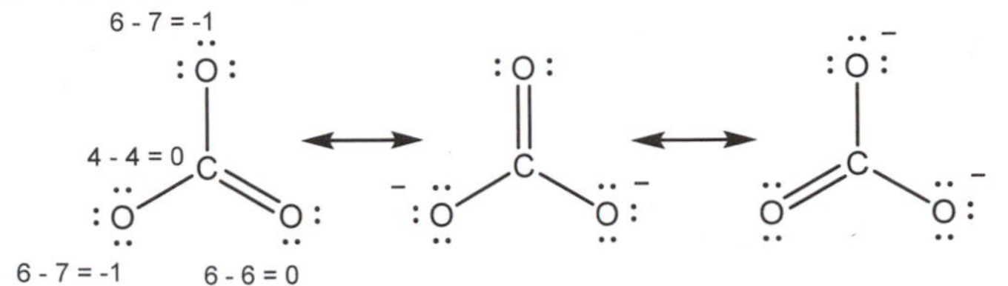

It is important to remember that the actual molecule or ion is not rapidly interconverting between these resonance structures. Instead, the actual molecule or ion is an average of the structures.

Using this criteria, you can eliminate the structure shown in choice **(C)** because the atoms have moved. You can also eliminate the structure shown in choice **(A)** because nitrogen cannot expand its octet and make 5 bonds; similarly carbon cannot expand its octet and make 5 bonds, so you also eliminate structure **(B)**. This leaves structure **(D)** for evaluation and comparison to the structure from the question:

Showing that **(D)** is a possible resonance structure for the original structure in the question.

Practice Questions Related to This: **PQ-11** and **PQ-12**

SQ-6. Using the bond energies, what is ΔH_{rxn} for the reaction below?

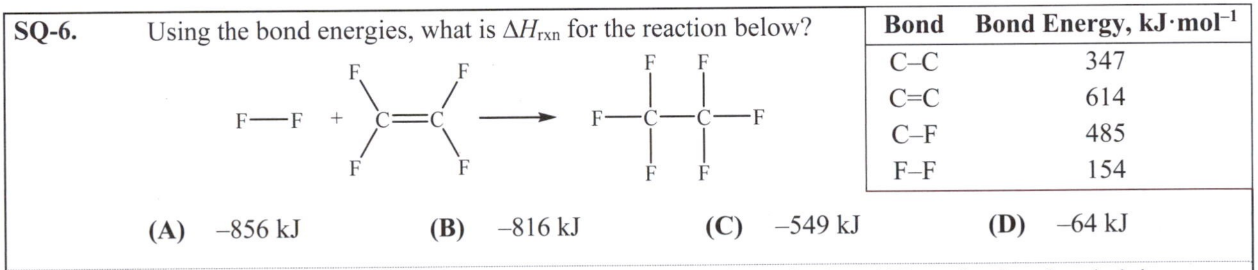

Bond	Bond Energy, $kJ \cdot mol^{-1}$
C–C	347
C=C	614
C–F	485
F–F	154

(A) –856 kJ **(B)** –816 kJ **(C)** –549 kJ **(D)** –64 kJ

Knowledge Required: (1) Definition of bond enthalpy. (2) Method to calculate ΔH_{rxn} using bond enthalpies.

Thinking it Through: You can see from reading the question that you will need to consider the bonds broken and the bonds formed to determine the overall change in enthalpy for the reaction. You decide to take this in parts and first determine how much energy is required to break the bond (reactant side):

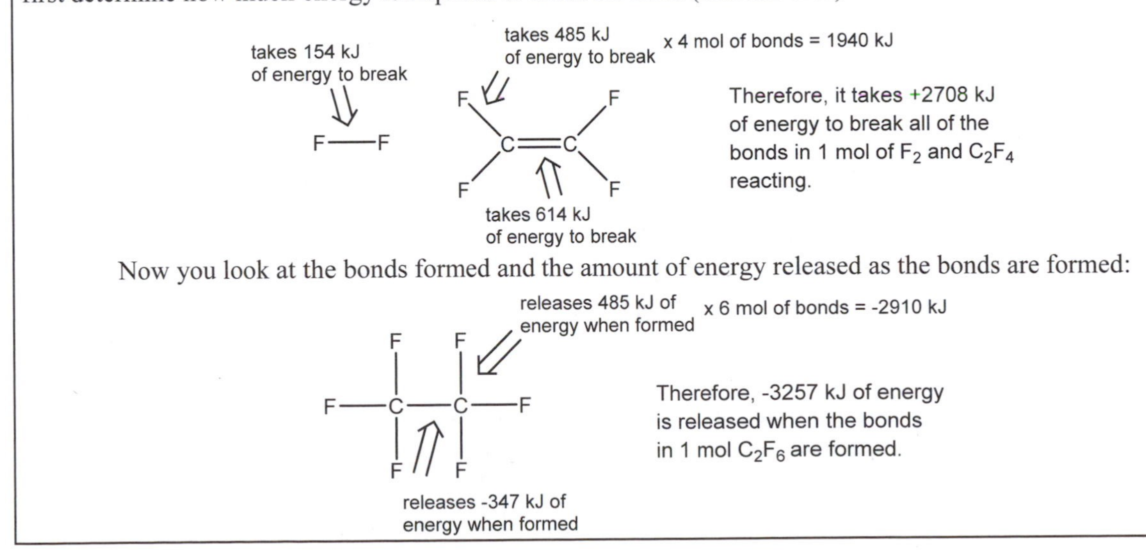

Now you look at the bonds formed and the amount of energy released as the bonds are formed:

Combining the two, the enthalpy of reaction (ΔH_{rxn}) is –3257 kJ plus +2708 kJ or overall –549 kJ or choice **(C)**.

Choice **(A)** was incorrectly determined by using half of the carbon/carbon double bond. Choice **(B)** was incorrectly determined by not considering the carbon-carbon bond (neither breaking the double bond nor making the single bond). Choice **(D)** was incorrectly determined by not including the number of bonds broken (so 154 kJ + 614 kJ) or number of bonds made (so –347 kJ + –485 kJ).

Practice Questions Related to This: **PQ-13** and **PQ-14**

SQ-7. An O=O bond is ____ than an O–O bond.

Conceptual

(A) longer and stronger
(B) longer and weaker
(C) shorter and stronger
(D) shorter and weaker

Knowledge Required: (1) Understanding of covalent bonds.

Thinking it Through: You could approach this by knowing facts about bonds, but you also know that you can approach this by considering a single versus a double bond in the context of valence bond theory. A single bond is often the overlap of two hybrid atomic orbitals, such as shown below for the O–O bond with sp hybrid atomic orbitals to form a σ bond:

A double bond is often the overlap of two unhybridized p orbitals to form a π bond in addition to the σ bond:

Therefore, now you can explain how a double bond is stronger than a single bond because there is additional overlap of unhybridized p orbitals (thus eliminating **(B)** and **(D)**). What is not immediately obvious from this is that double bonds are shorter than single bonds (and stronger bonds are typically shorter) or choice **(C)**.

Practice Questions Related to This: **PQ-15** and **PQ-16**

SQ-8. What is the molecular geometry of CO_3^{2-}?

(A) bent
(B) T-shaped
(C) trigonal bipyramidal
(D) trigonal planar

Knowledge Required: (1) Draw Lewis dot structures. (2) Molecular geometry.

Thinking it Through: The first thing you notice when you read this is that you are going to need to determine the shape, and the next thing you notice is that this is for an ion. Earlier, you walked through the process to determine the Lewis dot structure for a molecule, now you will walk through this with an ion (including using formal charge to "place" the charges). Again, you know that to begin the process to determine a Lewis dot structure, you will need to determine the number of valence electrons for the ion:

In one ion of CO_3^{2-}:		
	1 atom of C × 4 valence electrons for each C atom	= 4 electrons
	3 atoms of O × 6 valence electrons for each O atom	= 18 electrons
	–2 negative charge	= 2 electrons
	Total	= 24 electrons

Now you will need to place the atoms. For this simple ion (only one central atom), the central atom is listed first/the least electronegative atom and all other atoms are terminal. Following this, bond all of the terminal atoms to the central atom with single bonds.

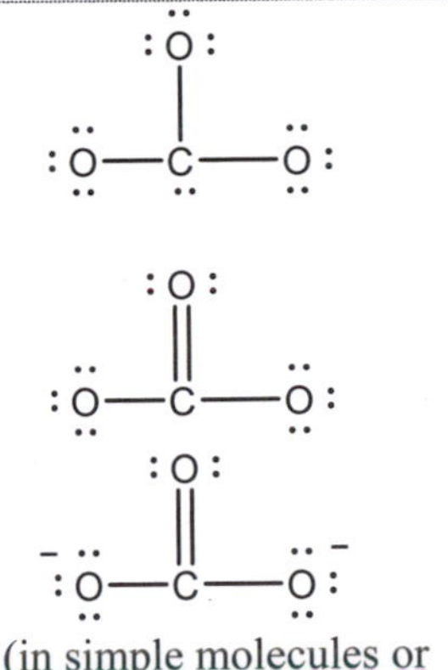

Next, complete the octets of each atom by adding lone pairs of electrons. For each oxygen atom, you will add three lone pairs and for carbon, you will add one.

Now, you count the number of electrons shown in your structure so far. For this one, you now have 26 which is more than 24.

To decrease to 24, you will need to make a multiple bond. You can add the double bond between carbon and any oxygen.

Then you will again complete the octets for all atoms.

Finally, to place the charges on the ions, you will use formal charge. The formal charges you determine will be the charges on the ion (here a net of –2, so for this ion, there are two atoms with a negative one formal charge).

Now, on to the shape of the ion. You start this by evaluating the central atom (in simple molecules or ions) or atom of focus (in more complex molecules or ions) and asking yourself how many atoms are bonded to the central atom and how many lone pairs are on the central atom. For the carbonate ion (CO_3^{2-}), this is three (three oxygen atoms, no lone pairs). You recall that it is not the type of bonding between the atoms, but only the number that matters (so no difference for the three oxygen atoms). For three electron groups (or electron families, regions or domains), the base geometry is trigonal planar with a bond angle of 120°. Because there are no lone pairs on the central atom, this is also the shape of the ion. Drawn more consistently with this shape and bond angle (shown to the right) which is choice **(D)**.

Choice **(A)** may be selected if the formula were read incorrectly. Choice **(B)** might be selected if the structure were incorrectly drawn with two lone pairs on the central atom while choice **(C)** may be selected if the structure was incorrectly drawn with one lone pair on the central atom.

Practice Questions Related to This: **PQ-17** and **PQ-18**

SQ-9. Which species contains a bond angle less than 120°?

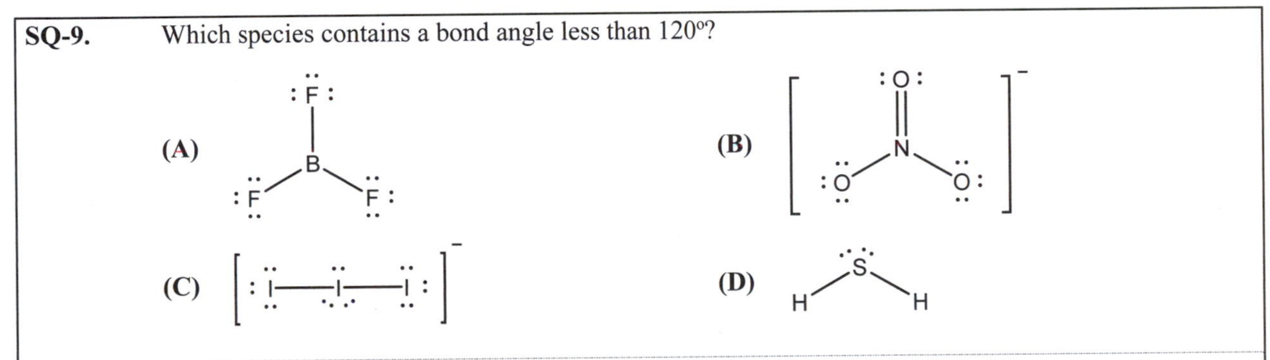

Knowledge Required: (1) Draw Lewis dot structures. (2) Molecular shape and corresponding bond angles.

Thinking it Through: Similar to the previous study question, you can see that you will need to consider the shape of each molecule or ion. You know to do this you will need to evaluate the environment of the central atom (in all of these simple molecules or ions).

Molecule or ion	# of bonded atoms	# of lone pairs	# of electron groups	Base geometry	Shape	Bond angle*
	3	0	3	trigonal planar	trigonal planar	120°
	3	0	3	trigonal planar	trigonal planar	120°

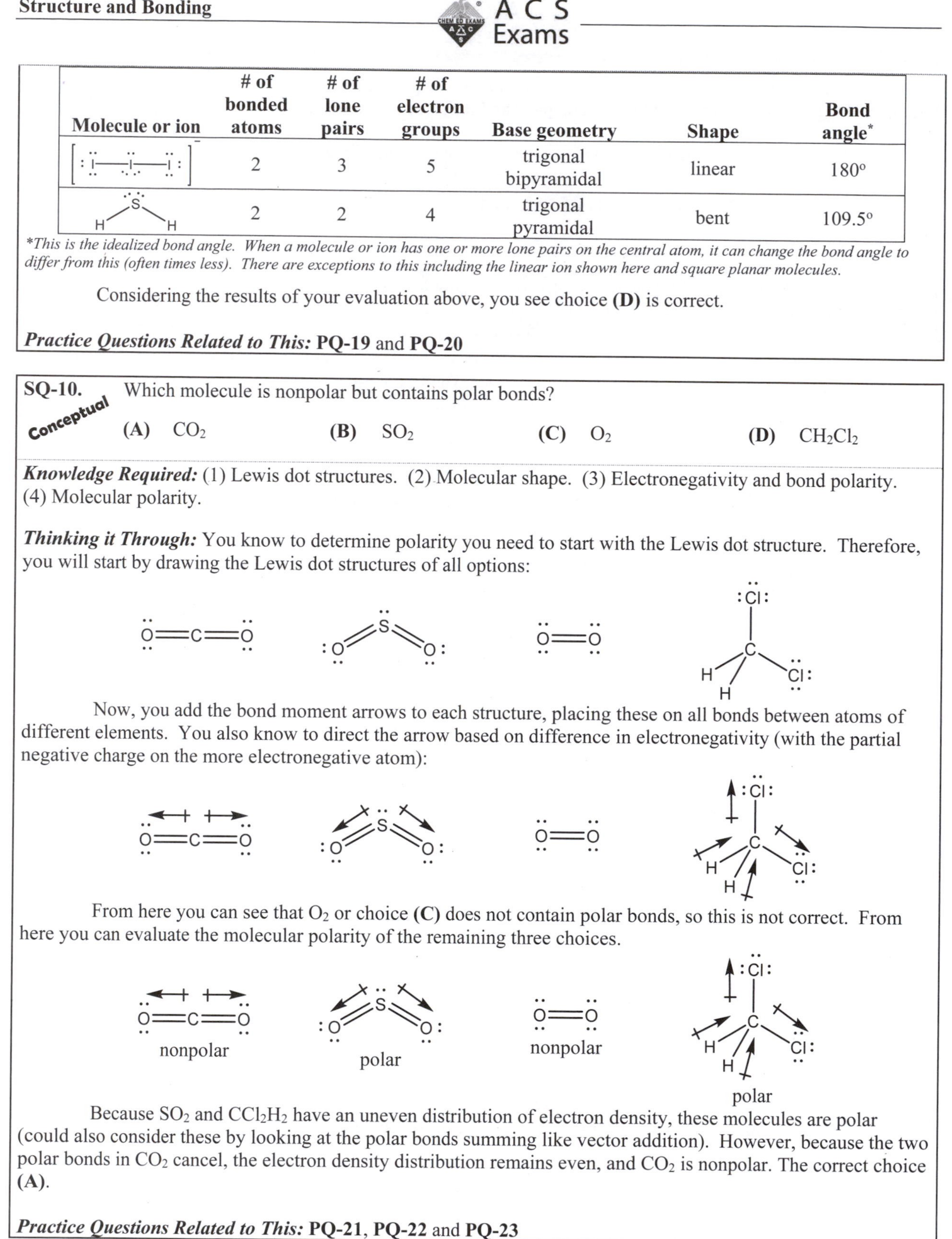

Molecule or ion	# of bonded atoms	# of lone pairs	# of electron groups	Base geometry	Shape	Bond angle*
$[I_3]^-$	2	3	5	trigonal bipyramidal	linear	180°
H_2S	2	2	4	trigonal pyramidal	bent	109.5°

**This is the idealized bond angle. When a molecule or ion has one or more lone pairs on the central atom, it can change the bond angle to differ from this (often times less). There are exceptions to this including the linear ion shown here and square planar molecules.*

Considering the results of your evaluation above, you see choice **(D)** is correct.

Practice Questions Related to This: **PQ-19** and **PQ-20**

SQ-10. Conceptual

Which molecule is nonpolar but contains polar bonds?

(A) CO_2 **(B)** SO_2 **(C)** O_2 **(D)** CH_2Cl_2

Knowledge Required: (1) Lewis dot structures. (2) Molecular shape. (3) Electronegativity and bond polarity. (4) Molecular polarity.

Thinking it Through: You know to determine polarity you need to start with the Lewis dot structure. Therefore, you will start by drawing the Lewis dot structures of all options:

Now, you add the bond moment arrows to each structure, placing these on all bonds between atoms of different elements. You also know to direct the arrow based on difference in electronegativity (with the partial negative charge on the more electronegative atom):

From here you can see that O_2 or choice **(C)** does not contain polar bonds, so this is not correct. From here you can evaluate the molecular polarity of the remaining three choices.

Because SO_2 and CCl_2H_2 have an uneven distribution of electron density, these molecules are polar (could also consider these by looking at the polar bonds summing like vector addition). However, because the two polar bonds in CO_2 cancel, the electron density distribution remains even, and CO_2 is nonpolar. The correct choice **(A)**.

Practice Questions Related to This: **PQ-21**, **PQ-22** and **PQ-23**

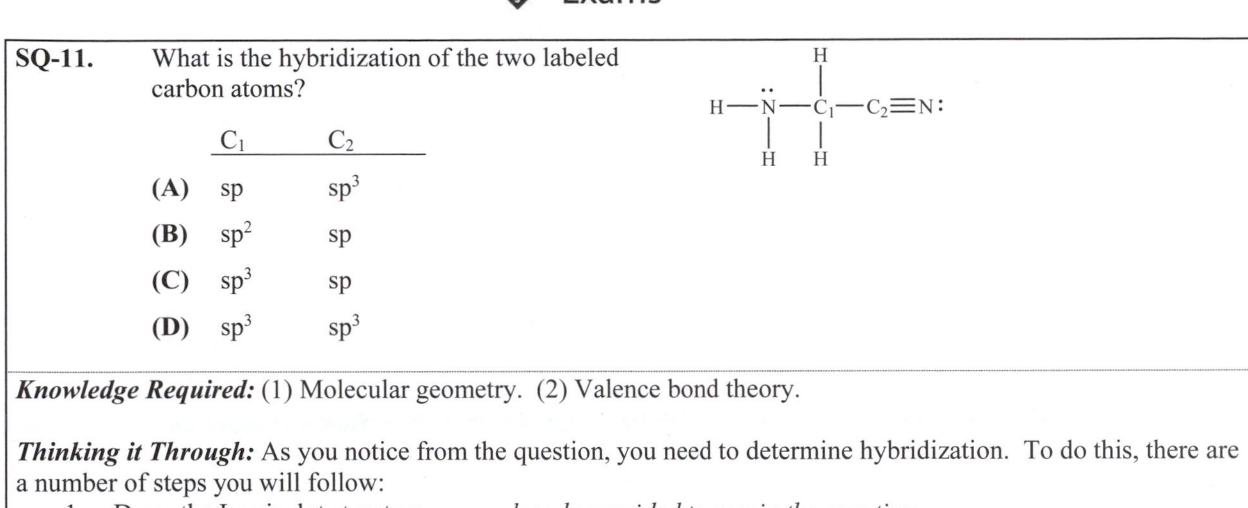

SQ-11. What is the hybridization of the two labeled carbon atoms?

	C_1	C_2
(A)	sp	sp^3
(B)	sp^2	sp
(C)	sp^3	sp
(D)	sp^3	sp^3

Knowledge Required: (1) Molecular geometry. (2) Valence bond theory.

Thinking it Through: As you notice from the question, you need to determine hybridization. To do this, there are a number of steps you will follow:

1. Draw the Lewis dot structure *already provided to you in the question*
2. Evaluate the types of bonds made by the targeted atom:

C_1	C_2
makes 4 single bonds	makes 1 single bond and 1 triple bond

3. Evaluate the bonds made in terms of overlap of atomic orbitals or hybrid atomic orbitals (you also know 1 single bond is often 1 σ bond, 1 double bond is often 1 σ bond and 1 π bond, and 1 triple bond is 1 σ bond and 2 π bonds):

C_1	C_2
makes 4 σ bonds	makes 2 σ bonds and 2 π bonds

4. Consider the need for hybrid atomic orbitals for each atom based on the bonds:

	C_1	C_2
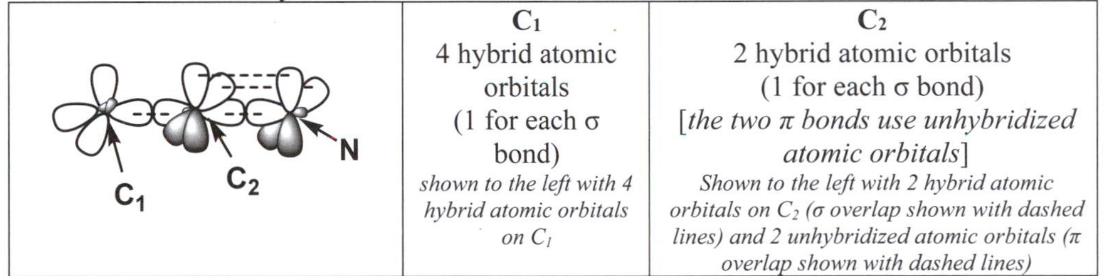	4 hybrid atomic orbitals (1 for each σ bond) *shown to the left with 4 hybrid atomic orbitals on C_1*	2 hybrid atomic orbitals (1 for each σ bond) [*the two π bonds use unhybridized atomic orbitals*] *Shown to the left with 2 hybrid atomic orbitals on C_2 (σ overlap shown with dashed lines) and 2 unhybridized atomic orbitals (π overlap shown with dashed lines)*

5. Because the hybrid atomic orbitals are made of s and p atomic orbitals, the number of hybrid atomic orbitals corresponds to the "hybridization" that you assign:
   2 hybrid atomic orbitals = 1 s and 1 p = "sp"
   3 hybrid atomic orbitals = 1 s and 2 p = "sp^2"
   4 hybrid atomic orbitals = 1 s and 3 p = "sp^3"

So you assign the hybridization of C_1 as sp^3 and C_2 as sp or choice **(C)**. Choice **(A)** reverses the assignment; choice **(B)** assigns C_1 incorrectly as sp^2; and choice **(D)** assigns C_2 incorrectly as sp^3.

Practice Questions Related to This: **PQ-24**, **PQ-25** and **PQ-26**

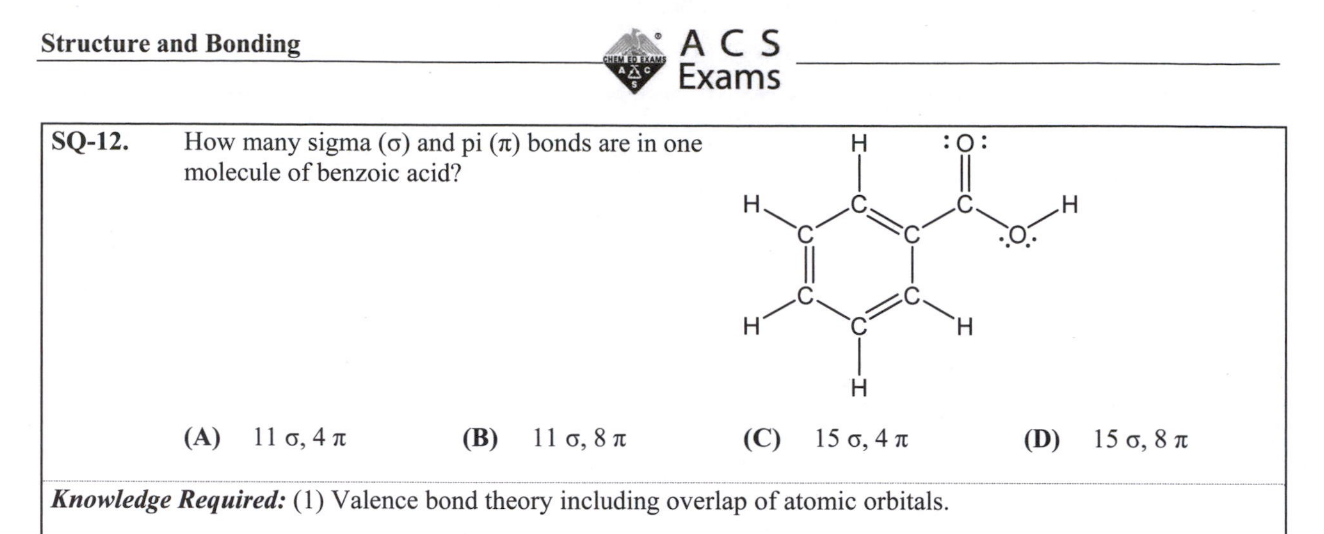

SQ-12. How many sigma (σ) and pi (π) bonds are in one molecule of benzoic acid?

(A) 11 σ, 4 π **(B)** 11 σ, 8 π **(C)** 15 σ, 4 π **(D)** 15 σ, 8 π

Knowledge Required: (1) Valence bond theory including overlap of atomic orbitals.

Thinking it Through: As discussed in the previous study question, you know that a single bond is often 1 sigma bond, a double bond is often 1 sigma and 1 pi bond, and a triple bond is 1 sigma bond and 2 pi bonds. Therefore, you will answer this question by counting the number of bonds and translating these into sigma and pi bonds:

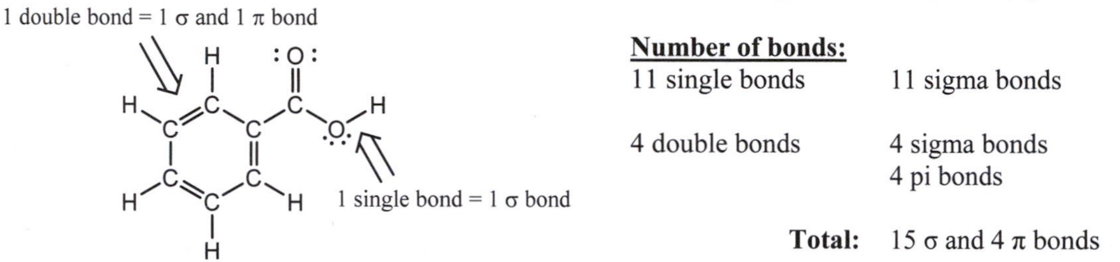

Number of bonds:	
11 single bonds	11 sigma bonds
4 double bonds	4 sigma bonds 4 pi bonds
Total:	15 σ and 4 π bonds

And you select the correct choice **(C)**. Choice **(A)** omits the sigma bonds from the double bonds. Choice **(B)** counts the double bonds as 2 pi bonds rather than 1 sigma and 1 pi bond. Choice **(D)** counts the double bonds as 2 sigma and 1 pi bond.

Practice Questions Related to This: **PQ-27** and **PQ-28**

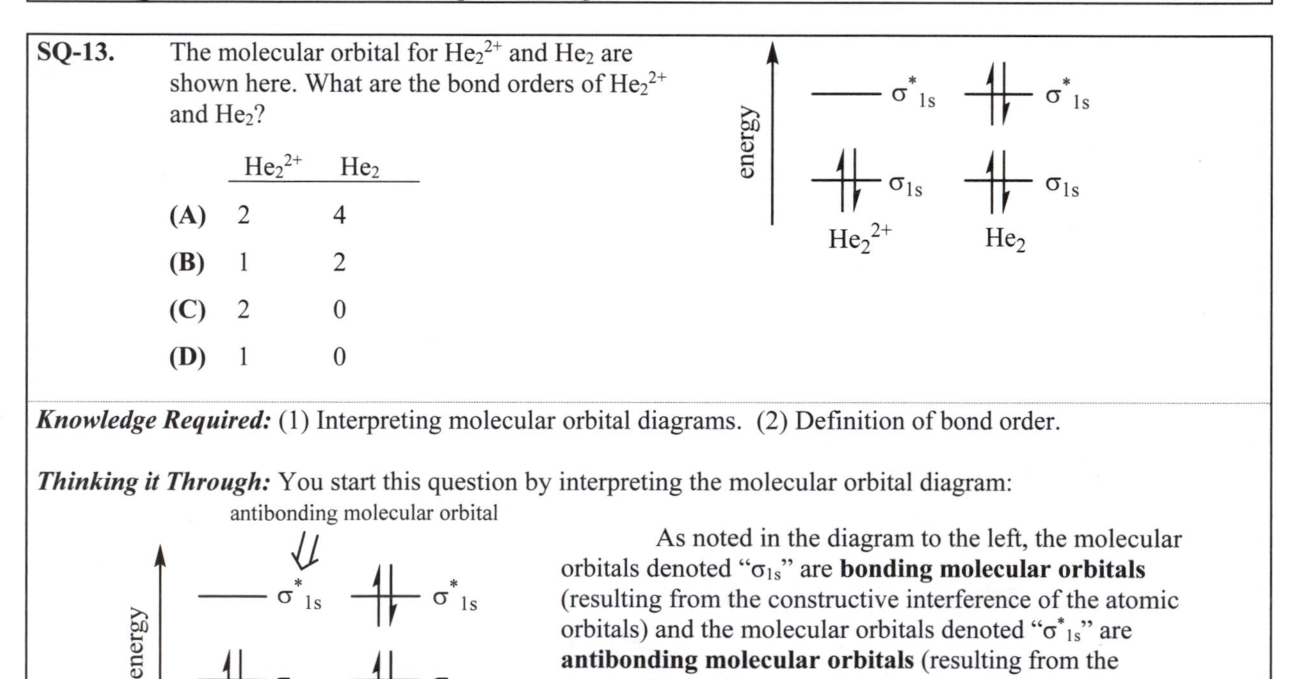

SQ-13. The molecular orbital for He_2^{2+} and He_2 are shown here. What are the bond orders of He_2^{2+} and He_2?

	He_2^{2+}	He_2
(A)	2	4
(B)	1	2
(C)	2	0
(D)	1	0

Knowledge Required: (1) Interpreting molecular orbital diagrams. (2) Definition of bond order.

Thinking it Through: You start this question by interpreting the molecular orbital diagram:

antibonding molecular orbital

energy σ^*_{1s} σ^*_{1s} σ_{1s} σ_{1s} He_2^{2+} He_2

electrons bonding molecular orbital

As noted in the diagram to the left, the molecular orbitals denoted "σ_{1s}" are **bonding molecular orbitals** (resulting from the constructive interference of the atomic orbitals) and the molecular orbitals denoted "σ^*_{1s}" are **antibonding molecular orbitals** (resulting from the destructive interference of the atomic orbitals).

The electrons fill these orbitals based on the Aufbau principle or lowest to highest energy (noted in the diagram).

Therefore, the first species, He_2^{2+} ion, has 2 electrons in the σ_{1s} bonding molecular orbital or can also be noted in the molecular orbital configuration of $(\sigma_{1s})^2$. The second species, He_2 molecular, has 2 electrons in the σ_{1s} bonding molecular orbital and 2 electrons in the σ^*_{1s} antibonding molecular orbital or can also be noted in the molecular orbital configuration of $(\sigma_{1s})^2(\sigma^*_{1s})^2$.

Bond order is determined as one-half the difference of antibonding electrons from bonding electrons:

$$\text{bond order} = \frac{1}{2}(\#\text{ of bonding electrons} - \#\text{ of antibonding electrons})$$

You can see from either the diagram or the configuration, He_2^{2+} has 2 bonding electrons, so a bond order of 1 while He_2 has 2 bonding electrons and 2 antibonding electrons or a bond order of 0 or choice **(D)**. Choice **(A)** reflects the electrons; choice **(B)** omits the antibonding electrons; and choice **(C)** omits the one-half.

Practice Questions Related to This: **PQ-29** and **PQ-30**

Practice Questions (PQ)

Conceptual **PQ-1.** Arrange CsBr, NaCl, and RbBr in increasing magnitude of lattice energy.

(A) CsBr < NaCl < RbBr **(B)** CsBr < RbBr < NaCl

(C) NaCl < RbBr < CsBr **(D)** RbBr < CsBr < NaCl

Conceptual **PQ-2.** Which ionic compound has the largest magnitude lattice energy?

(A) CaO **(B)** KBr **(C)** MgO **(D)** NaF

PQ-3. How many covalent bonds are represented in the formula NH_4Cl?

(A) 5 **(B)** 4 **(C)** 1 **(D)** 0

Conceptual **PQ-4.** Which statement best describes ionic bonding in lithium fluoride?

(A) The positive and negative charges of the ions cancel out.

(B) A lithium atom shares one electron with a fluorine atom.

(C) An electron is transferred from a lithium atom to a fluorine atom.

(D) An electrostatic attraction exists between lithium ions and fluoride ions.

PQ-5. What is the formal charge on the phosphorous atom in the phosphate ion as shown?

O
‖
O^-—P—O^-
|
O^-

(A) –3 **(B)** –1 **(C)** 0 **(D)** +5

PQ-6. Which molecule contains carbon with a negative formal charge?

(A) CO **(B)** CO_2 **(C)** H_2CO **(D)** CH_4

PQ-7. How many valence electrons are in the carbonate ion, CO_3^{2-}?

(A) 20 **(B)** 22 **(C)** 24 **(D)** 32

PQ-8. Which is the best Lewis structure for nitrogen monoxide?

(A) $:N\equiv O:$ **(B)** $:\dot{N}\equiv O\cdot$ **(C)** $:\dot{N}=\ddot{O}:$ **(D)** $\cdot\ddot{N}=\dot{O}:$

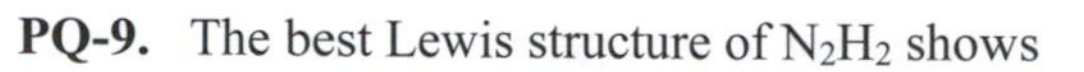

PQ-9. The best Lewis structure of N_2H_2 shows

(A) a nitrogen-nitrogen triple bond.

(B) a nitrogen-nitrogen single bond.

(C) each nitrogen with one nonbonding electron pair.

(D) each nitrogen with two nonbonding electron pairs.

PQ-10. What is the correct Lewis structure for PF_3?

(A) 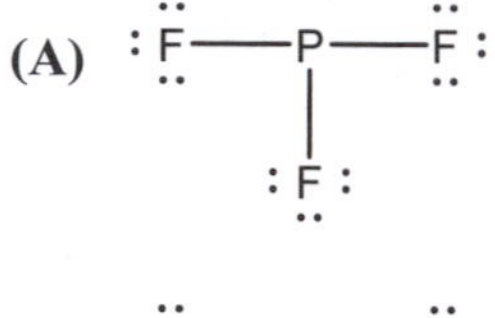

(B)

(C) 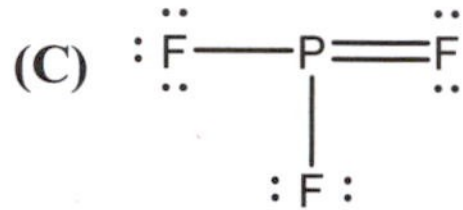

(D)

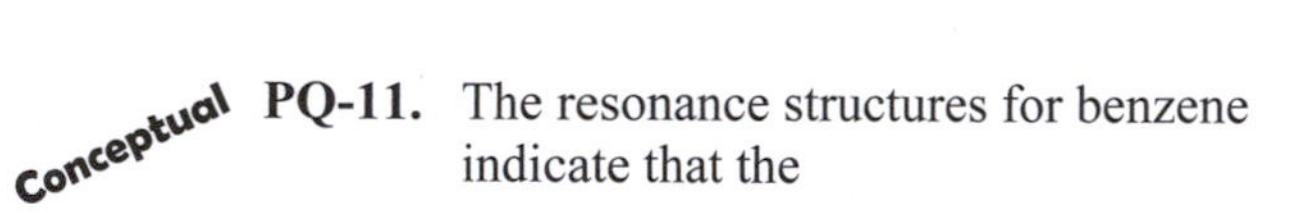

Conceptual **PQ-11.** The resonance structures for benzene indicate that the

I ⟷ II

(A) bonds between carbons in benzene are identical.

(B) benzene molecule exists as two unique structures.

(C) benzene molecule alternates between forms I and II.

(D) carbon-carbon single bonds are longer than the double bonds.

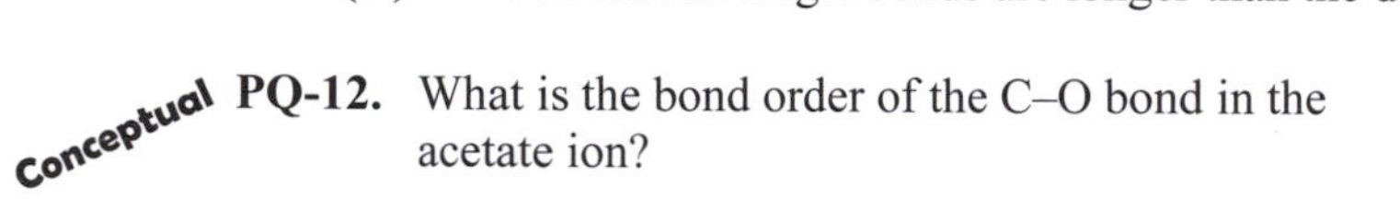

Conceptual **PQ-12.** What is the bond order of the C–O bond in the acetate ion?

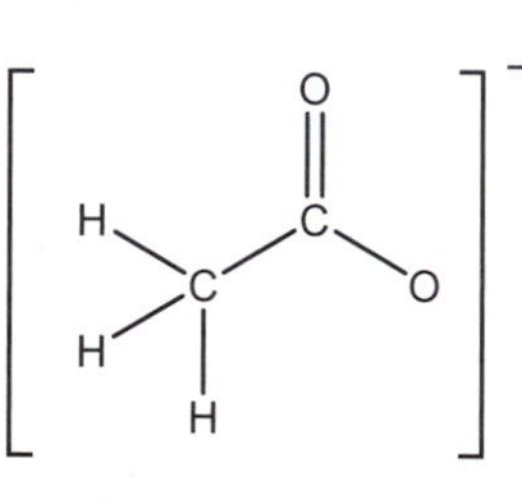

(A) 1 **(B)** 1.5 **(C)** 2

(D) switches between 1 and 2

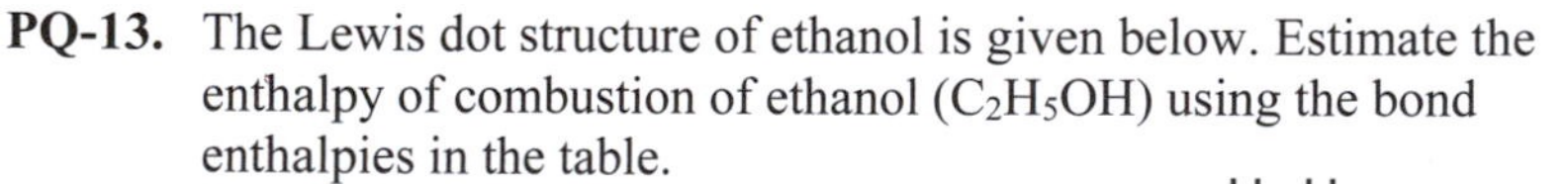

PQ-13. The Lewis dot structure of ethanol is given below. Estimate the enthalpy of combustion of ethanol (C_2H_5OH) using the bond enthalpies in the table.

$C_2H_5OH + 3O_2 \rightarrow 2CO_2 + 3H_2O$

Bond enthalpy / $kJ \cdot mol^{-1}$	
C–H	413
C–C	348
C–O	358
C=O	799
O–H	463
O=O	495

(A) $+1255\ kJ \cdot mol^{-1}$ **(B)** $-1255\ kJ \cdot mol^{-1}$ **(C)** $-1509\ kJ \cdot mol^{-1}$ **(D)** $-2044\ kJ \cdot mol^{-1}$

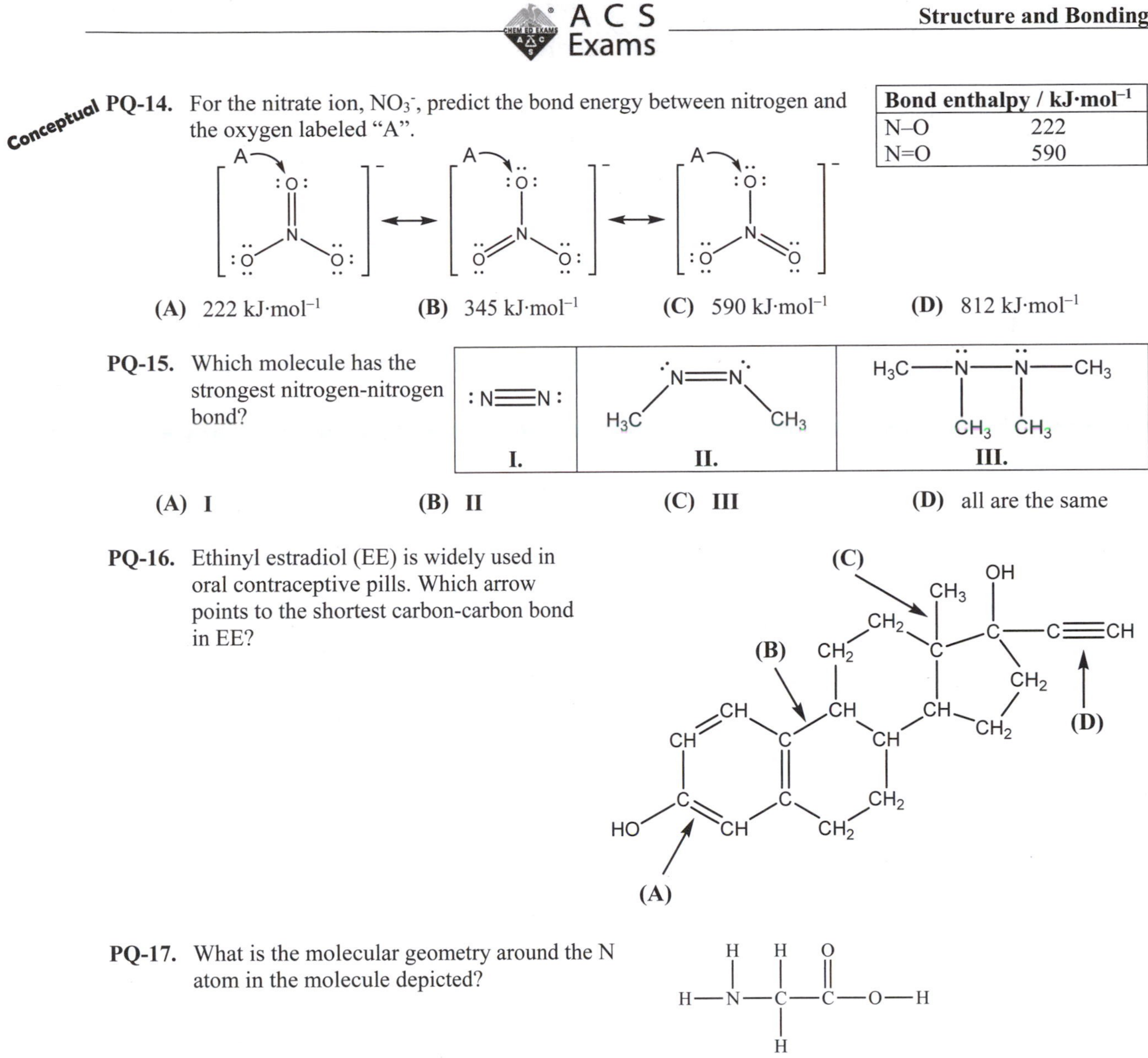

Conceptual **PQ-14.** For the nitrate ion, NO_3^-, predict the bond energy between nitrogen and the oxygen labeled "A".

Bond enthalpy / kJ·mol^{-1}	
N–O	222
N=O	590

(A) 222 kJ·mol^{-1} **(B)** 345 kJ·mol^{-1} **(C)** 590 kJ·mol^{-1} **(D)** 812 kJ·mol^{-1}

PQ-15. Which molecule has the strongest nitrogen-nitrogen bond?

:N≡N:	H_3C–N=N–CH_3	H_3C–N(CH_3)–N(CH_3)–CH_3
I.	**II.**	**III.**

(A) I **(B)** II **(C)** III **(D)** all are the same

PQ-16. Ethinyl estradiol (EE) is widely used in oral contraceptive pills. Which arrow points to the shortest carbon-carbon bond in EE?

PQ-17. What is the molecular geometry around the N atom in the molecule depicted?

H_2N–CH_2–C(=O)–O–H

(A) t-shaped
(B) tetrahedral
(C) trigonal planar
(D) trigonal pyramidal

PQ-18. A central atom has two lone pairs and three single bonds. What is the molecular geometry?

(A) T-shaped
(B) tetrahedral
(C) trigonal planar
(D) trigonal pyramidal

PQ-19. According to VSEPR, what are the approximate values of the bond angles X and Y?

H_3C–C(=O)–H (angle X at the CH_3 carbon, angle Y at the carbonyl carbon)

(A) X is 90° and Y is 180°
(B) X is 90° and Y is 120°
(C) X is 109.5° and Y is 180°
(D) X is 109.5° and Y is 120°

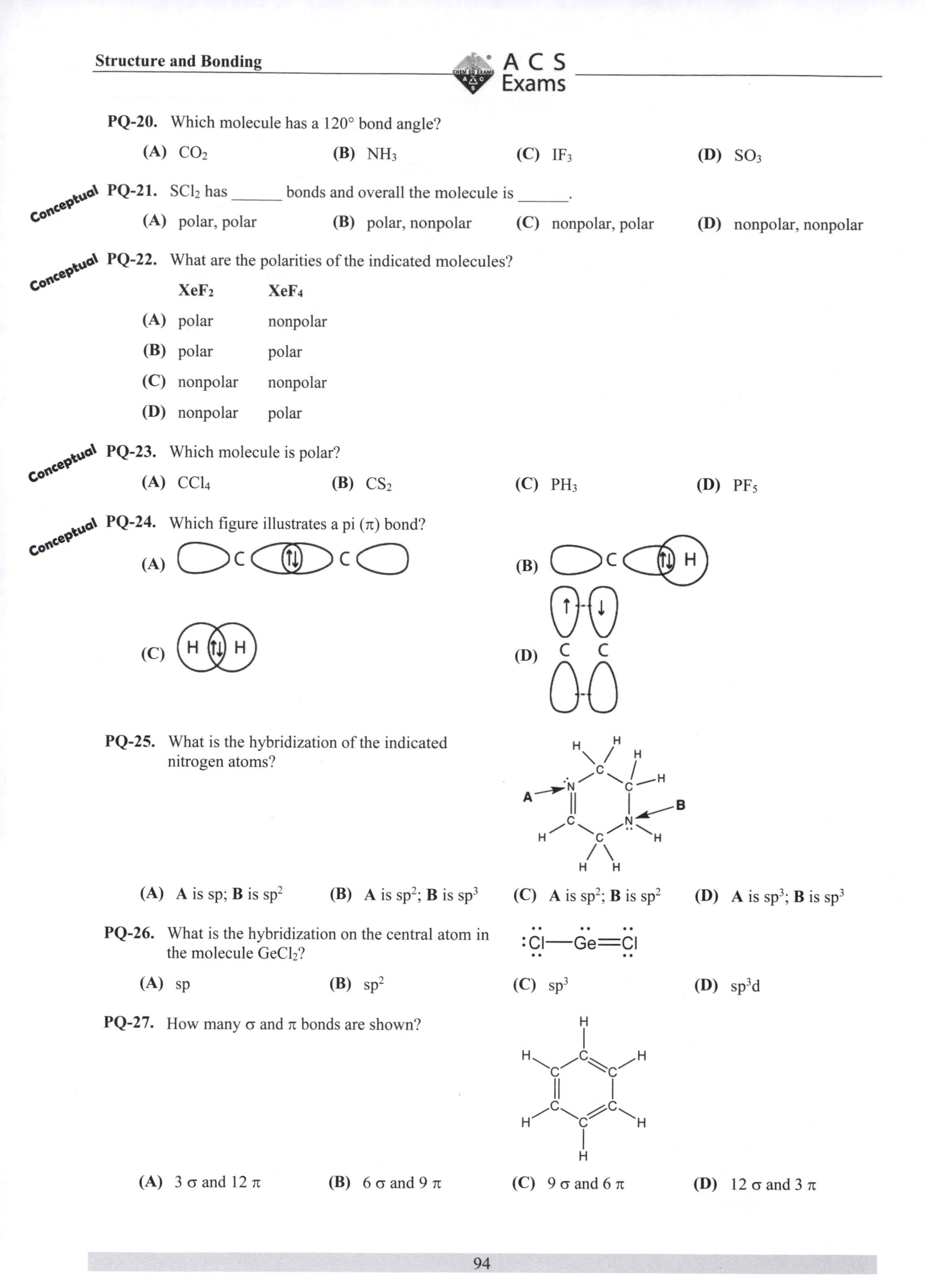

PQ-20. Which molecule has a 120° bond angle?

(A) CO_2 **(B)** NH_3 **(C)** IF_3 **(D)** SO_3

Conceptual **PQ-21.** SCl_2 has ______ bonds and overall the molecule is ______.

(A) polar, polar **(B)** polar, nonpolar **(C)** nonpolar, polar **(D)** nonpolar, nonpolar

Conceptual **PQ-22.** What are the polarities of the indicated molecules?

	XeF_2	XeF_4
(A)	polar	nonpolar
(B)	polar	polar
(C)	nonpolar	nonpolar
(D)	nonpolar	polar

Conceptual **PQ-23.** Which molecule is polar?

(A) CCl_4 **(B)** CS_2 **(C)** PH_3 **(D)** PF_5

Conceptual **PQ-24.** Which figure illustrates a pi (π) bond?

(A) **(B)**

(C) **(D)**

PQ-25. What is the hybridization of the indicated nitrogen atoms?

(A) **A** is sp; **B** is sp^2 **(B)** **A** is sp^2; **B** is sp^3 **(C)** **A** is sp^2; **B** is sp^2 **(D)** **A** is sp^3; **B** is sp^3

PQ-26. What is the hybridization on the central atom in the molecule $GeCl_2$?

(A) sp **(B)** sp^2 **(C)** sp^3 **(D)** sp^3d

PQ-27. How many σ and π bonds are shown?

(A) 3 σ and 12 π **(B)** 6 σ and 9 π **(C)** 9 σ and 6 π **(D)** 12 σ and 3 π

PQ-28. How many σ and π bonds are in the molecule shown?

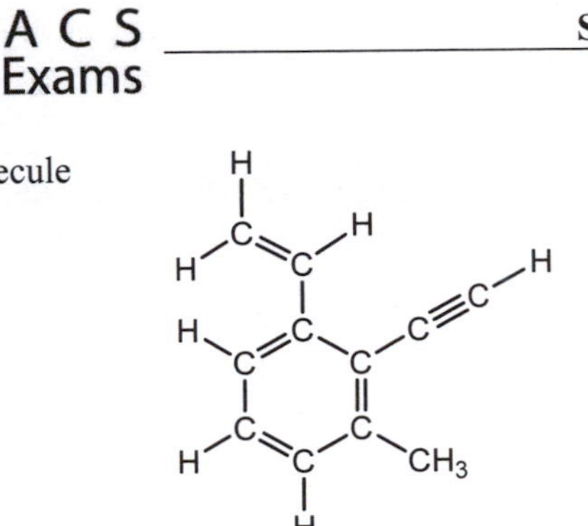

(A) 5 σ and 6 π **(B)** 13 σ and 5 π **(C)** 16 σ and 6 π **(D)** 21 σ and 6 π

Conceptual **PQ-29.** According to MO theory, F_2 would

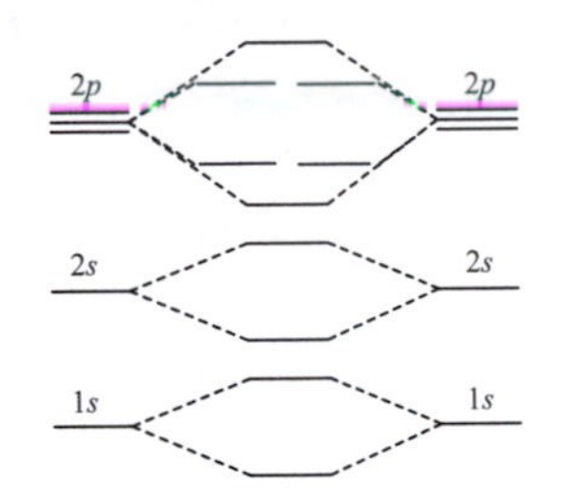

(A) have a bond order of 1 and be diamagnetic.

(B) have a bond order of 1 and be paramagnetic.

(C) have a bond order of 2 and be diamagnetic.

(D) have a bond order of 2 and be paramagnetic.

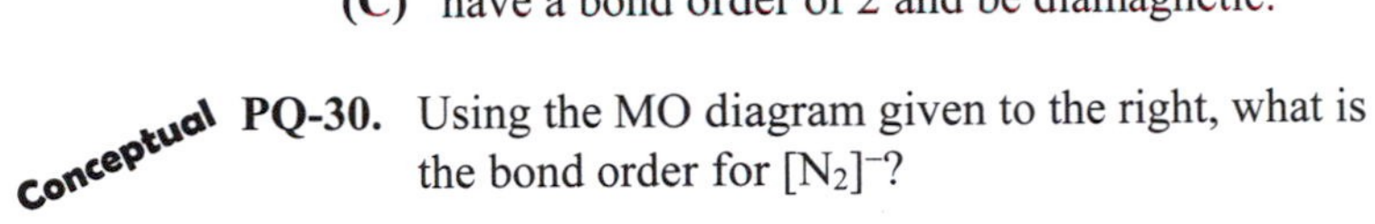

PQ-30. Using the MO diagram given to the right, what is the bond order for $[N_2]^-$?

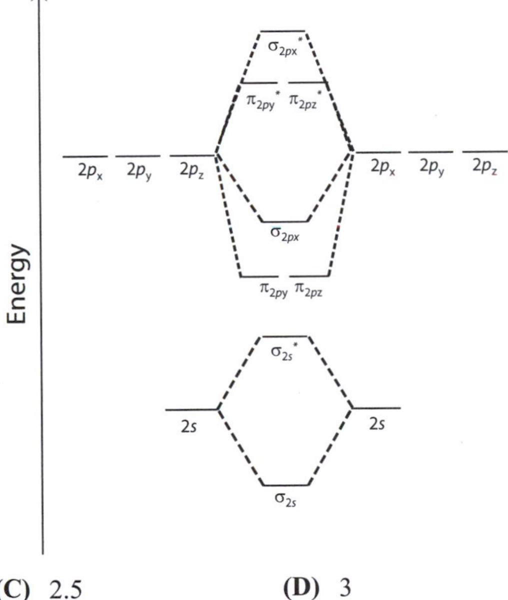

(A) 1.5 **(B)** 2 **(C)** 2.5 **(D)** 3

Answers to Study Questions

1. C
2. D
3. C
4. C
5. D
6. C
7. C
8. D
9. D
10. A
11. C
12. C
13. D

Answers to Practice Questions

1. B
2. C
3. B
4. D
5. C
6. A
7. C
8. C
9. C
10. B
11. A
12. B
13. B
14. B
15. A
16. D
17. D
18. A
19. D
20. D
21. A
22. C
23. C
24. D
25. B
26. B
27. D
28. D
29. A
30. C

Chapter 8 – States of Matter

Chapter Summary:

This chapter will focus on concepts related to gases including single gases, mixtures of gases and kinetic molecular theory. Additionally, this chapter includes concepts related to liquids and solids such as intermolecular forces and phase diagrams.

Specific topics covered in this chapter are:

- Properties of ideal gases
- Single ideal gases in one condition (ideal gas law)
- Single ideal gases in two conditions (combined gas law)
- Gases involved in reactions (stoichiometry of gases)
- Two or more ideal gases (gas mixtures)
- Kinetic molecular theory (including energy and speed of gases)
- Intermolecular forces of pure substances
- Vapor pressure
- Properties of solids
- Phase diagrams

Previous material that is relevant to your understanding of questions in this chapter include:

- Density and unit conversions (***Toolbox***)
- Formula calculations and molar mass (***Chapter 3***)
- Balancing equations and stoichiometry (***Chapter 4***)
- Lewis dot structures and polarity (***Chapter 7***)

Common representations used in questions related to this material:

Name	Example	Used in questions related to
Particulate representations	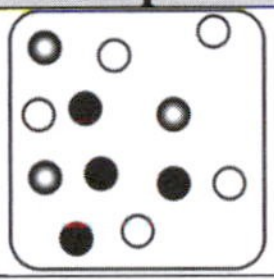	Gas conceptual problems or simple calculation problems Crystal structure problems
Structure	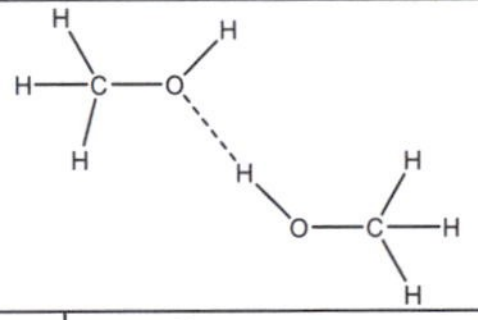	Intermolecular forces including identifying type of intermolecular force or how this interaction is represented
Graphs	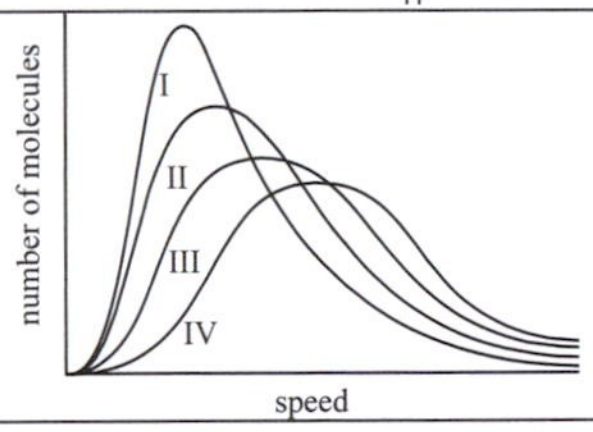	Kinetic molecular theory of gases Phase diagrams Vapor pressure diagrams

Where to find this in your textbook:

The material in this chapter typically aligns to two chapters (could be labeled as "Gases" and "Liquids and Solids") in your textbook. The names of your chapters may vary.

Practice exam:

There are practice exam questions aligned to the material in this chapter. Because there are a limited number of questions on the practice exam, a review of the breadth of the material in this chapter is advised in preparation for your exam.

How this fits into the big picture:

The material in this chapter aligns to the Big Idea of Intermolecular Forces (4) as listed on page 12 of this study guide.

Study Questions (SQ)

SQ-1. A gas behaves most ideally at

(A) high pressures and low temperatures. **(B)** high pressures and high temperatures.

(C) low pressures and low temperatures. **(D)** low pressures and high temperatures.

Knowledge Required: (1) Ideal gas properties.

Thinking it Through: You know you need to consider properties of ideal gases. One of the primary properties of an ideal gas is that there are no attractions between particles, and all interactions between particles are totally elastic. You know that these attractions are minimized by maintaning a sufficiently high temperature, which keeps the average energy of the particles sufficiently high. Coordinating with this you also know that ideal gases are assumed to have no volume. Therefore, to minimize the volume of the gas particles, the number of gas particles is sufficiently low bymaintaining either a low pressure or low density. Therefore, combining these, a gas behaves most ideally at low pressures and high temperatures or choice **(D)**. All other combinations are incorrect.

Finally, you also know that "sufficiently" high temperature or "sufficiently" low pressure or density varies depending on the gas. For a noble gas, this could be relatively low temperature and high pressure which would not be true for a heavier gas.

For most gases the ideal gas law will be adequate at or near conditions of Standard Temperature and Pressure (STP). These conditions are 0 °C and 1 bar (~1atm) of pressure.

Practice Questions Related to This: **PQ-1** and **PQ-2**

SQ-2. What is the density of Xe gas at 70 °C and 2.50 atm?

(A) 1.87 $g \cdot L^{-1}$ **(B)** 4.69 $g \cdot L^{-1}$ **(C)** 11.7 $g \cdot L^{-1}$ **(D)** 57.1 $g \cdot L^{-1}$

Knowledge Required: (1) Ideal gas law. (2) Temperature conversion to kelvin. (3) Definition of density.

Thinking it Through: You start this problem by defining density, which you know is mass per volume (or $g \cdot L^{-1}$). As you read the problem, you notice that you are not given either the mass or the volume of the gas. However, you also know that density is an intensive property, so you can use any sample size. Two logical sample sizes would be either 1.00 L of Xe or 1.00 mol of Xe.

You do the question first with 1.00 L of Xe. You now have temperature (70°C or 343 K), pressure (2.50 atm) and volume (1.00 L), so you can calculate n_{Xe} using the ideal gas law:

$$PV = nRT \qquad n = \frac{PV}{RT} = \frac{(2.50 \text{ atm})(1.0 \text{ L})}{(0.08206 \text{ L} \cdot \text{atm} \cdot \text{mol}^{-1} \cdot \text{K}^{-1})(343 \text{ K})} = 0.0889 \text{ mol}$$

$$(0.0889 \text{ mol Xe})\left(\frac{131.29 \text{ g Xe}}{\text{mol Xe}}\right) = 11.7 \text{ g Xe per 1.00 L or the density is } 11.7 \text{ g} \cdot \text{L}^{-1}$$

Alternatively, you know you can also use a sample size of 1.00 mol. You now have temperature (70°C or 343 K), pressure (2.50 atm) and number of moles (1.00 mol), so you can calculate V_{Xe} using the ideal gas law:

$$PV = nRT \qquad V = \frac{nRT}{P} = \frac{(0.08206 \text{ L} \cdot \text{atm} \cdot \text{mol}^{-1} \cdot \text{K}^{-1})(343 \text{ K})(1.0 \text{ mol})}{(2.50 \text{ atm})} = 11.3 \text{ L}$$

1.00 mol Xe is 131.29 g Xe. Calculating the density: $\frac{131.29 \text{ g Xe}}{11.3 \text{ L}} = 11.7 \text{ g} \cdot \text{L}^{-1}$ which is choice **(C)**.

Choice **(A)** would be determined when using the density equation ($\text{density} = \frac{MM \times P}{RT}$) but solving incorrectly for density and dividing by P while choice **(B)** uses the equation but omits the gas constant and uses 70°C instead of 343 K. Choice **(D)** is incorrectly determined by 70 °C rather 343 K.

Practice Questions Related to This: **PQ-3**, **PQ-4**, **PQ-5**, and **PQ-6**

SQ-3. A 10.0 L sample of an ideal gas at 25.0 °C is heated at constant pressure until the volume has doubled. What is the final temperature of the gas?

(A) 50.0 °C **(B)** 149 °C **(C)** 323 °C **(D)** 596 °C

Knowledge Required: (1) Combined gas law. (2) Temperature conversion to kelvin.

Thinking it Through: The first thing you notice for this question is that you have one ideal gas in two conditions, which means you will use the combined gas law: $\frac{P_1V_1}{T_1} = \frac{P_2V_2}{T_2}$

For this question, pressure is held constant: $\frac{\not{P_1}V_1}{T_1} = \frac{\not{P_2}V_2}{T_2}$ $\frac{V_1}{T_1} = \frac{V_2}{T_2}$ which is Charles Law.

Finally, you know that the relationship is for absolute temperature, so the temperature must be converted to kelvin: 25 °C = 298 K: $T_2 = \frac{V_2T_1}{V_1} = \frac{(20.0\text{ L})(298\text{ K})}{(10.0\text{ L})} = 596\text{ K} = 323\,^\circ\text{C}$ which is choice **(C)**.

Choice **(A)** is incorrectly calculated by forgetting to convert the temperature to kelvin. Choice **(B)** is calculated by inverting the equation and using $T_2 = \frac{V_1T_1}{V_2}$ instead as well as not converting back to °C. Choice **(D)** is the temperature in kelvin.

Practice Questions Related to This: **PQ-7** and **PQ-8**

SQ-4. What volume of $CO_2(g)$, measured at 0.993 atm and 28 °C, must react in order to completely consume 59.0 g of $Al_2O_3(s)$?

Molar mass / g·mol^{-1}	
Al_2O_3	101.96

$$Al_2O_3(s) + 3CO_2(g) \rightarrow Al_2(CO_3)_3(s)$$

(A) 1.34 L **(B)** 4.03 L **(C)** 14.4 L **(D)** 43.3 L

Knowledge Required: (1) Ideal gas law. (2) Stoichiometry and mole ratios.

Thinking it Through: You start this problem by considering what you are given: a mass (59.0 g) of one of the reactants (Al_2O_3) and pressure and temperature of the other reactant (CO_2). Secondly, what you need to find is the volume of the other reactant (CO_2). So, your approach will be:

$$\text{mass of Al}_2\text{O}_3 \xrightarrow{\text{use molar mass}} n_{\text{Al}_2\text{O}_3} \xrightarrow{\text{use the mol ratio}} n_{\text{CO}_2} \xrightarrow{\text{use P, T}} V_{\text{CO}_2}$$

$$(59.0\text{ g Al}_2\text{O}_3)\left(\frac{\text{mol Al}_2\text{O}_3}{101.96\text{ g Al}_2\text{O}_3}\right)\left(\frac{3\text{ mol CO}_2}{1\text{ mol Al}_2\text{O}_3}\right) = 1.74\text{ mol CO}_2$$

$$PV = nRT \qquad V_{\text{CO}_2} = \frac{n_{\text{CO}_2}RT}{P} = \frac{(1.74\text{ mol CO}_2)(0.08206\text{ L}\cdot\text{atm}\cdot\text{mol}^{-1}\cdot\text{K}^{-1})(301\text{ K})}{0.993\text{ atm}} = 43.3\text{ L}$$

which is choice **(D)**.

Choice **(C)** is incorrectly determined by omitting the mole ratio of 1 mol Al_2O_3 to 3 mol CO_2. Choice **(B)** uses 28 °C rather than 301 K and Choice **(A)** uses 28 °C as well as omits the mole ratio.

Practice Questions Related to This: **PQ-9**, **PQ-10**, and **PQ-11**

SQ-5. *Conceptual*

If the diagram shown is representative of the particles in a sample of ideal gases, which statement is true?

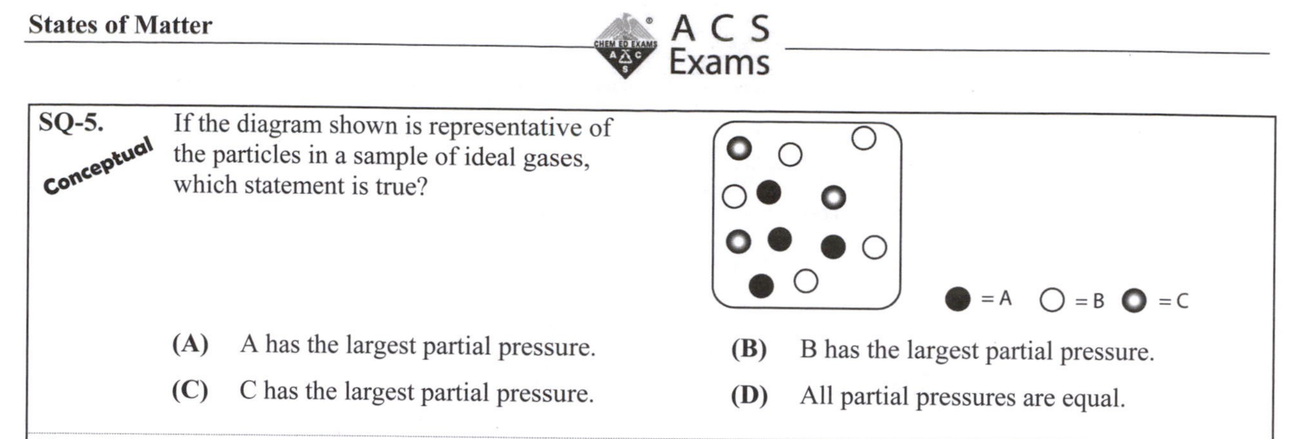

(A) A has the largest partial pressure.
(B) B has the largest partial pressure.
(C) C has the largest partial pressure.
(D) All partial pressures are equal.

Knowledge Required: (1) Particulate representations. (2) Definition of partial pressure.

Thinking it Through: For this question, you see that you first are going to interpret the particulate diagram. For this problem, there are 4 particles of A(g), 5 particles of B(g) and 3 particles of C(g) which you can equate to number of moles (so 4 mol A(g), 5 mol B(g) and 3 mol C(g)). Then you realize that you need to determine which gas has the highest partial pressure. Partial pressure is defined as the pressure an ideal gas would exert if it occupied the container alone. Using Avogadro's law, $\frac{P_1}{n_1} = \frac{P_2}{n_2}$, you know that pressure and amount (number of moles) are directly proportional. Therefore, the greatest number of moles corresponds to the largest partial pressure (or gas B), which is choice **(B)**.

Choices **(A)** and **(C)** correspond to selecting smaller number of moles of gas (possibly inverting Avogadro's law). Choice **(D)** would be true only if there were equivalent number of particles of all gases. Finally, it is important to remember that it is either the number of moles of particles or number of particles which can be atomic gases (like what is shown in the figure) or molecules (such as N_2 or CH_4).

Practice Questions Related to This: **PQ-12** and **PQ-13**

SQ-6. *Conceptual*

The graph to the right represents four different gases at the same temperature. Which are the correct label for curves I–IV?

	Curve I	Curve II	Curve III	Curve IV
(A)	CO_2	F_2	Cl_2	Kr
(B)	Kr	CO_2	Cl_2	F_2
(C)	Kr	Cl_2	CO_2	F_2
(D)	F_2	CO_2	Cl_2	Kr

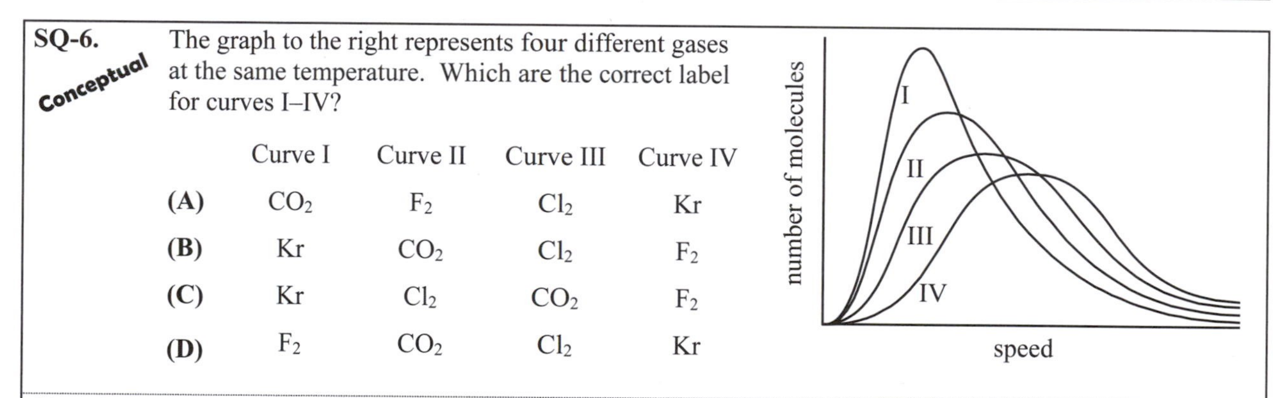

Knowledge Required: (1) Kinetic molecular theory. (2) Maxwell-Boltzmann distributions. (2) Molar masses.

Thinking it Through: You start this question by interpreting the Maxwell-Boltzmann distributions shown in the figure. The frequency distributions are showing the fraction of particles with a particular speed. Simply selecting one distribution, you can see that distribution is close to Gaussian. Further, you can also see that the maximum (or the mode) for each distribution varies depending on the curve. Finally, you also know the equation for molecular speed (for root-mean-square speed) is $u_{rms} = \sqrt{\frac{3RT}{MM}}$ where "*MM*" is the molar mass. You can see that the molecular speed is inversely proportional to the square root of the molar mass. Therefore, the higher the molar mass, the lower the average speed as shown in the table. Therefore, the correct choice is **(C)**. Choice **(D)** reverses this; choice **(A)** uses the number of atoms in the molecule rather the molar mass; and choice **(B)** reverses the responses without diatomic molar masses.

Gas	Molar mass	Curve
Kr	84	I
Cl_2	71	II
CO_2	44	III
F_2	38	IV

Practice Questions Related to This: **PQ-14**, **PQ-15**, **PQ-16**, and **PQ-17**

SQ-7. *Conceptual*

Which correctly shows a hydrogen bond between two molecules of methanol? The hydrogen bond is represented as a dashed line.

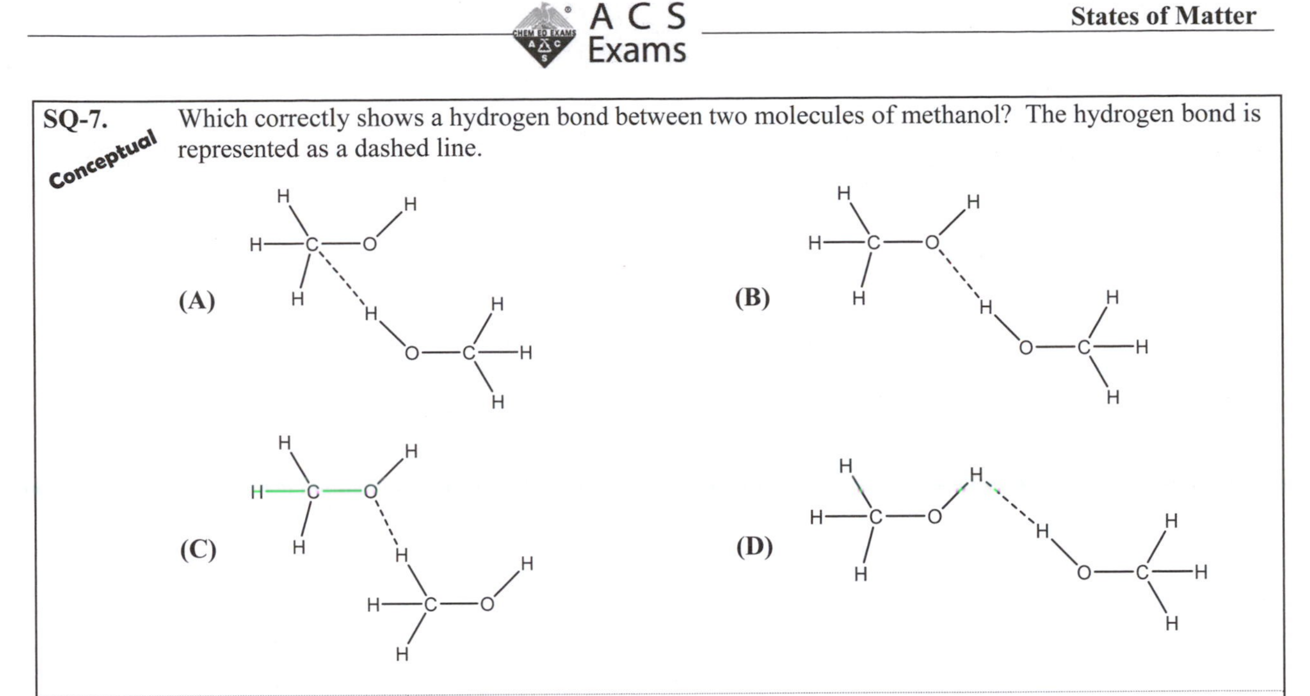

Knowledge Required: (1) Definition of hydrogen bonding. (2) Interpretation of structures.

Thinking it Through: You start this question by recalling the definition of hydrogen bonding. You know that hydrogen bonding is an intermolecular force, which means that it is an interaction *between* molecules and is not a covalent bond *within* a molecule. Therefore, you know you are considering an interaction due to opposite charges on each molecule. You know that hydrogen bonding specifically is due to the presence of N, O, or F bonded to hydrogen with a lone pair of electrons on the N, O, or F. The interaction in a pure substance (where all of the molecules are the same) is then between the N, O or F on one molecule and the H bonded to the N, O, F on another molecule:

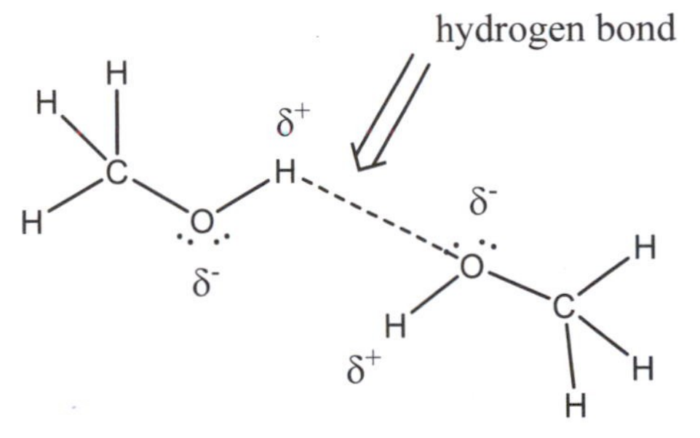

The corresponding partial charges are included above to show the attraction of hydrogen bonding. When comparing this structure to the choices, you can see the correct choice is **(B)**.

Choice **(A)** has the interaction between the correct hydrogen and carbon. Choice **(C)** has the interaction between the oxygen correctly but the wrong hydrogen (bonded to carbon and not oxygen on the neighboring molecule). Choice **(D)** has the interaction between two similar hydrogen atoms on neighboring molecules.

Practice Questions Related to This: **PQ-18**, **PQ-19**, and **PQ-20**

SQ-8. *Conceptual*

What is the correct order of increasing normal boiling point of NaCl, Br_2 and ICl?

(A) Br_2 < ICl < NaCl **(B)** ICl < Br_2 < NaCl **(C)** NaCl < Br_2 < ICl **(D)** NaCl < ICl < Br_2

Knowledge Required: (1) Structures of simple substances. (2) Intermolecular forces. (3) Relationship between strength of intermolecular forces and boiling point.

Thinking it Through: As you start this, you see that you are asked to rank substances by boiling point. Therefore, you know that as intermolecular forces increase, boiling point increases (because the interactions between the

particles are stronger and it takes more energy to overcome the interactions). Now, you can evaluate the different substances and intermolecular forces. To do this, you will generate the structures and assign the intermolecular forces remembering that all substances have dispersion forces and polar substances have dipole forces (Br_2 is nonpolar; ICl is polar):

Formula	Structure	Intermolecular forces	Relative strength
Br_2	:Br—Br:	dispersion	low
ICl	:I—Cl:	dispersion, dipole	medium
NaCl	Na^+ $:Cl:^-$	ionic	high

As shown in the table, NaCl is assigned as the highest boiling point as this is ionic and to melt (and to boil), ionic bonds must be broken. Following that ICl is the next highest as this substance has two types of intermolecular forces; Br_2 has the lowest strength of intermolecular forces as it has only one type of intermolecular force. It is important to note that these rules can be followed only when comparing similar molecules (particularly similarly sized molecules). Therefore, this corresponds to choice **(A)**.

Choice **(B)** reverses Br_2 and ICl likely due to using molar mass rather than number of intermolecular forces (or missing that ICl is polar and has dipole forces). Choices **(C)** and **(D)** incorrectly place NaCl as the lowest boiling point (with either the correct ordering of Br_2 and ICl, choice **(C)** or again inverting Br_2 and ICl, choice **(D)**).

Practice Questions Related to This: **PQ-21**, **PQ-22**, and **PQ-23**

SQ-9. Conceptual

Two vapor pressure curves, A and B are shown. One is for methanol, CH_3OH, and the other is for 1-hexanol, $CH_3CH_2CH_2CH_2CH_2CH_2OH$. Which statement correctly completes the sentence?
At T = 35 °C,

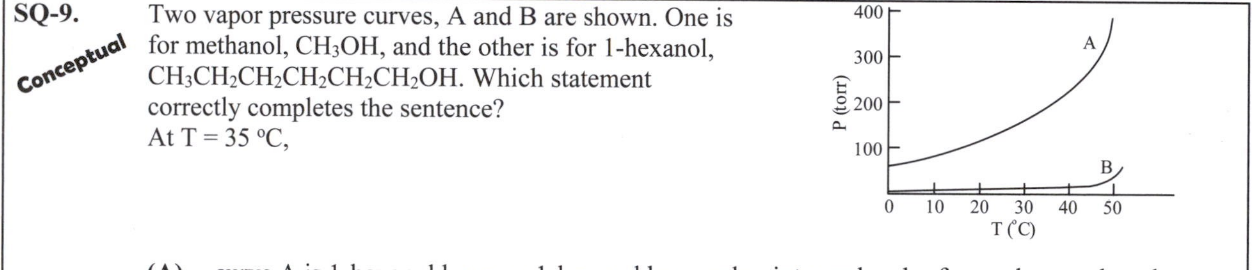

(A) curve A is 1-hexanol because 1-hexanol has weaker intermolecular forces than methanol.

(B) curve A is 1-hexanol because 1-hexanol has stronger intermolecular forces than methanol.

(C) curve A is methanol because methanol has weaker intermolecular forces than 1-hexanol.

(D) curve A is methanol because methanol has stronger intermolecular forces than 1-hexanol.

Knowledge Required: (1) Intermolecular forces. (2) Definition of vapor pressure. (3) Relationship between vapor pressure and intermolecular forces.

Thinking it Through: As described in the previous study question (**SQ-8**), in order to assess intermolecular forces, you start with the structure of each of the molecules and determine the intermolecular forces:

Formula	Structure	Intermolecular forces
CH_3OH	H_3C–Ö–H	dispersion, dipole, hydrogen bonding
$CH_3CH_2CH_2CH_2CH_2CH_2OH$	H_3C–CH_2–CH_2–CH_2–CH_2–CH_2–Ö–H	dispersion, dipole, hydrogen bonding

Because both molecules have the same intermolecular forces, you would then consider the molar masses (or total number of electrons or size) of the molecules. The greater the molar mass, the stronger the intermolecular forces.

Therefore, you then know that as the intermolecular forces increase, the number of particles in the vapor phase above the liquid decreases (because it takes more energy to overcome the intermolecular forces and

transition to the vapor state). Because the number of particles in the vapor phase is lower, the vapor pressure is lower as shown below with a particulate diagram for each:

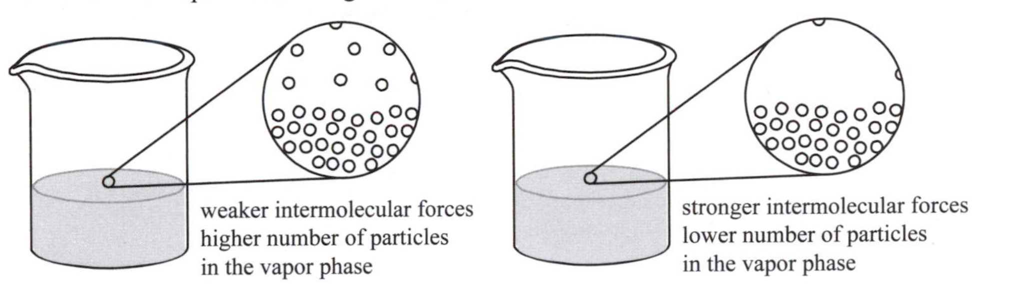

Therefore, you know that as the intermolecular forces increases, the vapor pressure decreases, and 1-hexanol which has the highest intermolecular forces would have the lowest vapor pressure.

Finally, comparing this to the graph, you can see that, at 35 °C, curve A has a higher vapor pressure and curve B has a lower vapor pressure. You then assign curve A as methanol because it has lower intermolecular forces and curve B as 1-hexanol because it has higher intermolecular forces. This then corresponds to choice **(C)**.

Choice **(A)** is incorrect because 1-hexanol is assigned as having weaker intermolecular forces than methanol. Curve A is incorrectly assigned to the substance with stronger intermolecular forces in Choice **(B)**. Choice **(D)** combines both errors from Choice **(A)** and **(B)**.

Practice Questions Related to This: **PQ-24** and **PQ-25**

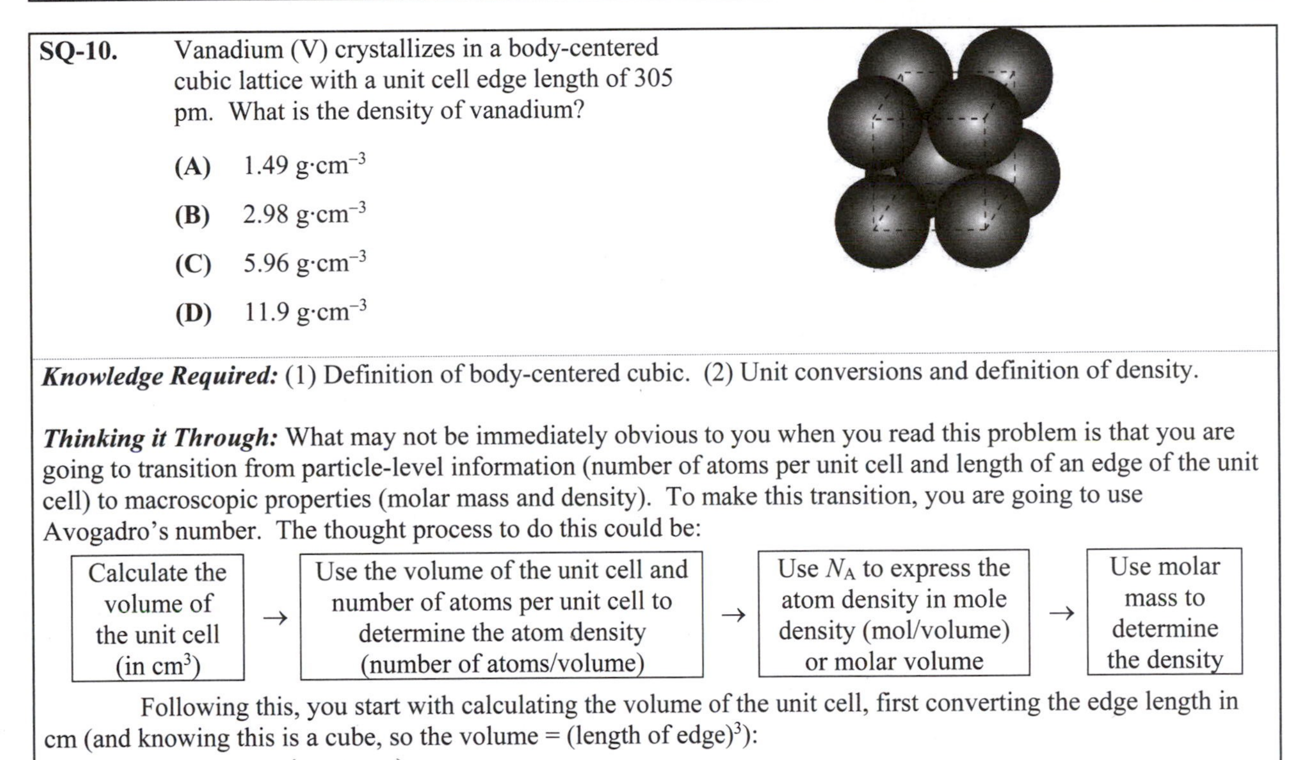

SQ-10. Vanadium (V) crystallizes in a body-centered cubic lattice with a unit cell edge length of 305 pm. What is the density of vanadium?

(A) 1.49 g·cm^{-3}

(B) 2.98 g·cm^{-3}

(C) 5.96 g·cm^{-3}

(D) 11.9 g·cm^{-3}

Knowledge Required: (1) Definition of body-centered cubic. (2) Unit conversions and definition of density.

Thinking it Through: What may not be immediately obvious to you when you read this problem is that you are going to transition from particle-level information (number of atoms per unit cell and length of an edge of the unit cell) to macroscopic properties (molar mass and density). To make this transition, you are going to use Avogadro's number. The thought process to do this could be:

Calculate the volume of the unit cell (in cm^3)	→	Use the volume of the unit cell and number of atoms per unit cell to determine the atom density (number of atoms/volume)	→	Use N_A to express the atom density in mole density (mol/volume) or molar volume	→	Use molar mass to determine the density

Following this, you start with calculating the volume of the unit cell, first converting the edge length in cm (and knowing this is a cube, so the volume = (length of edge)^3):

$$(305\ \text{pm})\left(\frac{10^{-10}\ \text{cm}}{1\ \text{pm}}\right) = 3.05\times10^{-8}\ \text{cm} \qquad V_{\text{unit cell}} = \left(3.05\times10^{-8}\ \text{cm}\right)^3 = 2.84\times10^{-23}\ \text{cm}^3$$

Now, you continue with the calculation, knowing that a body-centered cubic lattice has two atoms per unit cell (given the edge length as shown in the figure); one full atom for the center and $^1/_8$ atom per corner × 8 corners:

$$\left(\frac{2\ \text{atoms}}{\text{unit cell}}\right)\left(\frac{\text{unit cell}}{2.84\times10^{-23}\ \text{cm}^3}\right)\left(\frac{1\ \text{mol atoms V}}{6.022\times10^{23}\ \text{atoms}}\right)\left(\frac{50.94\ \text{g V}}{1\ \text{mol V}}\right) = 5.96\ \text{g}\cdot\text{cm}^{-3}$$ which is choice **(C)**.

If the number of atoms is inverted, one would calculate Choice **(A)**. Choice **(B)** would be selected if the number of atoms per unit cell were incorrectly determined to be one or entirely omitted. Choice **(D)** similarly would be selected if the number of atoms per unit cell were incorrectly determined to be four (such as for a face-centered cubic).

Practice Questions Related to This: **PQ-26** and **PQ-27**

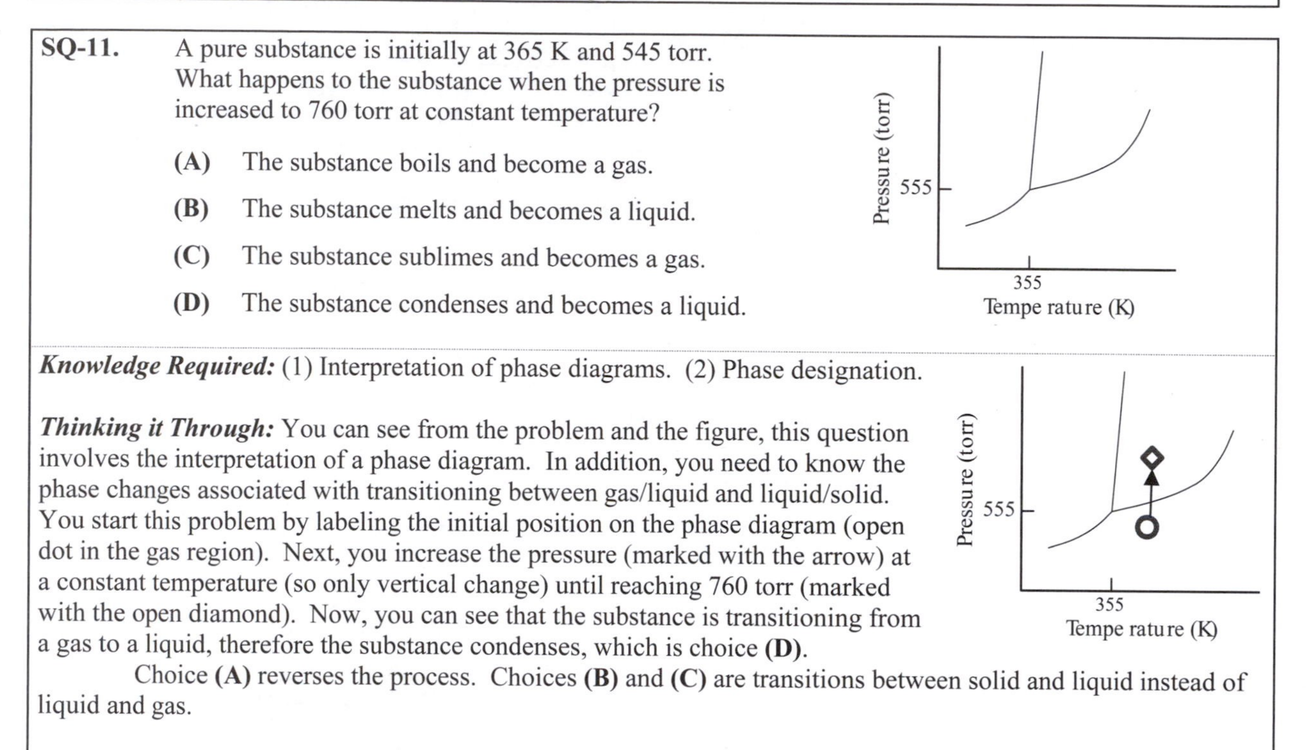

SQ-11. A pure substance is initially at 365 K and 545 torr. What happens to the substance when the pressure is increased to 760 torr at constant temperature?

(A) The substance boils and become a gas.

(B) The substance melts and becomes a liquid.

(C) The substance sublimes and becomes a gas.

(D) The substance condenses and becomes a liquid.

Knowledge Required: (1) Interpretation of phase diagrams. (2) Phase designation.

Thinking it Through: You can see from the problem and the figure, this question involves the interpretation of a phase diagram. In addition, you need to know the phase changes associated with transitioning between gas/liquid and liquid/solid. You start this problem by labeling the initial position on the phase diagram (open dot in the gas region). Next, you increase the pressure (marked with the arrow) at a constant temperature (so only vertical change) until reaching 760 torr (marked with the open diamond). Now, you can see that the substance is transitioning from a gas to a liquid, therefore the substance condenses, which is choice **(D)**.

Choice **(A)** reverses the process. Choices **(B)** and **(C)** are transitions between solid and liquid instead of liquid and gas.

Practice Questions Related to This: **PQ-28**, **PQ-29**, and **PQ-30**

Practice Questions (PQ)

Conceptual **PQ-1.** Which sample deviates most from ideal gas behavior?

(A) Xe at 1 atm and −50 °C

(B) Xe at 200 atm and −50 °C

(C) He at 200 atm and −50 °C

(D) Xe at 200 atm and 1000 °C

PQ-2. All gases approach ideal behavior under conditions of

(A) low density and high temperature.

(B) low density and low temperature.

(C) high density and high temperature.

(D) high density and low temperature.

PQ-3. What is the density of $F_2(g)$ at 1.00 atm and 25.0 °C?

(A) 0.776 $g{\cdot}L^{-1}$ **(B)** 0.850 $g{\cdot}L^{-1}$ **(C)** 1.55 $g{\cdot}L^{-1}$ **(D)** 1.70 $g{\cdot}L^{-1}$

PQ-4. A 14.0 g sample of $N_2(g)$ occupies what volume at 0 °C and 1.00 atm?

(A) 5.60 L **(B)** 11.2 L **(C)** 14.0 L **(D)** 22.4 L

Conceptual **PQ-5.** Which gas is most dense at 2 atm and 400K?

(A) F_2 **(B)** N_2 **(C)** Ne **(D)** O_2

PQ-6. In a 10.0 L vessel at 100.0 °C, 10.0 grams of an unknown gas exert a pressure of 1.13 atm. What is the gas?

	Gas	Molar mass ($g \cdot mol^{-1}$)
(A)	NH_3	17
(B)	HCN	27
(C)	NO	39
(D)	NO_2	46

Conceptual **PQ-7.** What happens to the volume of a fixed amount of gas if both the pressure and absolute temperature are doubled?

(A) halved **(B)** doubles **(C)** quadruples **(D)** no change

PQ-8. A sample of gas occupies 3.00 L at 1.00 atm. What volume will it occupy at 1.45 atm and the same temperature?

(A) 2.07 L **(B)** 4.35 L **(C)** 2280 L **(D)** 1572.4 L

PQ-9. What volume of oxygen gas, measured at 27 °C and 0.987 atm, is produced from the decomposition of 67.5 g of HgO(s)?

Molar mass / $g \cdot mol^{-1}$	
HgO	216.59

$$2HgO(s) \rightarrow 2Hg(l) + O_2(g)$$

(A) 3.49 L **(B)** 3.89 L **(C)** 6.98 L **(D)** 7.77 L

PQ-10. What mass of solid aluminum is needed to react with excess $H_2SO_4(aq)$ to produce 20.0 L of $H_2(g)$ at 1.0 atm and 298 K?

Molar mass / $g \cdot mol^{-1}$	
Al	26.98

$$2Al(s) + 3H_2SO_4(aq) \rightarrow Al_2(SO_4)_3(aq) + 3H_2(g)$$

(A) 0.545 g **(B)** 14.7 g **(C)** 22.0 g **(D)** 33.1 g

PQ-11. What is the percent yield when 1.72 g of H_2O_2 decomposes and produces 375 mL of O_2 gas measured at 42 °C and 1.52 atm?

Molar mass / $g \cdot mol^{-1}$	
H_2O_2	34.02

$$2H_2O_2(aq) \rightarrow 2H_2O(l) + O_2(g)$$

(A) 15.3% **(B)** 30.7% **(C)** 43.7% **(D)** 87.2%

Conceptual **PQ-12.** Suppose a gas mixture contains equal moles of He(g) and $O_2(g)$. Which is true?

(A) The partial pressure of each gas is the same.

(B) The partial pressure of He(g) is four times the partial pressure of $O_2(g)$.

(C) The partial pressure $O_2(g)$ is two times the partial pressure of He(g).

(D) The partial pressure of $O_2(g)$ is eight times the partial pressure of He(g).

PQ-13. A mixture of 3.25 moles of oxygen gas and 2.75 moles of nitrogen gas exert a total pressure of 22.4 atm. What is the partial pressure of oxygen?

(A) 72.8 atm **(B)** 12.1 atm **(C)** 11.2 atm **(D)** 10.3 atm

Conceptual **PQ-14.** According to kinetic-molecular theory, why does pressure increase as the temperature of an ideal gas increases?

I	The gas molecules collide more frequently with the wall.
II	The gas molecules collide more energetically with the wall.

(A) Only **I** **(B)** Only **II** **(C)** Both **I** and **II** **(D)** Neither **I** nor **II**

Conceptual **PQ-15.** For a gas sample containing equimolar amounts of nitrogen and hydrogen at 300 K, hydrogen has ______ average speed and ______ average kinetic energy compared to nitrogen.

(A) lower; the same **(B)** the same; the same **(C)** higher; the same **(D)** higher; higher

Conceptual **PQ-16.** The gases, F_2, H_2, N_2, and O_2 at STP are represented in the plot. Which gas corresponds to curve III?

(A) O_2

(B) N_2

(C) H_2

(D) F_2

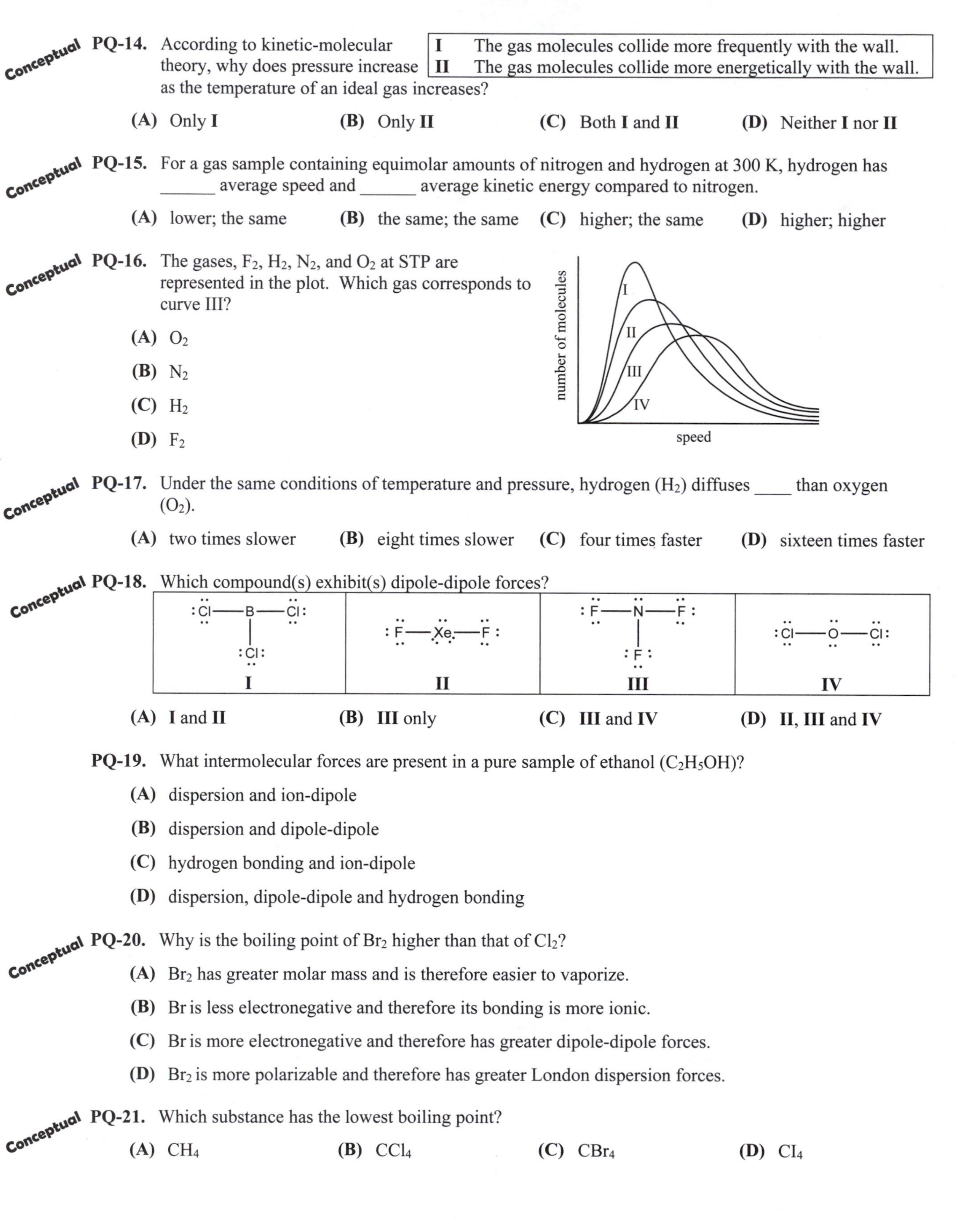

Conceptual **PQ-17.** Under the same conditions of temperature and pressure, hydrogen (H_2) diffuses ____ than oxygen (O_2).

(A) two times slower **(B)** eight times slower **(C)** four times faster **(D)** sixteen times faster

Conceptual **PQ-18.** Which compound(s) exhibit(s) dipole-dipole forces?

BCl_3 (Cl–B(–Cl)–Cl)	F–Xe–F	F–N(–F)–F	Cl–O–Cl
I	**II**	**III**	**IV**

(A) **I** and **II** **(B)** **III** only **(C)** **III** and **IV** **(D)** **II, III** and **IV**

PQ-19. What intermolecular forces are present in a pure sample of ethanol (C_2H_5OH)?

(A) dispersion and ion-dipole

(B) dispersion and dipole-dipole

(C) hydrogen bonding and ion-dipole

(D) dispersion, dipole-dipole and hydrogen bonding

Conceptual **PQ-20.** Why is the boiling point of Br_2 higher than that of Cl_2?

(A) Br_2 has greater molar mass and is therefore easier to vaporize.

(B) Br is less electronegative and therefore its bonding is more ionic.

(C) Br is more electronegative and therefore has greater dipole-dipole forces.

(D) Br_2 is more polarizable and therefore has greater London dispersion forces.

Conceptual **PQ-21.** Which substance has the lowest boiling point?

(A) CH_4 **(B)** CCl_4 **(C)** CBr_4 **(D)** CI_4

Conceptual **PQ-22.** Arrange LiF, HCl, HF, and F_2 in order of increasing normal boiling point.

(A) $F_2 < HF < HCl < LiF$

(B) $F_2 < HCl < HF < LiF$

(C) $F_2 < HCl < LiF < HF$

(D) $HF < LiF < HCl < F_2$

Conceptual **PQ-23.** Which pure substance has the lowest boiling point?

(A) $N(CH_3)_3$ — 59 g/mol

(B) $N(C_2H_5)_3$ — 101 g/mol

(C) $CH_3CH_2CH_2NH_2$ — 59 g/mol

(D) $H_2NCH_2CH_2NH_2$ — 60 g/mol

Conceptual **PQ-24.** Which pure substance has the lowest vapor pressure at 25 °C?

(A) CO_2

(B) H_2S

(C) H_2O

(D) O_2

PQ-25. As intermolecular forces increase, enthalpy of vaporization ________ and vapor pressure ________.

(A) increases; increases

(B) increases; decreases

(C) decreases; increases

(D) decreases; decreases

PQ-26. Aluminum metal crystallizes in a face-centered cubic unit cell. How many aluminum atoms are in one unit cell?

(A) 14

(B) 7

(C) 4

(D) 2

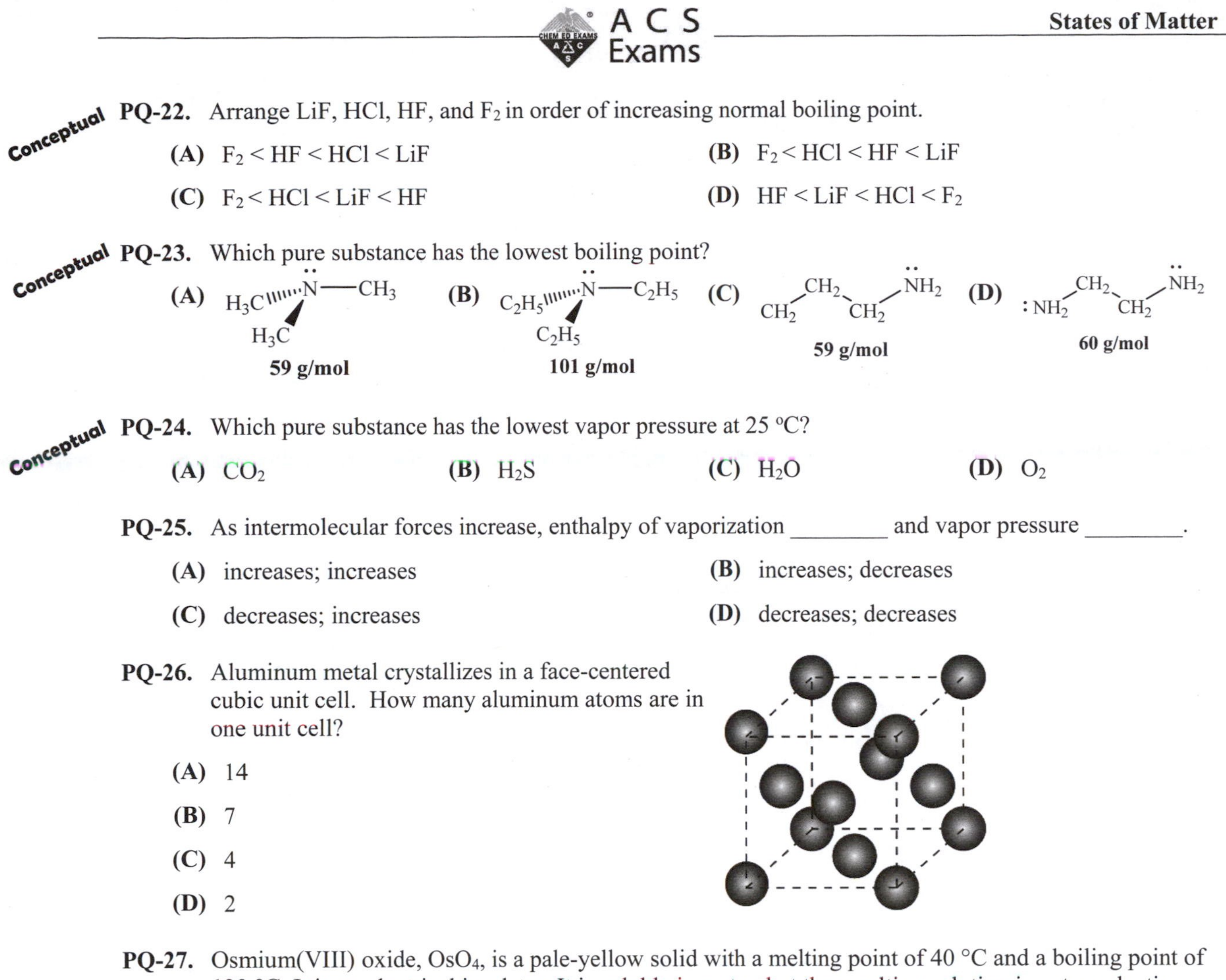

PQ-27. Osmium(VIII) oxide, OsO_4, is a pale-yellow solid with a melting point of 40 °C and a boiling point of 130 °C. It is an electrical insulator. It is soluble in water, but the resulting solution is not conducting. Based on these observations what is the most appropriate classification of OsO_4?

(A) covalent-network solid

(B) ionic solid

(C) metallic solid

(D) molecular solid

PQ-28. Which point corresponds to the normal boiling point of a substance?

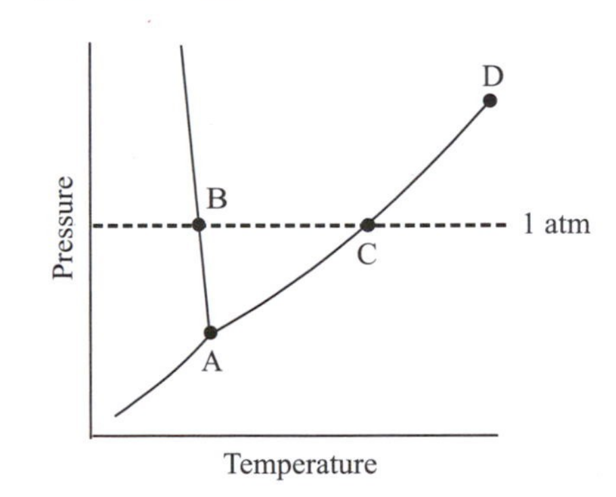

PQ-29. The process described by changing the system from point A to point B is ____.

(A) condensation

(B) evaporation

(C) melting

(D) sublimation

Conceptual **PQ-30.** Which arrow on the phase diagram corresponds to the particle-level phase change shown?

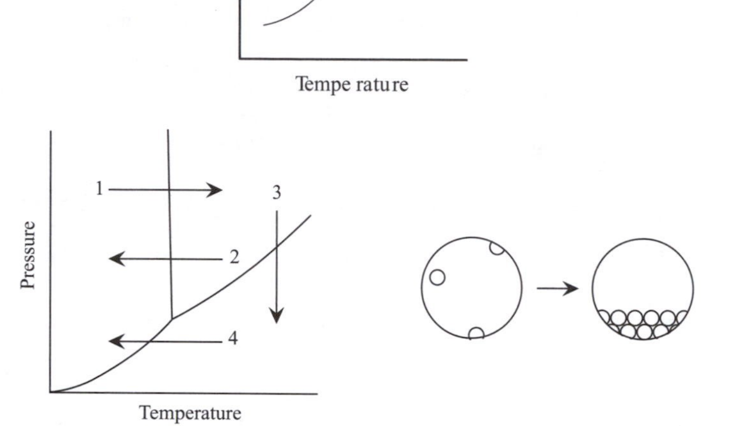

(A) 1

(B) 2

(C) 3

(D) 4

Answers to Study Questions

1. D
2. C
3. C
4. D
5. B
6. C
7. B
8. A
9. C
10. C
11. D

Answers to Practice Questions

1. B
2. A
3. C
4. B
5. A
6. B
7. D
8. A
9. B
10. B
11. D
12. A
13. B
14. C
15. C
16. B
17. C
18. C
19. D
20. D
21. A
22. B
23. A
24. C
25. B
26. C
27. D
28. C
29. C
30. D

General Chemistry II material begins here

Sample datasheet and periodic table provided as reference on the next page.

ABBREVIATIONS AND SYMBOLS					
amount of substance	*n*	gas constant	*R*	molar mass	*M*
atmosphere	atm	gram	g	mole	mol
atomic mass unit	u	hour	h	Planck's constant	*h*
atomic molar mass	*A*	joule	J	pressure	*P*
Avogadro constant	N_A	kelvin	K	second	s
Celsius temperature	°C	kilo– prefix	k	speed of light	*c*
centi– prefix	c	liter	L	temperature, K	*T*
energy of activation	E_a	measure of pressure	mmHg	time	*t*
enthalpy	*H*	milli– prefix	m	volume	*V*
frequency	ν	molar	M		

CONSTANTS & CONVERSIONS

$R = 8.314\ \text{J·mol}^{-1}\text{·K}^{-1}$

$R = 0.0821\ \text{L·atm·mol}^{-1}\text{·K}^{-1}$

$N_A = 6.022 \times 10^{23}\ \text{mol}^{-1}$

$h = 6.626 \times 10^{-34}\ \text{J·s}$

$c = 2.998 \times 10^{8}\ \text{m·s}^{-1}$

0 °C = 273.15 K

1 atm = 760 mmHg

1 atm = 760 torr

EQUATIONS

Equations may be included in this position of your exam.

These vary depending on the type of exam you are taking.

For example, if you are expected to do a calculation involving an integrated rate law, you may be provided with a series of equations in this position.

PERIODIC TABLE OF THE ELEMENTS

1	2	3	4	5	6	7	8	9	10	11	12	13	14	15	16	17	18
1 **H** 1.008																	2 **He** 4.003
3 **Li** 6.941	4 **Be** 9.012											5 **B** 10.81	6 **C** 12.01	7 **N** 14.01	8 **O** 16.00	9 **F** 19.00	10 **Ne** 20.18
11 **Na** 22.99	12 **Mg** 24.31											13 **Al** 26.98	14 **Si** 28.09	15 **P** 30.97	16 **S** 32.07	17 **Cl** 35.45	18 **Ar** 39.95
19 **K** 39.10	20 **Ca** 40.08	21 **Sc** 44.96	22 **Ti** 47.88	23 **V** 50.94	24 **Cr** 52.00	25 **Mn** 54.94	26 **Fe** 55.85	27 **Co** 58.93	28 **Ni** 58.69	29 **Cu** 63.55	30 **Zn** 65.39	31 **Ga** 69.72	32 **Ge** 72.61	33 **As** 74.92	34 **Se** 78.96	35 **Br** 79.90	36 **Kr** 83.80
37 **Rb** 85.47	38 **Sr** 87.62	39 **Y** 88.91	40 **Zr** 91.22	41 **Nb** 92.91	42 **Mo** 95.94	43 **Tc**	44 **Ru** 101.1	45 **Rh** 102.9	46 **Pd** 106.4	47 **Ag** 107.9	48 **Cd** 112.4	49 **In** 114.8	50 **Sn** 118.7	51 **Sb** 121.8	52 **Te** 127.6	53 **I** 126.9	54 **Xe** 131.3
55 **Cs** 132.9	56 **Ba** 137.3	57 **La** 138.9	72 **Hf** 178.5	73 **Ta** 180.9	74 **W** 183.8	75 **Re** 186.2	76 **Os** 190.2	77 **Ir** 192.2	78 **Pt** 195.1	79 **Au** 197.0	80 **Hg** 200.6	81 **Tl** 204.4	82 **Pb** 207.2	83 **Bi** 209.0	84 **Po**	85 **At**	86 **Rn**
87 **Fr**	88 **Ra**	89 **Ac**	104 **Rf**	105 **Db**	106 **Sg**	107 **Bh**	108 **Hs**	109 **Mt**	110 **Ds**	111 **Rg**	112 **Cn**	113 **Nh**	114 **Fl**	115 **Mc**	116 **Lv**	117 **Ts**	118 **Og**

58 **Ce** 140.1	59 **Pr** 140.9	60 **Nd** 144.2	61 **Pm**	62 **Sm** 150.4	63 **Eu** 152.0	64 **Gd** 157.3	65 **Tb** 158.9	66 **Dy** 162.5	67 **Ho** 164.9	68 **Er** 167.3	69 **Tm** 168.9	70 **Yb** 173.0	71 **Lu** 175.0
90 **Th** 232.0	91 **Pa** 231.0	92 **U** 238.0	93 **Np**	94 **Pu**	95 **Am**	96 **Cm**	97 **Bk**	98 **Cf**	99 **Es**	100 **Fm**	101 **Md**	102 **No**	103 **Lr**

Please note that the periodic table changes to keep current with recent discoveries.
The periodic table you use may vary from the table shown here.

Chapter 9 – Solutions and Aqueous Reactions, Part 2

Chapter Summary:

This chapter will focus on the additional aspects of solutions, usually found in the second semester of a standard general chemistry course (whereas the first part of solutions is usually found in the first semester). Concepts discussed here and explained in example items go into more detail related to intermolecular forces associated with solution formation and properties of solutions. Reactions of solutions are discussed separately in chapters related to equilibrium.

Specific topics covered in this chapter are:

- Solution formation including intermolecular forces in solutions
- Solubility as it relates to intermolecular forces
- Concentration units and conversions between
- Solubility of solids and gases, particularly in aqueous solution
- Colligative properties including freezing point depression, boiling point elevation and osmotic pressure

Previous material that is relevant to your understanding of questions in this chapter include:

- The mole and formula calculations (***Chapter 3***)
- Stoichiometry and balancing equations (***Chapter 4***)
- Molar concentrations and solution stoichiometry (***Chapter 5***)
- Lewis dot structures and polarity (***Chapter 7***)
- Intermolecular forces (***Chapter 8***)

Common representations used in questions related to this material:

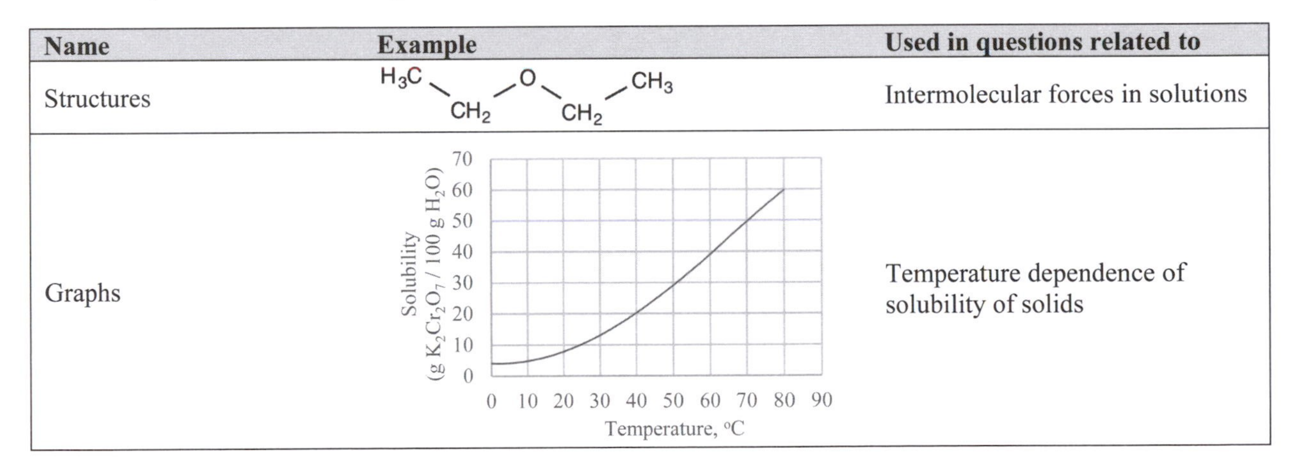

Name	Example	Used in questions related to
Structures	H_3C, CH_2, O, CH_2, CH_3	Intermolecular forces in solutions
Graphs	(graph)	Temperature dependence of solubility of solids

Where to find this in your textbook:

The material in this chapter typically aligns to a solutions chapter (could be labeled as "Physical Properties of Solutions") in your textbook. The name of your chapter may vary.

Practice exam:

There are practice exam questions aligned to the material in this chapter. Because there are a limited number of questions on the practice exam, a review of the breadth of the material in this chapter is advised in preparation for your exam.

How this fits into the big picture:

The material in this chapter aligns to Intermolecular Interactions (4) as listed on page 12 of this study guide.

Study Questions (SQ)

SQ-1. *Conceptual*

Where is the hydrophilic (attracted to water) region of the molecule?

(A) Region 1

(B) Region 2

(C) Region 3

(D) The three regions are equally hydrophilic

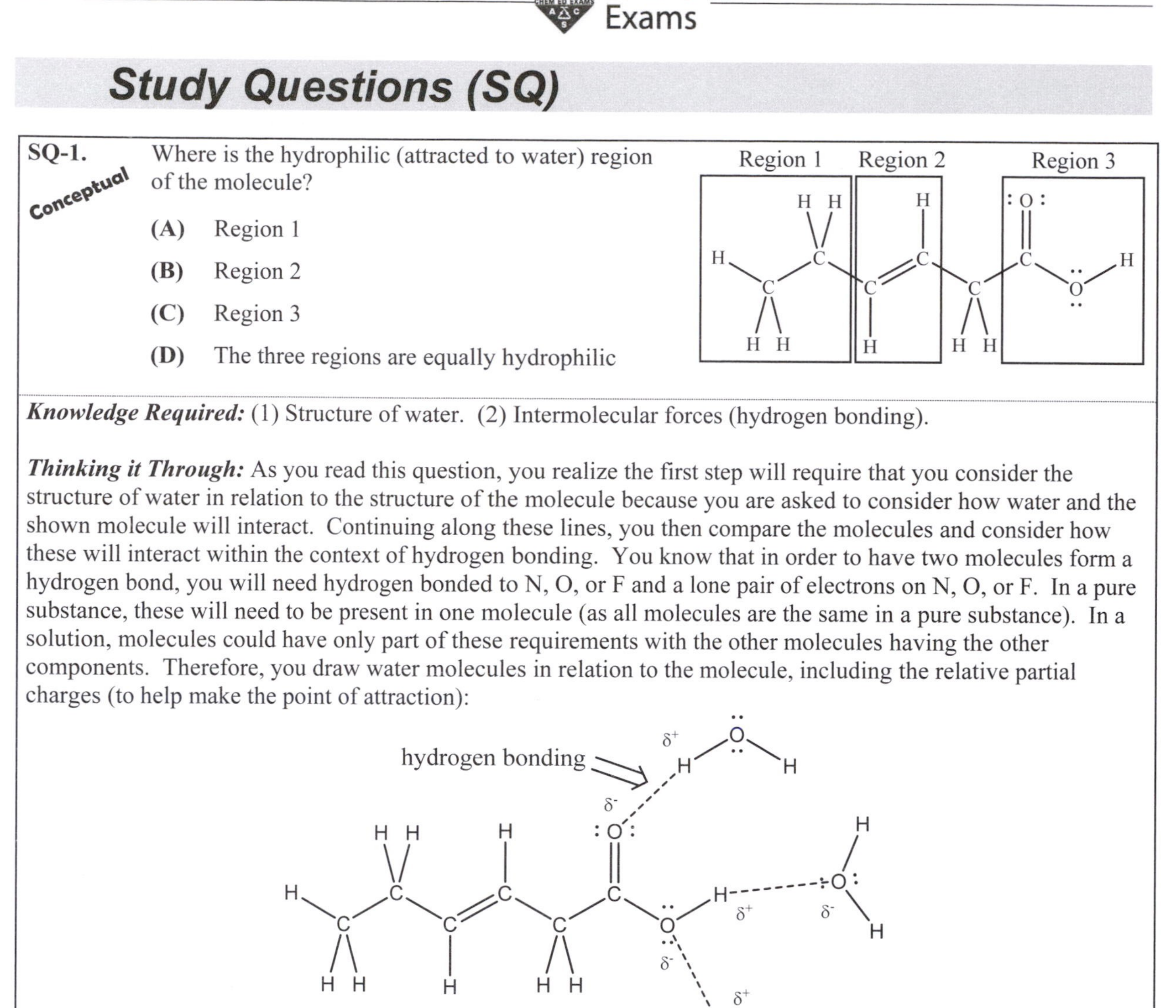

Knowledge Required: (1) Structure of water. (2) Intermolecular forces (hydrogen bonding).

Thinking it Through: As you read this question, you realize the first step will require that you consider the structure of water in relation to the structure of the molecule because you are asked to consider how water and the shown molecule will interact. Continuing along these lines, you then compare the molecules and consider how these will interact within the context of hydrogen bonding. You know that in order to have two molecules form a hydrogen bond, you will need hydrogen bonded to N, O, or F and a lone pair of electrons on N, O, or F. In a pure substance, these will need to be present in one molecule (as all molecules are the same in a pure substance). In a solution, molecules could have only part of these requirements with the other molecules having the other components. Therefore, you draw water molecules in relation to the molecule, including the relative partial charges (to help make the point of attraction):

Therefore, water is attracted to region 3 (Choice **(C)**) and not region 1 (Choice **(A)**) or region 2 (Choice **(B)**) or all three equally (Choice **(D)**).

NOTE: Drawing your own representations is often an effective problem-solving strategy in chemistry.

Practice Questions Related to This: **PQ-1** and **PQ-2**

SQ-2. *Conceptual*

Which molecule is most soluble in water?

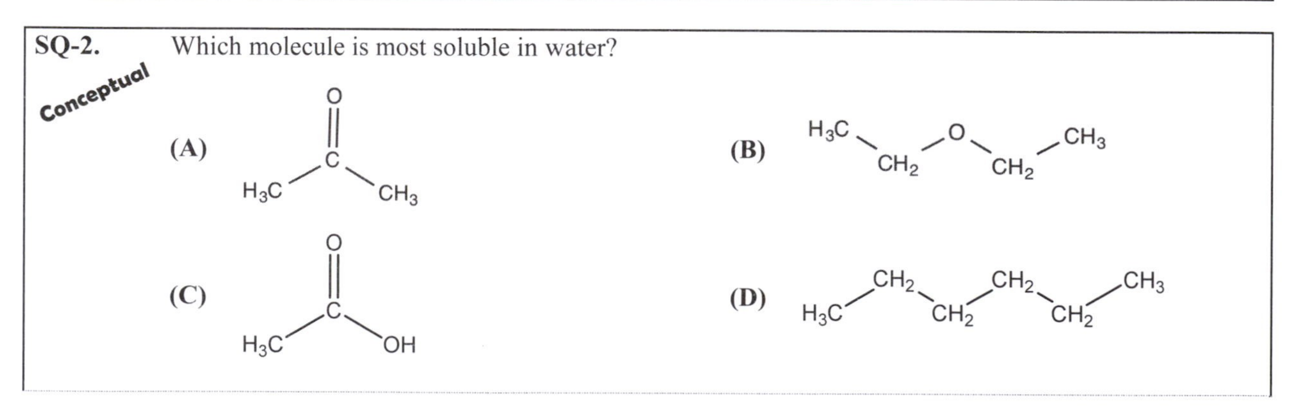

Knowledge Required: (1) Structure of water. (2) Intermolecular forces (hydrogen bonding).

Thinking it Through: Similar to the previous study question, to solve this you are going to consider each structure interacting with the structure of water, paying close attention to components that would support greater attraction due to hydrogen bonding. As a reminder again, in a solution these are:

1. Hydrogen bonded to N, O, or F
2. A lone pair of electrons on N, O, or F

Molecules interacting with water can have one or more locations for this attraction to occur. Considering each molecule with water, including relevant partial charges and hydrogen bonding:

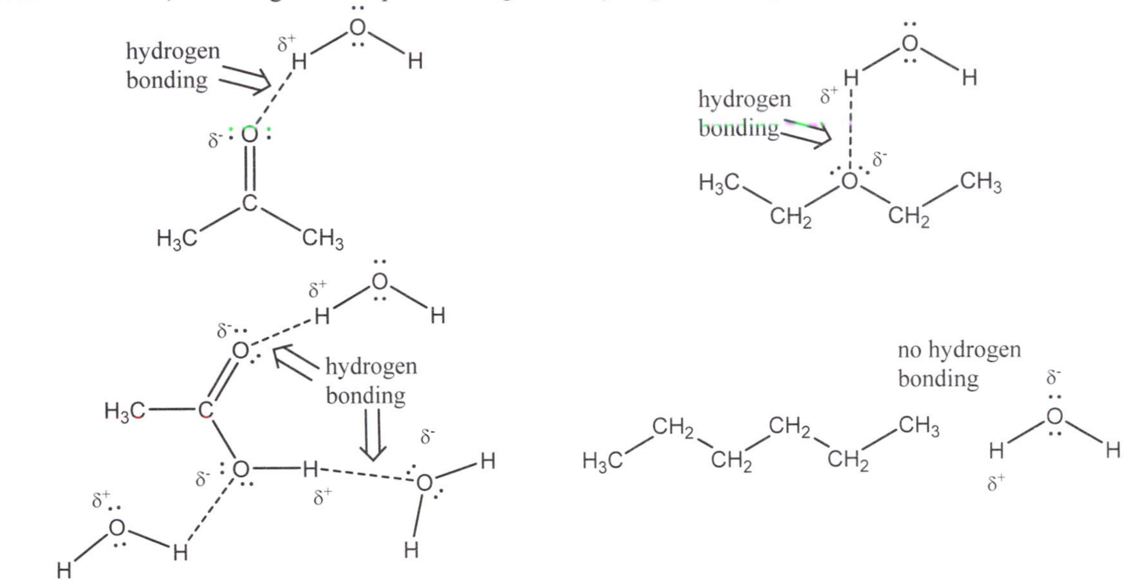

From these diagrams, you see that the greater the number of attractions, the more soluble the molecule will be in water. Adding to this decision, you consider the relative sizes of the molecules (as solubility can be conceptualized as a "substitution" process of the solute molecules into the water molecule's location, thus needing the same relative attractions and size). You see that the molecules are reasonably the same size.

Therefore, you can see that one molecule will not hydrogen bond with water at all (Choice **(D)**), therefore this molecule will be the least soluble in water. You also see that molecules in Choices **(A)** and **(B)** have limited attractions to water making these incorrect as well. Choice **(C)** is ideal for substitution, as it has multiple opportunities for hydrogen bonding, making **(C)** or acetic acid, the correct choice.

Practice Questions Related to This: **PQ-3**, **PQ-4**, and **PQ-5**

SQ-3. A solution of NaCl in water has a concentration of 20.5% by mass. What is the molal concentration of the solution?

Molar mass / $g \cdot mol^{-1}$	
NaCl	58.44

(A) 0.205 *m* **(B)** 0.258 *m* **(C)** 3.51 *m* **(D)** 4.41 *m*

Knowledge Required: (1) Concentration unit of molality. (2) Concentration unit of percent by mass.

Thinking it Through: You see this question starts by giving you percent by mass and asks you to determine molal concentration. So, to solve this, you first define both concentration units. First for percent by mass, you define this with a general definition, then for a solution and finally further define the mass of the solution as the sum of the masses of the solvent and solute:

$$\text{Percent by mass} = \frac{\text{mass of one component}}{\text{total mass}} \times 100\% = \frac{\text{mass solute}}{\text{mass solution}} \times 100\% = \frac{\text{mass solute}}{\text{mass solute} + \text{mass solvent}} \times 100\%$$

Now you define molal concentration: $$\text{molal concentration} = \frac{\text{mol solute}}{\text{mass solvent (kg)}}$$

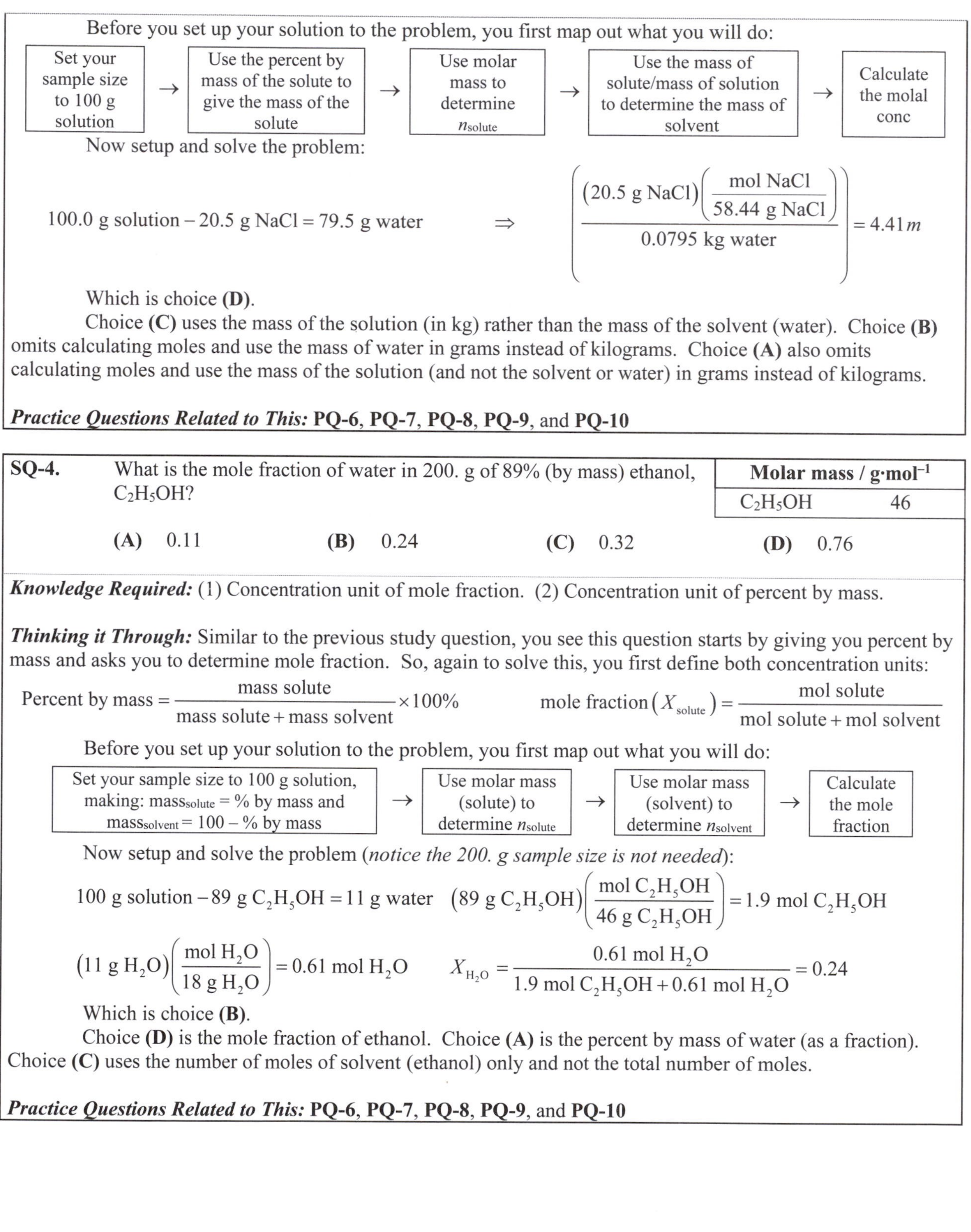

Before you set up your solution to the problem, you first map out what you will do:

Set your sample size to 100 g solution	→	Use the percent by mass of the solute to give the mass of the solute	→	Use molar mass to determine n_{solute}	→	Use the mass of solute/mass of solution to determine the mass of solvent	→	Calculate the molal conc

Now setup and solve the problem:

$$100.0\text{ g solution} - 20.5\text{ g NaCl} = 79.5\text{ g water} \quad \Rightarrow \quad \left(\frac{(20.5\text{ g NaCl})\left(\frac{\text{mol NaCl}}{58.44\text{ g NaCl}}\right)}{0.0795\text{ kg water}}\right) = 4.41m$$

Which is choice **(D)**.

Choice **(C)** uses the mass of the solution (in kg) rather than the mass of the solvent (water). Choice **(B)** omits calculating moles and use the mass of water in grams instead of kilograms. Choice **(A)** also omits calculating moles and use the mass of the solution (and not the solvent or water) in grams instead of kilograms.

Practice Questions Related to This: **PQ-6**, **PQ-7**, **PQ-8**, **PQ-9**, and **PQ-10**

SQ-4. What is the mole fraction of water in 200. g of 89% (by mass) ethanol, C_2H_5OH?

Molar mass / g·mol^{-1}	
C_2H_5OH	46

(A) 0.11 **(B)** 0.24 **(C)** 0.32 **(D)** 0.76

Knowledge Required: (1) Concentration unit of mole fraction. (2) Concentration unit of percent by mass.

Thinking it Through: Similar to the previous study question, you see this question starts by giving you percent by mass and asks you to determine mole fraction. So, again to solve this, you first define both concentration units:

$$\text{Percent by mass} = \frac{\text{mass solute}}{\text{mass solute} + \text{mass solvent}} \times 100\% \qquad \text{mole fraction}\left(X_{\text{solute}}\right) = \frac{\text{mol solute}}{\text{mol solute} + \text{mol solvent}}$$

Before you set up your solution to the problem, you first map out what you will do:

Set your sample size to 100 g solution, making: $mass_{solute}$ = % by mass and $mass_{solvent}$ = 100 – % by mass	→	Use molar mass (solute) to determine n_{solute}	→	Use molar mass (solvent) to determine $n_{solvent}$	→	Calculate the mole fraction

Now setup and solve the problem (*notice the 200. g sample size is not needed*):

$$100\text{ g solution} - 89\text{ g }C_2H_5OH = 11\text{ g water} \quad (89\text{ g }C_2H_5OH)\left(\frac{\text{mol }C_2H_5OH}{46\text{ g }C_2H_5OH}\right) = 1.9\text{ mol }C_2H_5OH$$

$$(11\text{ g }H_2O)\left(\frac{\text{mol }H_2O}{18\text{ g }H_2O}\right) = 0.61\text{ mol }H_2O \qquad X_{H_2O} = \frac{0.61\text{ mol }H_2O}{1.9\text{ mol }C_2H_5OH + 0.61\text{ mol }H_2O} = 0.24$$

Which is choice **(B)**.

Choice **(D)** is the mole fraction of ethanol. Choice **(A)** is the percent by mass of water (as a fraction). Choice **(C)** uses the number of moles of solvent (ethanol) only and not the total number of moles.

Practice Questions Related to This: **PQ-6**, **PQ-7**, **PQ-8**, **PQ-9**, and **PQ-10**

SQ-5. *Conceptual*

A mixture of 100 g of $K_2Cr_2O_7$ and 200 g of water is stirred at 60 °C until no more of the salt dissolves. The resulting solution is poured off into a separate beaker, leaving the undissolved solid behind. The solution is now cooled to 20 °C. What mass of $K_2Cr_2O_7$ crystallizes from the solution during the cooling?

(A) 9 g

(B) 18 g

(C) 31 g

(D) 62 g

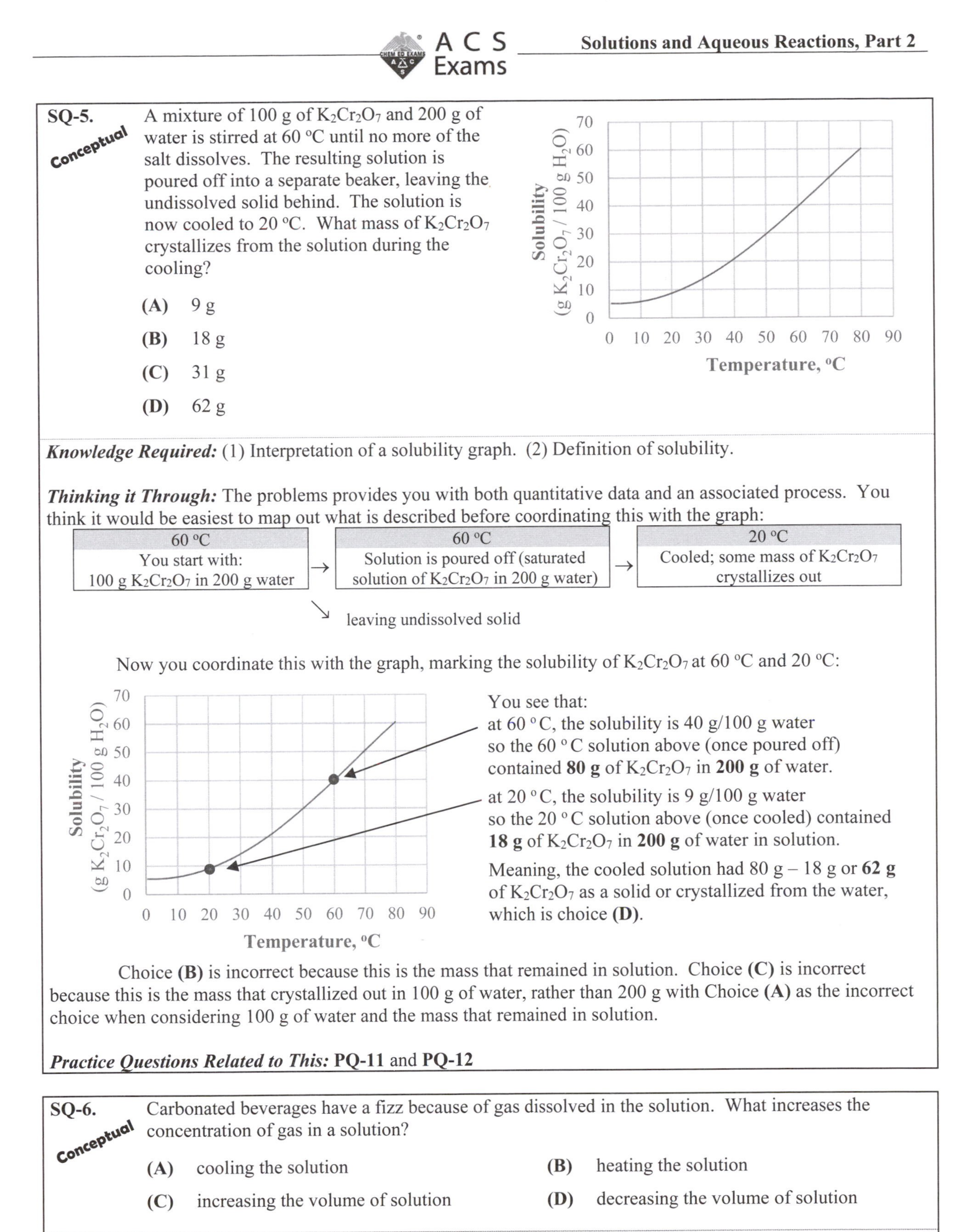

Knowledge Required: (1) Interpretation of a solubility graph. (2) Definition of solubility.

Thinking it Through: The problems provides you with both quantitative data and an associated process. You think it would be easiest to map out what is described before coordinating this with the graph:

60 °C		60 °C		20 °C
You start with: 100 g $K_2Cr_2O_7$ in 200 g water	→	Solution is poured off (saturated solution of $K_2Cr_2O_7$ in 200 g water)	→	Cooled; some mass of $K_2Cr_2O_7$ crystallizes out

↘ leaving undissolved solid

Now you coordinate this with the graph, marking the solubility of $K_2Cr_2O_7$ at 60 °C and 20 °C:

You see that:
at 60 °C, the solubility is 40 g/100 g water so the 60 °C solution above (once poured off) contained **80 g** of $K_2Cr_2O_7$ in **200 g** of water.

at 20 °C, the solubility is 9 g/100 g water so the 20 °C solution above (once cooled) contained **18 g** of $K_2Cr_2O_7$ in **200 g** of water in solution.

Meaning, the cooled solution had 80 g – 18 g or **62 g** of $K_2Cr_2O_7$ as a solid or crystallized from the water, which is choice **(D)**.

Choice **(B)** is incorrect because this is the mass that remained in solution. Choice **(C)** is incorrect because this is the mass that crystallized out in 100 g of water, rather than 200 g with Choice **(A)** as the incorrect choice when considering 100 g of water and the mass that remained in solution.

Practice Questions Related to This: **PQ-11** and **PQ-12**

SQ-6. *Conceptual*

Carbonated beverages have a fizz because of gas dissolved in the solution. What increases the concentration of gas in a solution?

(A) cooling the solution

(B) heating the solution

(C) increasing the volume of solution

(D) decreasing the volume of solution

Knowledge Required: (1) Solubility of gases. (2) Properties of concentration.

Thinking it Through: You recall that solubility of gases can be expressed quantitatively using Henry's law which expresses relationship of the concentration of the gas (in the solution, c), the pressure of the gas above the solution (P), and a constant for the gas (k): $c = kP$ (also written as $c = K_HP$). From this, you see that as pressure of the gas

increases, the concentration of the gas increases. You also suspect that solubility of gases vary based on their structure, and this would be expressed in the Henry's law constant, *k*. However, from this you also see that the volume of the solution will not change the concentration of the gas in the solution, so neither choice **(C)** or **(D)** are correct.

You do know that the Henry's law constant is a constant for a gas at a given temperature and as temperature increases, solubility of the gas decreases or *k* decreases. Therefore, heating the beverage will decrease the concentration of the gas (Choice **(B)**), but cooling the beverage will increase the concentration or fizziness, meaning Choice **(A)** is correct.

Practice Questions Related to This: **PQ-13** and **PQ-14**

SQ-7. When 100. g of an unknown compound was dissolved in 1.00 kg of water, the freezing point was lowered by 6.36 °C. What is the identity of this unknown compound? (K_f for water = 1.86°C·m^{-1})

(A) CsCl **(B)** KCl **(C)** LiF **(D)** NaCl

Knowledge Required: (1) Colligative properties, freezing point depression. (2) Ionic compounds and van't Hoff coefficients. (2) Molal concentration and molar mass.

Thinking it Through: As you read the question, you approach this problem by first evaluating the equation for freezing point depression (which you know is the appropriate colligative property because of the cue to this in the question): $\Delta T = K_f m$. You further examine the responses and see that all of the substances are ionic with the same 1:1 ratio of cation to anion, meaning that 2 ions form for each formula unit of the compound. Therefore, you refine the equation for freezing point depression for the van't Hoff coefficient of 2 ($i = 2$): $\Delta T = iK_f m = 2K_f m$:

$$\Delta T_f = iK_f m \qquad m = \frac{\Delta T_f}{iK_f} = \frac{6.36\ ^\circ\text{C}}{2\left(1.86\ ^\circ\text{C}\cdot m^{-1}\right)} = 1.71\ m$$

From here, you consider molal concentration and the other information provided in the problem to determine first the number of moles of solute and then (using the mass of solute), the molar mass of the solute:

$$\text{molal concentration} = \frac{\text{mol solute}}{\text{mass solvent}} \qquad \left(\frac{1.71\ \text{mol}}{\text{kg solvent}}\right)(1.00\ \text{kg water}) = 1.71\ \text{mol solute}$$

$$\left(\frac{100.\ \text{g solute}}{1.71\ \text{mol solute}}\right) = 58.5\ \text{g}\cdot\text{mol}^{-1}$$ which is NaCl or response **(D)**.

Choice **(A)** would likely be selected if the calculation of molal concentration was inverted (calculated molar mass of 170 g·mol^{-1}). Choice **(B)** may be selected if number of moles are calculated incorrectly by using the mass of solution rather than mass of solvent. Choice **(C)** may be selected if the van't Hoff coefficient was omitted (calculated molar mass of 29 g·mol^{-1}).

Practice Questions Related to This: **PQ-15**, **PQ-16**, and **PQ-17**

SQ-8. *Conceptual* Which 0.1 molal aqueous solution will have the lowest freezing point?

(A) $Al(NO_3)_3$ **(B)** $CaCl_2$ **(C)** C_2H_5OH **(D)** NaCl

Knowledge Required: (1) Colligative properties, freezing point depression. (2) Ionic compounds and van't Hoff coefficients.

Thinking it Through: You know that colligative properties depend on the number of particles in solution (the molal concentration) but not the identity of the solute. You also know this is reflected in the equation for freezing point depression as noted in the previous study question. Because the concentrations are all the same (0.1 *m*) and the solvent is the same (water with the same freezing point depression constant, K_f), the only component that will vary will be the van't Hoff coefficient, *i*. So, you start by evaluating the number of particles in solution for each solute:

hydration equation	i_{calc}	hydration equation	i_{calc}
$Al(NO_3)_3(s) \rightarrow Al^{3+}(aq) + 3NO_3^-(aq)$	4	$CaCl_2(s) \rightarrow Ca^{2+}(aq) + 2Cl^-(aq)$	3
$C_2H_5OH(l) \rightarrow C_2H_5OH(aq)$	1	$NaCl(s) \rightarrow Na^+(aq) + Cl^-(aq)$	2

Note: i_{calc} is used here because you don't have the measured value (i_{meas}); in calculations, i_{meas} would be better than i_{calc}.

You are asked for the lowest freezing point or largest ΔT_f. Therefore, choice **(A)** is correct with a calculated van't Hoff coefficient of 4. The other choices are incorrect because the coefficients are all smaller than the value for $Al(NO_3)_3$.

Practice Questions Related to This: **PQ-18**, **PQ-19**, and **PQ-20**

Practice Questions (PQ)

Conceptual **PQ-1.** Which pair of compounds will form hydrogen bonds with one another?

(A) CH_4 and H_2O **(B)** CH_4 and NH_3 **(C)** HF and CH_4 **(D)** H_2O and NH_3

Conceptual **PQ-2.** Which pair of compounds will form hydrogen bonds with one another?

(A) cyclohexane (H_2C, CH_2 ring) and H—Cl

(B) CCl_4 and CH_4

(C) CH_3OH and CH_4

(D) CH_3OH and CH_3NH_2

Conceptual **PQ-3.** What would be expected to be the most soluble in ethanol (CH_3CH_2OH)?

(A) $CH_3CH_2CH_3$ **(B)** CO_2 **(C)** NH_3 **(D)** SF_6

Conceptual **PQ-4.** Which substance is most soluble in water?

(A) ethane, CH_3CH_3 **(B)** ethanol, CH_3CH_2OH

(C) *n*-butane, $CH_3CH_2CH_2CH_3$ **(D)** 1-butanol, $CH_3CH_2CH_2CH_2OH$

Conceptual **PQ-5.** Consider the structure of Vitamin A_1 ($C_{20}H_{30}O$). Vitamin A_1 should be

(A) insoluble in water and in oil.

(B) less soluble in water than in oil.

(C) equally soluble in water and in oil.

(D) more soluble in water than in oil.

PQ-6. The mass percent of acetic acid (CH_3COOH) in a bottle of vinegar is 5.45% in water. What is the molar concentration of acetic acid in vinegar, assuming the density of vinegar is 1.005 $g \cdot mL^{-1}$?

Molar mass / $g \cdot mol^{-1}$	
CH_3COOH	60.06

(A) 0.545 M **(B)** 0.903 M **(C)** 0.908 M **(D)** 0.912 M

PQ-7. A 250-mL bottle of a sports drink solution contains 4.50% by mass of sodium chloride. What is the molal concentration of sodium chloride in this bottle of sports drink?

Molar mass / g·mol^{-1}	
NaCl	58.44

(A) 0.806 *m* **(B)** 0.770 *m* **(C)** 0.180 *m* **(D)** 0.0113 *m*

PQ-8. A 1400 g sample of stream water contains 12.2 ppm of mercury. What is the mass of Hg in the sample?

(A) 0.11 mg **(B)** 2.4 mg **(C)** 8.7 mg **(D)** 17 mg

PQ-9. What mass of water is needed to dissolve 292.5 g of NaCl to produce a 0.25 *m* aqueous solution?

Molar mass / g·mol^{-1}	
NaCl	58.44

(A) 20 kg **(B)** 5.0 kg **(C)** 0.80 kg **(D)** 0.050 kg

PQ-10. The density of a 3.539 M HNO_3 aqueous solution is 1.150 g·mL^{-1} at 20 °C. What is the molal concentration?

Molar mass / g·mol^{-1}	
HNO_3	63.02

(A) 3.077 *m* **(B)** 3.818 *m* **(C)** 3.946 *m* **(D)** 5.252 *m*

Conceptual **PQ-11.** The solubility of a substance is 60 g per 100 mL of water at 15 °C. A solution of this substance is prepared by dissolving 75 g in 100 mL of water at 75 °C. The solution is then cooled slowly to 15 °C without any solid separating. The solution is

(A) supersaturated at 75°C. **(B)** supersaturated at 15°C.

(C) unsaturated at 15°C. **(D)** saturated at 15°C.

Conceptual **PQ-12.** A student mixes 20.0 g of a salt in 100.0 g of water at 20 °C and obtains a homogeneous solution. Which salt could this be and why?

Solubility at 20 °C in 100.0 g of water	
KCl	10.0 g
KNO_3	30.0 g

(A) KCl because it is saturated. **(B)** KCl because it is unsaturated.

(C) KNO_3 because it is saturated. **(D)** KNO_3 because it is unsaturated.

PQ-13. The solubility of carbon dioxide in water is very low in air (1.05×10^{-5} M at 25 °C) because the partial pressure of carbon dioxide in air is only 0.00030 atm. What partial pressure of carbon dioxide is needed to dissolve 100.0 mg of carbon dioxide in 1.00 L of water?

(A) 0.0649 atm **(B)** 2.86 atm **(C)** 28.6 atm **(D)** 64.9 atm

Conceptual **PQ-14.** How will you increase the solubility of oxygen in water? The partial pressure of oxygen (P_{O_2}) is 0.21 atm in air at 1 atm (P_{ext}).

(A) increase P_{O_2} but keep P_{ext} constant **(B)** decrease P_{O_2} but keep P_{ext} constant

(C) increase P_{ext} but keep P_{O_2} constant **(D)** decrease P_{ext} but keep P_{O_2} constant

PQ-15. A 2.50 g sample of naphthalene, $C_{10}H_8$, was dissolved in 100 g of benzene. What is the freezing point of the benzene solution? The freezing point of pure benzene is 5.45 °C; K_f = 5.07 °C·m^{-1}.

Molar mass / g·mol^{-1}	
$C_{10}H_8$	128.17

(A) –0.989 °C **(B)** 0.989 °C **(C)** 4.46 °C **(D)** 6.44 °C

PQ-16. A technician has the measured osmotic pressure of a solution to determine the molar mass of a covalent solute. Which other information would need to be measured in order to determine the molar mass?

I.	Temperature
II.	Volume of solution
III.	Mass of solute

(A) Only **I** **(B)** Only **III** **(C)** Only **I** and **II** **(D)** **I**, **II**, and **III**

PQ-17. A solution was made by adding 800 g of ethanol, C_2H_5OH, to 8.0×10^3 g of water. How much would this lower the freezing point

K_f / °C·m^{-1}	
H_2O	1.86

(A) 3.2 °C **(B)** 4.1 °C **(C)** 8.2 °C **(D)** 16 °C

Conceptual **PQ-18.** Which aqueous solution will have the lowest osmotic pressure?

(A) 0.04 M $MgCl_2$ **(B)** 0.05 M KBr **(C)** 0.1 M LiCl **(D)** 0.2 M Na_2SO_4

Conceptual **PQ-19.** Consider pure water separated from an aqueous sugar solution by a semipermeable membrane. After some time has passed, what (if anything) will happen to the concentration of the sugar solution?

(A) It will decrease. **(B)** It will increase.

(C) It will remain the same. **(D)** It cannot be determined.

Conceptual **PQ-20.** Which solution has the higher boiling point?

Solution:	**Molar mass / g·mol^{-1}**
(1) 200.0 g glucose dissolved in 1.00 kg of water	glucose, 180
(2) 200.0 g sucrose dissolved in 1.00 kg of water	sucrose, 342

(A) solution **1**

(B) solution **2**

(C) Both would boil at the same temperature as pure water.

(D) Both would boil at the same temperature, but above that of pure water.

Answers to Study Questions

1. C
2. C
3. D
4. B
5. D
6. A
7. D
8. A

Answers to Practice Questions

1. D
2. D
3. C
4. B
5. B
6. D
7. A
8. D
9. A
10. B
11. B
12. D
13. A
14. A
15. C
16. D
17. B
18. B
19. A
20. A

Chapter 10 – Kinetics

Chapter Summary:

This chapter will focus on chemical kinetics. This includes the topics of reaction rate and rate laws. Methods of determining the rate laws will be presented and calculations with integrated rate laws will be reviewed. The collision theory of chemical kinetics and reaction mechanisms will be reviewed.

Specific topics covered in this chapter are:

- Determining a rate law from plots of data
- Method of initial rates
- Solving problems using integrated rate laws
- Half-life problems
- Collision theory and reaction energy diagrams
- Calculation of activation energy for a reaction
- Reaction mechanisms
- Relationship of kinetics to equilibrium

Previous material that is relevant to your understanding of questions in this chapter include:

- Significant figures (***Toolbox***)
- Scientific notation (***Toolbox***)
- Equilibrium ***(Chapter 11)***
- Enthalpy ***(Chapter 6)***

Common representations used in questions related to this material:

Name	Example	Used in questions related to
Compound units	$M{\cdot}s^{-1}$; $L{\cdot}mol^{-1}{\cdot}s^{-1}$	rates and rate constants
Energy diagrams	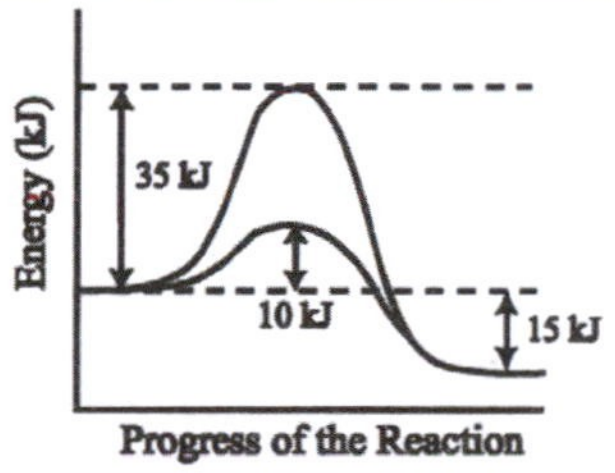 Energy (kJ); 35 kJ; 10 kJ; 15 kJ; Progress of the Reaction	activation energy

Where to find this in your textbook:

The material in this chapter typically aligns to "Kinetics" or "Chemical Kinetics" in your textbook. The name of your chapter may vary.

Practice exam:

There are practice exam questions aligned to the material in this chapter. Because there are a limited number of questions on the practice exam, a review of the breadth of the material in this chapter is advised in preparation for your exam.

How this fits into the big picture:

The material in this chapter aligns to the Big Idea of Kinetics (7) as listed on page 12 of this study guide.

Study Questions (SQ)

SQ-1. What is the rate law of this reaction?

$2H_2(g) + 2NO(g) \rightarrow N_2(g) + 2H_2O(g)$

Trial #	$[H_2]_0$ /M	$[NO]_0$ / M	Initial rate / $M{\cdot}s^{-1}$
1	$1.0{\times}10^{-3}$	$1.0{\times}10^{-3}$	$2.0{\times}10^{3}$
2	$2.0{\times}10^{-3}$	$2.0{\times}10^{-3}$	$1.60{\times}10^{4}$
3	$2.0{\times}10^{-3}$	$1.0{\times}10^{-3}$	$4.0{\times}10^{3}$

(A) rate = $k[H_2]^2[NO]$

(B) rate = $k[H_2][NO]^2$

(C) rate = $k[H_2]^2[NO]^2$

(D) rate = $k[N_2][H_2O]^2/[H_2]^2[NO]^2$

Knowledge Required: (1) Definition of rate law. (2) How to use the method of initial rates to determine the rate law.

Thinking it Through: You are given data with varying starting amounts (initial concentration noted as the "0" subscript, i.e. $[H_2]_0$) of each reactant and the reaction rate that was measured. You write the generic rate law for the reaction as:

$$\text{rate} = k[H_2]^x[NO]^y$$

To find the value of each exponent, you select a pair of trials that holds one concentration constant and allows you to see the effect on the rate by changing the other concentration. You select Trials 1 and 3 to determine the value of x. Taking the ratio of the trials you find: (NOTE: The k will cancel out.)

$$\frac{\text{rate}_3}{\text{rate}_1} = \frac{k[H_2]^x[NO]^y}{k[H_2]^x[NO]^y}$$

$$\frac{4.0\times10^3}{2.0\times10^3} = \frac{k(2.0\times10^{-3})^x(1.0\times10^{-3})^y}{k(1.0\times10^{-3})^x(1.0\times10^{-3})^y} \qquad 2 = \frac{(2.0\times10^{-3})^x(1.0\times10^{-3})^y}{(1.0\times10^{-3})^x(1.0\times10^{-3})^y} = 2^x$$

$$x = 1$$

The order of the reaction with respect H_2 is 1.

Repeating this process for Trials 2 and 3 to find the value of y.

$$\frac{\text{rate}_2}{\text{rate}_3} = \frac{k[H_2]^x[NO]^y}{k[H_2]^x[NO]^y}$$

$$\frac{1.6\times10^4}{4.0\times10^3} = \frac{k(2.0\times10^{-3})^x(2.0\times10^{-3})^y}{k(2.0\times10^{-3})^x(1.0\times10^{-3})^y} \qquad 4 = \frac{(2.0\times10^{-3})^x(2.0\times10^{-3})^y}{(2.0\times10^{-3})^x(1.0\times10^{-3})^y} = 2^y$$

$$y = 2$$

The order of the reaction with respect NO is 2.
The rate law for this reaction is rate = $k[H_2][NO]^2$. This is choice **(B)**.
Choice **(A)** is incorrect because it reverses the exponents.
Choice **(C)** is incorrect because it uses the coefficients from the balanced equation.
Choice **(D)** is incorrect because it is the equilibrium constant expression for the reaction.

Practice Questions Related to This: **PQ-1, PQ-2,** and **PQ-3**

SQ-2. *Conceptual*

The rate law for the reaction

$$A + B \rightarrow C + D$$

is first order in [A] and second order in [B]. If [A] is halved and [B] is doubled, the rate of the reaction will

(A) remain the same.

(B) be increased by a factor of 2.

(C) be increased by a factor of 4.

(D) be increased by a factor of 8.

Knowledge Required: (1) The meanings of the terms rate law, first order, and second order. (2) The ability to use the rate law expression.

Thinking it Through: You are told the order of the reaction with respect to each reactant. You recognize that this is telling you the exponents in the rate expression. The exponent for [A] is 1 and the exponent for [B] is 2. You write the rate expression for the reaction:

$$\text{rate} = k[\text{A}][\text{B}]^2$$

You realize the problem is asking you to determine the effect of changing the initial concentrations of both reactants on the rate of the reaction. This is a similar problem to **SQ-1**, where you were given the initial amounts and the rates but were asked to find the exponents. Here you know the exponents and the amounts and are being asked to find the ratio of the rates. As before you set this up as a ratio of the two rates (for two reaction trials noted by "1"and "2" subscripts):

$$\frac{\text{rate}_2}{\text{rate}_1} = \frac{k[\text{A}]_2[\text{B}]_2^2}{k[\text{A}]_1[\text{B}]_1^2}$$

Where rate_1, $[\text{A}]_1$, and $[\text{B}]_1$ are the rate, concentration of A, concentration of B for trial #1 and rate_2, $[\text{A}]_2$, and $[\text{B}]_2$ are the rate, concentration of A, concentration of B for trial #2. The ratio of the rates is what you need to find.

You also know the relationship between $[\text{A}]_1$ and $[\text{A}]_2$ and $[\text{B}]_1$ and $[\text{B}]_2$ from the information in the question: $[\text{A}]_2 = \frac{1}{2}[\text{A}]_1$ (because "[A] is halved")
and $[\text{B}]_2 = 2[\text{B}]_1$ (because "[B] is doubled")

Substituting these into the expression you get:

$$\frac{\text{rate}_2}{\text{rate}_1} = \frac{k[\text{A}]_2[\text{B}]_2^2}{k[\text{A}]_1[\text{B}]_1^2} = \frac{k\left(\frac{1}{2}[\text{A}]_1\right)\left(2[B]_1\right)^2}{k[\text{A}]_1[\text{B}]_1^2} = \frac{k\left(\frac{1}{2}[\text{A}]_1\right)4[\text{B}]_1^2}{k[\text{A}]_1[\text{B}]_1^2}$$

$$\frac{\text{rate}_2}{\text{rate}_1} = \frac{\left(\frac{1}{2}\right)4}{(1)(1)} = 2$$

The correct answer is choice **(B).**

Choice **(A)** is incorrect because the 2 was not squared in the expression $(2[\text{B}])^2$. Choice **(C)** is incorrect because it omitted the ½ term. Choice **(D)** is incorrect because the ½ term was mistakenly put in the denominator rather than the numerator.

Practice Questions Related to This: **PQ-4** and **PQ-5**

SQ-3. The half-life for the first order conversion of cyclobutene to ethylene,

$$C_4H_8(g) \rightarrow 2C_2H_4(g)$$

is 22.7 s at a particular temperature. How many seconds are needed for the partial pressure of cyclobutane to decrease from 100 mmHg to 10 mmHg?

(A) 0.101 s

(B) 52.0 s

(C) 75.5 s

(D) 5233 s

Knowledge Required: (1) The relationship between half-life and the rate constant for first-order reaction. (2) The ability to use the half-life equation and the integrated first-law equation.

Thinking it Through: You are given a reaction and told that it is first-order. You need to select and use the first-order integrated rate law to find the amount of time needed for the pressure to fall from 100 mmHg to 10 mmHg. You realize that you are not given a value for the rate constant, *k*. You recognize that because you know the half-life, you can find the value of the rate constant. Because the reaction is first-order you use the half-life expression for a first-order reaction:

$$t_{1/2} = \frac{\ln 2}{k}$$

Rearranging and substituting the given half-life you calculate the value of *k*.

$$k = \frac{\ln 2}{t_{1/2}} = \frac{\ln 2}{22.7\text{ s}} = 0.0305\text{ s}^{-1}$$

You then use this value of *k* in the first-order integrated rate law.

$$\ln[C_4H_8] = -kt + \ln[C_4H_8]_o$$

$$t = -\frac{\ln[C_4H_8] - \ln[C_4H_8]_o}{k} = -\frac{\ln(10) - \ln(100)}{0.0305\text{ s}^{-1}} = 75.5\text{ s}$$ which is choice **(C)**.

Choice **(A)** is incorrect because it used the value of the half-life as the rate constant.
Choice **(B)** is incorrect because it used 1/half-life as the rate constant, instead of ln2/half-life.
Choice **(D)** is incorrect because it used the expression for the second order half-life to find the value of *k*.

Practice Questions Related to This: **PQ-6** and **PQ-7**

SQ-4. The half-life for the first-order radioactive decay of ^{32}P is 14.2 days. How many days would be required for a sample of a radiopharmaceutical containing ^{32}P to decrease to 20.0% of its initial activity?

(A) 33.0 d **(B)** 49.2 d **(C)** 71.0 d **(D)** 286 d

Knowledge Required: (1) The knowledge that radioactive decay is first-order. (2) The ability to use the integrated form of the first-order rate law. (3) Relationship between half-life and rate constant.

Thinking it Through: You are being asked to find the time for a process to occur. You know that you have to use the integrated rate law. You also know that nuclear decay is a first-order process. To use the first-order integrated rate law you need a value for the rate constant. You use the relationship for the first-order half-life:

$$t_{1/2} = \frac{\ln 2}{k}$$

Solving for *k* you get: $k = \frac{\ln 2}{t_{1/2}} = \frac{\ln 2}{14.2\text{ d}} = 0.0488\text{ d}^{-1}$

The question is asking you for the time required for the sample to decrease to 20.0% of its original activity. This is different from other problems that give you specific amounts of initial and final concentrations. To solve the problem, you realize that you can pick any initial amount and the final amount will be 20.0% of that initial value. To keep the math simple, you assume you started with 100.0%. You then need to find the time required for the amount to decrease to 20.0% of 100%, or 0.200*1.000 = 0.200. You now know:

$$[^{32}P]_o = 1.00;\quad [^{32}P] = 0.20;\quad k = 0.0488\text{ d}^{-1}$$

You then use the integrated first-order rate law to find the time needed.

$$\ln[^{32}P] = -kt + \ln[^{32}P]_o$$

$$t = -\frac{\ln[^{32}P] - \ln[^{32}P]_o}{k} = -\frac{\ln(0.200) - \ln(1.00)}{0.0488\,d^{-1}} = 33.0\,d$$ which is choice **(A)**.

Choice **(B)** is incorrect because it represents 5 half-lives multiplied by ln(2).
Choice **(C)** is incorrect because it assumes it took 5 half-lives to get to 20.0% left.
Choice **(D)** is incorrect because it is 20 half-lives.

Practice Questions Related to This: **PQ-8** and **PQ-9**

SQ-5. Conceptual

Plots are shown for the reaction $NO_2(g) \rightarrow NO(g) + \frac{1}{2} O_2(g)$.
What is the rate law for the reaction?

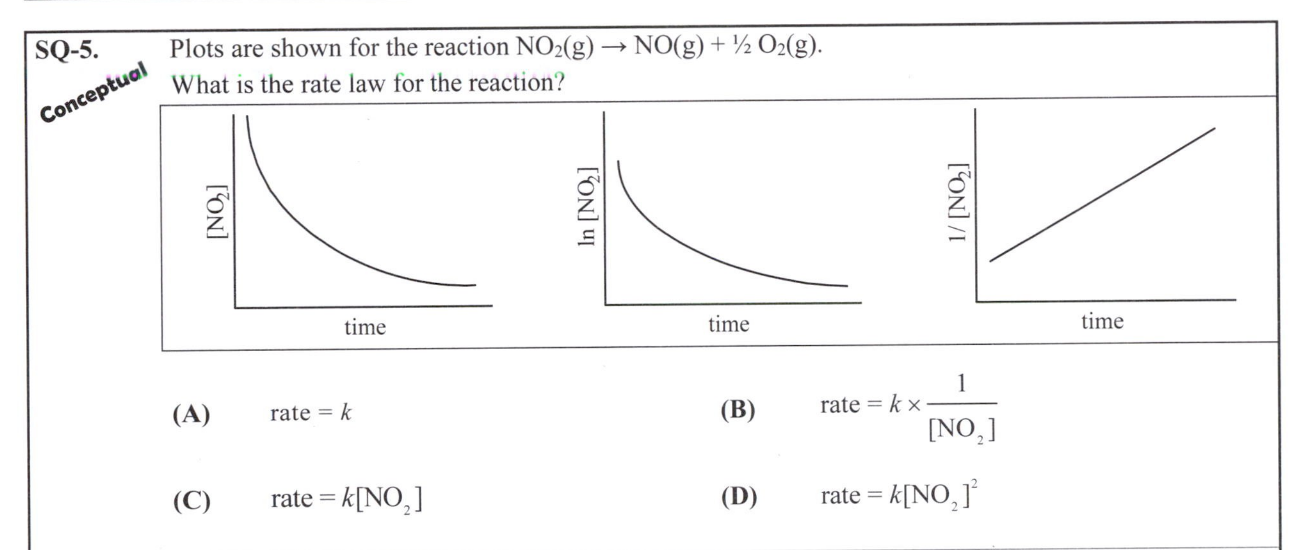

(A) rate = k

(B) rate = $k \times \frac{1}{[NO_2]}$

(C) rate = $k[NO_2]$

(D) rate = $k[NO_2]^2$

Knowledge Required: (1) Definition of rate law. (2) Ability to interpret integrated rate law equations.

Thinking it Through: You are given three plots of the concentration versus time data for a reaction. Each plot differs in how it plots the concentration of NO_2. You recall the three common integrated rate law expressions for reactions with a single reactant: $X \rightarrow P$

Order	Rate Law	Integrated Rate Law	Linear Plot
zero	rate = k	$[X]_t = -kt + [X]_o$	[X] vs time
first	rate = $k[X]^1$	$\ln[X]_t = -kt + \ln[X]_o$	ln[X] vs time
second	rate = $k[X]^2$	$1/[X]_t = kt + 1/[X]_o$	1/[X] vs time

Each of the integrated rate laws has the form of a y = mx + b equation. When you examine the plots provided, you notice that the plot of $1/[NO_2]$ vs time is linear and the others are nonlinear. You decide that this means the rate law is a second order rate law: rate = $k[NO_2]^2$ which is choice **(D).**

Choice **(A)** is incorrect because this is the zero-order rate law and the plot of $[NO_2]$ is not linear.
Choice **(B)** is incorrect because this is taken directly from the plot with $1/[NO_2]$ on the y-axis.
Choice **(C)** is incorrect because this is a first-order rate law and the plot of $\ln[NO_2]$ is not linear.

Practice Questions Related to This: **PQ-10** and **PQ-11**

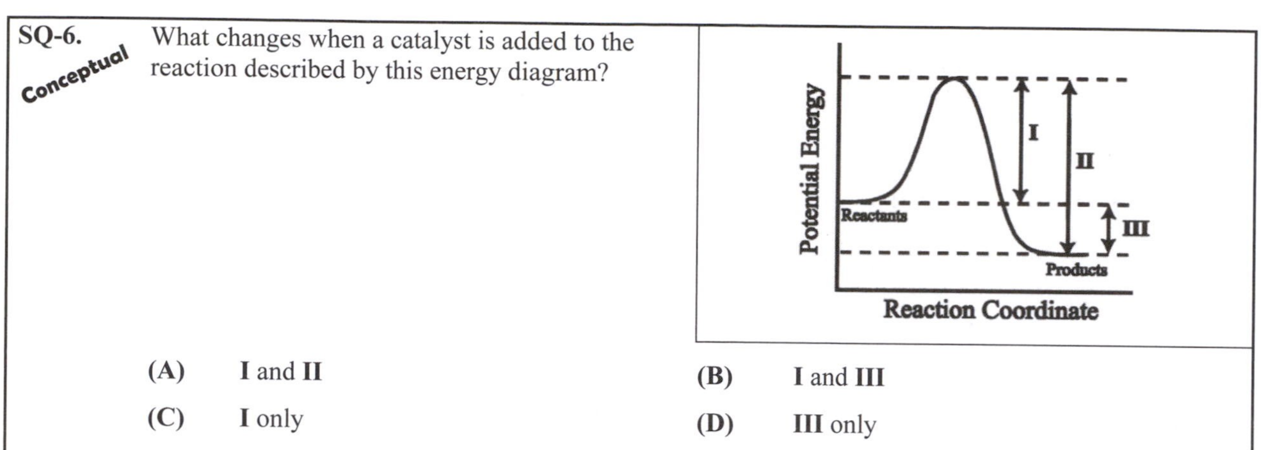

SQ-6. *Conceptual* What changes when a catalyst is added to the reaction described by this energy diagram?

(A) **I** and **II**
(B) **I** and **III**
(C) **I** only
(D) **III** only

Knowledge Required: (1) Interpretation of reaction energy diagrams. (2) Meaning of catalyst and activation energy.

Thinking it Through: You are being asked to interpret a reaction energy diagram. Several parts of the diagram have been labeled. Examining the diagram, you recall what each labeled portion represents.

I is the activation energy for the reaction in the forward direction.
II is the activation energy for the reaction in the reverse direction.
III is the difference in energy of the reactants and products, which is the ΔH for the reaction.

The question asks you to determine which of these is affected by the addition of the catalyst. You remember that a catalyst provides an alternative pathway for the reaction. This alternative pathway has a lower activation energy. The presence of a catalyst does not affect the value of ΔH for the reaction. Because arrows **I** and **II** represent activation energies, these will be affected by the addition of a catalyst. The correct answer is choice **(A)**.

Choice **(B)** and **(D)** are incorrect because a catalyst does not affect the value of the enthalpy change, **III**. Choice **(C)** is incorrect because a catalyst will affect the activation energy for both the forward and reverse reactions.

Practice Questions Related to This: **PQ-12, PQ-13,** and **PQ-14**

SQ-7. *Conceptual* Which statement regarding chemical reactions is true according to collision theory?

(A) All molecular collisions result in chemical reactions.
(B) Catalysts make individual collisions more effective, increasing reaction rates.
(C) Proper orientation of molecules is required for collision to result in chemical reactions.
(D) Increasing the temperature of a reaction decreases the kinetic energy of molecules, making collisions more effective.

Knowledge Required: (1) Understanding of the collision theory of chemical reactions. (2) Knowledge of what factors influence the rate of chemical reactions.

Thinking it Through: The question asks you to evaluate a series of statements about chemical reactions and collision theory. You decide to evaluate each option individually.

You know that for a collision to lead to a chemical reaction it must satisfy two criteria. The collision must have sufficient energy for the reaction to occur. Also, the molecules must be in the proper orientation for the collision to be effective. Therefore, not all collisions result in a chemical reaction. Therefore choice **(A)** is incorrect.

The presence of a catalyst does increase the rate of a reaction. However, the way a catalyst increases the rate of a reaction is not by making the collisions more effective. Instead a catalyst increases the rate of a reaction by providing an alternative route from reactants to products that has a smaller activation energy. Processes with smaller activation energies result in a faster reaction rate. Therefore choice **(B)** is incorrect.

You know that one of the two requirements for a collision to lead to a chemical reaction is that the reactants must collide with the proper orientation. Therefore choice **(C)** is correct.

Increasing the temperature **often** leads to an increase in the rate of a reaction. However, the reason is because as the temperature increases the average kinetic energy of the reactants increases. This can increase the rate of a reaction in two ways. First, there are more collisions that have the minimum energy needed for a reaction to occur. Second, the higher kinetic energy increases the number of collisions, which means the odds of a collision having the proper orientation increases. This increases the rate of a reaction. Therefore choice **(D)** is incorrect.

Practice Questions Related to This: **PQ-15, PQ-16, PQ-17, PQ-18, PQ-19, PQ-20, PQ-21, PQ-22, PQ-23,** and **PQ-24**

SQ-8. The activation energy for a particular reaction is 83.1 kJ·mol^{-1}. By what factor will the rate constant increase when the temperature is increased from 50.0 °C to 60.0 °C?

(A) 2.53 **(B)** 1.00 **(C)** 0.927 **(D)** 0.395

Knowledge Required: (1) How to use the Arrhenius equation. (2) The relationship between activation energy and the value of the rate constant.

Thinking it Through: You are being asked to determine the relative change in the value of the rate constant when the temperature changes. You recall the Arrhenius equation which relates the value of the activation energy, the temperature, and the rate constant. You also remember that all the temperature values must by in kelvin:

$$k = Ae^{-E_a/RT}$$

An alternative form of the equation is: $\ln k = \left(\frac{-E_a}{R}\right)\left(\frac{1}{T}\right) + \ln(A)$

This form of the equation is useful if you have measured values of k at different temperatures. You recognize that this equation is in the form: $y = mx + b$. If you plot $\ln(k)$ versus $1/T$ the slope of this plot is equal to $(-E_a/R)$.

Another form of this equation, often called the two-point form, is used when you only have two values of k or temperature values. You get this form of the equation by subtracting the equation at the first temperature, T_1, from that at the second temperature, T_2:

$$\ln k_2 - \ln k_1 = \left(\frac{-E_a}{R}\right)\left(\frac{1}{T_2}\right) - \left(\frac{-E_a}{R}\right)\left(\frac{1}{T_1}\right) + \ln(A) - \ln(A)$$

$$\ln k_2 - \ln k_1 = \left(\frac{-E_a}{R}\right)\left(\frac{1}{T_2} - \frac{1}{T_1}\right) \qquad \ln\left(\frac{k_2}{k_1}\right) = \left(\frac{-E_a}{R}\right)\left(\frac{1}{T_2} - \frac{1}{T_1}\right)$$

You recognize that the ratio of k_2/k_1 is what the problem is asking for and you substitute the given values into the equation. **Note**: you are careful to make sure the units of R and E_a match and the temperature is in kelvin:

$$\ln\left(\frac{k_2}{k_1}\right) = \left(\frac{-83.1\,\text{kJ}\cdot\text{mol}^{-1}}{8.314\,\text{J}\cdot\text{mol}^{-1}\cdot\text{K}^{-1}}\right)\left(\frac{1000\text{ J}}{1\text{ kJ}}\right)\left(\frac{1}{333.15\text{ K}} - \frac{1}{323.15\text{ K}}\right)$$

$$\ln\left(\frac{k_2}{k_1}\right) = 0.927 \qquad \frac{k_2}{k_1} = e^{0.927} = 2.53$$

The correct answer is choice **(A)**.

Choice **(B)** is not correct because it is the answer if the units of E_a are not converted to J. Choice **(C)** is not correct because it is the natural log (ln) of the ratio of the rate constants. Choice **(D)** is not correct because it is the ratio k_1/k_2.

Practice Questions Related to This: **PQ-25** and **PQ-26**

SQ-9. Consider the reaction

$$2NO_2(g) + F_2(g) \rightleftharpoons 2NO_2F(g)$$

A proposed mechanism for the reaction is

$NO_2 + F_2 \rightleftharpoons NO_2F + F$ (slow)

$NO_2 + F \rightleftharpoons NO_2F$ (fast)

What is the rate law for this mechanism?

(A) $\text{rate} = k\dfrac{[NO_2F]}{[NO_2]^2[F_2]}$

(B) $\text{rate} = k[NO_2]^2[F_2]$

(C) $\text{rate} = k[NO_2][F_2]$

(D) $\text{rate} = k[NO_2][F]$

Knowledge Required: (1) The fact that the slowest step in the reaction mechanism determines the rate law for the reaction. (2) Being able to write a rate law expression from an elementary step.

Thinking it Through: You are given a mechanism and are asked to write the rate law for the proposed mechanism. The slowest step in a reaction mechanism is the step with the highest activation energy and the species involved in this step determine the rate law for the reaction. You note that you are told the first step is the slowest. You know that you can write the rate law for this elementary step using the coefficients from the elementary step as the exponents in the rate law.

You write the rate law for the slow step as: $rate = k[NO_2][F_2]$ which is choice **(C)**.

Choice **(A)** is not correct because it is the equilibrium expression for the reaction, not a rate law. Choice **(B)** is not correct because it uses the stoichiometry of the overall reaction. Choice **(D)** is not correct because it is not the derived from the slowest step.

Practice Questions Related to This: **PQ-27, PQ-28,** and **PQ-29**

SQ-10. *Conceptual*

Consider this equilibrium:

$$2SO_2(g) + O_2(g) \rightleftharpoons 2SO_3(g)$$

The forward reaction is proceeding at a certain rate at some temperature and pressure. When the pressure is increased, what might we expect for the forward reaction?

(A) a greater rate of reaction and a greater yield of SO_3 at equilibrium

(B) a greater rate of reaction and a same yield of SO_3 at equilibrium

(C) a lesser rate of reaction and a smaller yield of SO_3 at equilibrium

(D) a lesser rate of reaction and a greater yield of SO_3 at equilibrium

Knowledge Required: (1) How to use LeChatlier's principle. (2) Qualitative understanding of rate laws.

Thinking it Through: You need to predict what will happen to an equilibrium system when it is subjected to a change. You know that a change in pressure may shift the equilibrium and depends on the moles of gaseous substances as reactants and products. An increase in pressure will shift the equilibrium in the direction that minimizes the increase in pressure. This is the side that has the fewest moles of gas. For this reaction, the side with fewer moles of gas is the product side. You realize that this means that an increase in pressure will cause the rate of the reaction producing SO_3 to increase. The correct answer is choice **(A)**.

Choices **(B)** and **(C)** is incorrect because SO_3 will increase. Choice **(D)** is incorrect because the rate of the reaction will increase.

Practice Questions Related to This: **PQ-30**

Practice Questions (PQ)

Conceptual **PQ-1.** When the reaction:

$CH_3Cl(g) + H_2O(g) \rightarrow CH_3OH(g) + HCl(g)$

was studied, the tabulated data were obtained. Based on these data, what are the reaction orders?

Exp	$[CH_3Cl]_0$ / M	$[H_2O]_0$ / M	Initial Rate / $M{\cdot}s^{-1}$
1	0.100	0.100	0.182
2	0.200	0.200	1.45
3	0.200	0.400	5.81

(A) CH_3Cl: first order H_2O: first order
(B) CH_3Cl: first order H_2O: second order
(C) CH_3Cl: second order H_2O: first order
(D) CH_3Cl: second order H_2O: second order

PQ-2. The reaction between acetone and bromine in acidic solution is represented by the equation:
$CH_3COCH_3(aq) + Br_2(aq) + H_3O^+(aq) \rightarrow$ products
The tabulated kinetic data were gathered. Based on these data, the experimental rate law is

Exp	$[CH_3COCH_3]_0$ / M	$[Br_2]_0$ / M	$[H_3O^+]_0$ / M	Initial Rate / $M{\cdot}s^{-1}$
1	0.30	0.050	0.050	5.8×10^{-5}
2	0.30	0.100	0.050	5.8×10^{-5}
3	0.30	0.050	0.100	1.2×10^{-4}
4	0.40	0.050	0.200	3.2×10^{-4}

(A) rate = $k[CH_3COCH_3][Br_2][H_3O^+]$
(B) rate = $k[CH_3COCH_3][H_3O^+]$
(C) rate = $k[H_3O^+]^2$
(D) rate = $k[CH_3COCH_3][Br_2]$

PQ-3. Initial rate data for the reaction

$$2H_2(g) + Cl_2(g) \rightarrow 2HCl(g)$$

are given in the table. What is the rate law for the reaction?

Exp	$[H_2]_0$ / M	$[Cl_2]_0$ / M	Initial Rate / $M{\cdot}s^{-1}$
1	0.0020	0.0050	2.5×10^{-3}
2	0.0020	0.0025	1.3×10^{-3}
3	0.0015	0.0025	1.3×10^{-3}
4	0.0050	0.0010	0.5×10^{-3}

(A) rate = $k[Cl_2]^2$
(B) rate = $k[Cl_2]$
(C) rate = $k[H_2]$
(D) rate=$k[H_2][Cl_2]$

Conceptual **PQ-4.** The gas phase reaction, $A_2 + B_2 \rightarrow 2AB$, proceeds by bimolecular collisions between A_2 and B_2 molecules. The rate law is rate =$k[A_2][B_2]$
If the concentrations of both A_2 and B_2 are doubled, the reaction rate will change by a factor of

(A) ½.
(B) $\sqrt{2}$.
(C) 2.
(D) 4.

Conceptual **PQ-5.** For the reaction of chorine and nitric oxide,

$$2NO(g) + Cl_2(g) \rightarrow 2NOCl(g)$$

doubling the concentration of chlorine doubles the rate of reaction. Doubling the concentration of both reactants increases the rate of reaction by a factor of eight. The reaction is

(A) first order in both NO and Cl_2.
(B) first order on NO and second order in Cl_2.
(C) second order in NO and first order in Cl_2.
(D) second order in both NO and Cl_2.

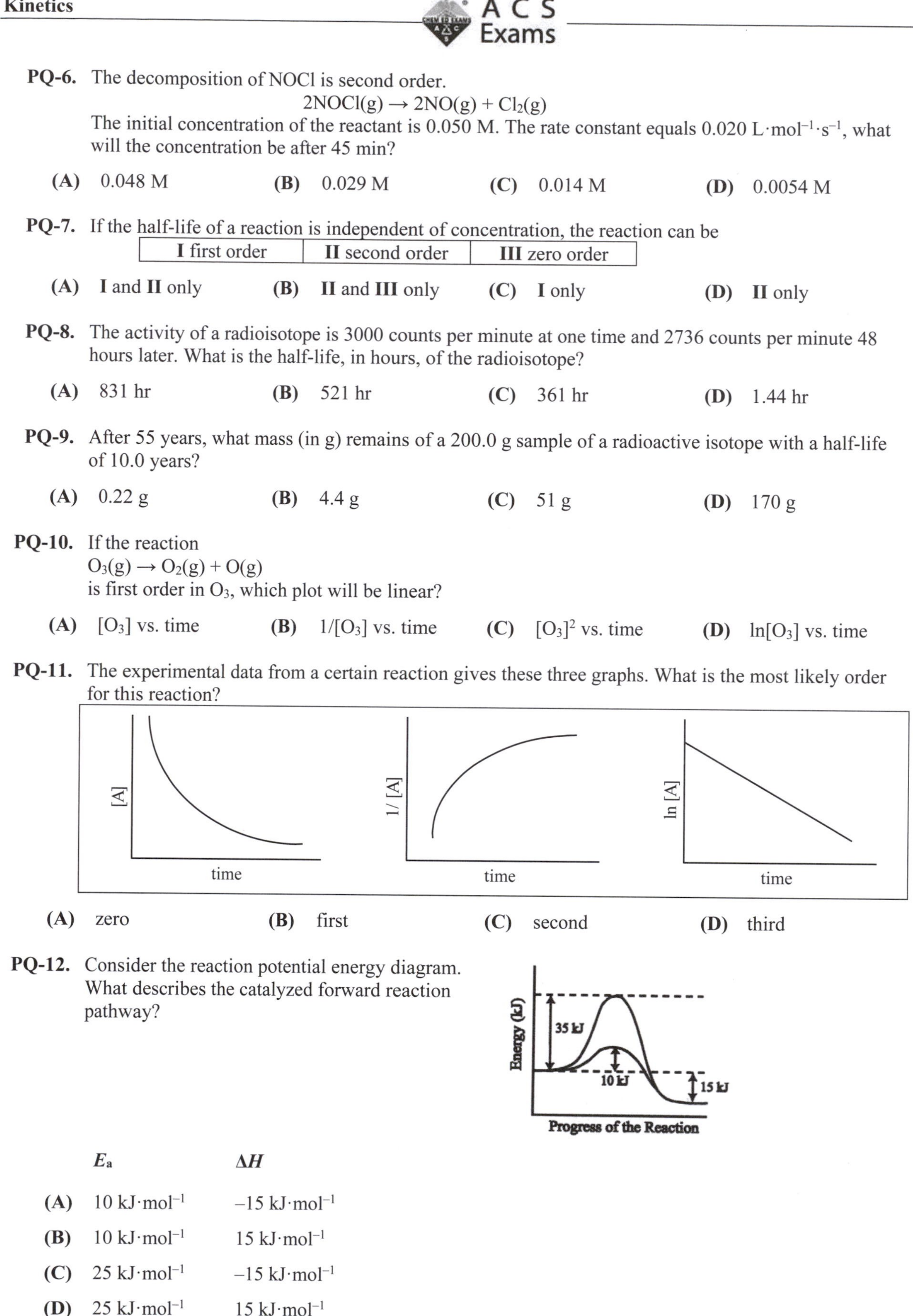

PQ-6. The decomposition of NOCl is second order.

$$2NOCl(g) \rightarrow 2NO(g) + Cl_2(g)$$

The initial concentration of the reactant is 0.050 M. The rate constant equals 0.020 $L \cdot mol^{-1} \cdot s^{-1}$, what will the concentration be after 45 min?

(A) 0.048 M **(B)** 0.029 M **(C)** 0.014 M **(D)** 0.0054 M

PQ-7. If the half-life of a reaction is independent of concentration, the reaction can be

I first order	**II** second order	**III** zero order

(A) **I** and **II** only **(B)** **II** and **III** only **(C)** **I** only **(D)** **II** only

PQ-8. The activity of a radioisotope is 3000 counts per minute at one time and 2736 counts per minute 48 hours later. What is the half-life, in hours, of the radioisotope?

(A) 831 hr **(B)** 521 hr **(C)** 361 hr **(D)** 1.44 hr

PQ-9. After 55 years, what mass (in g) remains of a 200.0 g sample of a radioactive isotope with a half-life of 10.0 years?

(A) 0.22 g **(B)** 4.4 g **(C)** 51 g **(D)** 170 g

PQ-10. If the reaction

$O_3(g) \rightarrow O_2(g) + O(g)$

is first order in O_3, which plot will be linear?

(A) $[O_3]$ vs. time **(B)** $1/[O_3]$ vs. time **(C)** $[O_3]^2$ vs. time **(D)** $\ln[O_3]$ vs. time

PQ-11. The experimental data from a certain reaction gives these three graphs. What is the most likely order for this reaction?

(A) zero **(B)** first **(C)** second **(D)** third

PQ-12. Consider the reaction potential energy diagram. What describes the catalyzed forward reaction pathway?

	E_a	ΔH
(A)	10 $kJ \cdot mol^{-1}$	–15 $kJ \cdot mol^{-1}$
(B)	10 $kJ \cdot mol^{-1}$	15 $kJ \cdot mol^{-1}$
(C)	25 $kJ \cdot mol^{-1}$	–15 $kJ \cdot mol^{-1}$
(D)	25 $kJ \cdot mol^{-1}$	15 $kJ \cdot mol^{-1}$

PQ-13. Which line segment represents the activation energy for the reaction between **C** and **D** to form **A** and **B**?

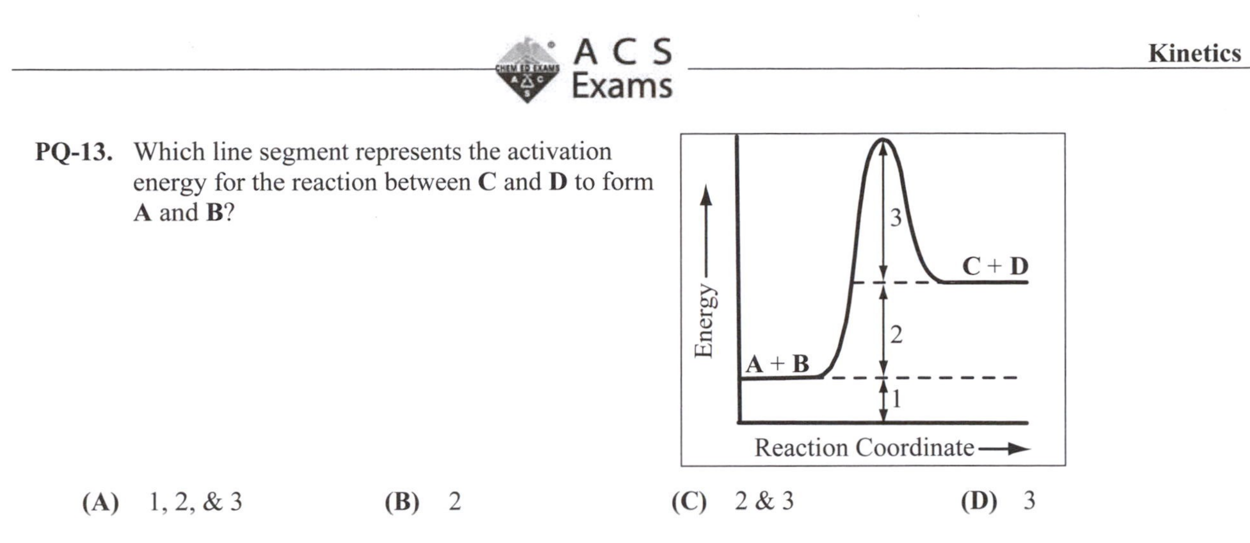

(A) 1, 2, & 3 **(B)** 2 **(C)** 2 & 3 **(D)** 3

PQ-14. A certain reaction has $\Delta H = -75\ \text{kJ·mol}^{-1}$ and an activation energy of $40\ \text{kJ·mol}^{-1}$. A catalyst is found that lowers the activation energy of the forward reaction by $15\ \text{kJ·mol}^{-1}$. What is the activation energy of the reverse reaction in the presence of the same catalyst?

(A) $25\ \text{kJ·mol}^{-1}$ **(B)** $60\ \text{kJ·mol}^{-1}$ **(C)** $90\ \text{kJ·mol}^{-1}$ **(D)** $100\ \text{kJ·mol}^{-1}$

Conceptual **PQ-15.** Which statement best explains why the activation energy is changed by adding a catalyst?

(A) The catalyst changes the reaction mechanism.

(B) The catalyst changes the free energy of reaction.

(C) The catalyst increases the kinetic energy of the reactants.

(D) The catalyst lowers the reaction volume, increasing the concentration of reactants.

Conceptual **PQ-16.** Which reaction rate is more affected by a change in temperature?

	Reaction	E_a / kJ·mol^{-1}
I	$H_2(g) + I_2(g) \rightarrow 2HI(g)$	173
II	$CH_3CHO(g) \rightarrow CH_4(g) + CO(g)$	356

(A) reaction **I** because the activation energy is lower

(B) reaction **I** because the number of moles of gas stays the same

(C) reaction **II** because number of moles of gases increases as reaction goes forward

(D) reaction **II** because the activation energy is higher

PQ-17. What will increase the value of the rate constant for an elementary step?

(A) adding a catalyst

(B) raising the temperature

(C) increasing the concentration of products

(D) increasing the concentration of reactants

Conceptual **PQ-18.** The Arrhenius equation describes the relationship between the rate constant, *k*, and the energy of activation, E_a.

$$k = Ae^{-E_a/RT}$$

In this equation, *A* is an empirical constant, *R* is the ideal-gas constant, *e* is the base of natural logarithms, and *T* is the absolute temperature. According to the Arrhenius equation,

(A) at constant temperature, reactions with lower activation energies proceed more rapidly.

(B) at constant temperature, reactions with lower activation energies proceed less rapidly.

(C) at constant energy of activation, reactions at lower temperatures proceed more rapidly.

(D) at constant energy of activation, reactions with smaller values of *A* proceed more rapidly.

Conceptual **PQ-19.** What will be the effect of increasing the temperature of reactants that are known to undergo an endothermic reaction?

(A) Both the rate of reaction and the value of the equilibrium constant increases.

(B) The rate of reaction increases and the value of the equilibrium constant decreases.

(C) The rate of reaction decreases and the value of the equilibrium constant increases.

(D) The rate of reaction increases and the value of the equilibrium constant is unchanged.

Conceptual **PQ-20.** The change in temperature from 10 °C to 20 °C is found to double the rate of a particular chemical reaction. How did the change in temperature affect the reaction?

(A) The number of molecules with sufficient energy to react increased.

(B) The number of molecules with sufficient energy to react decreased.

(C) The activation energy increased.

(D) The activation energy decreased.

Conceptual **PQ-21.** The value for the rate constant of a reaction can generally be expected to

(A) decrease with increasing temperature.

(B) increase with increasing temperature.

(C) decrease with increasing temperature only when the reaction is exothermic.

(D) increase with increasing temperature only when the reaction is exothermic.

Conceptual **PQ-22.** Two reactions with different activation energies have the same rate at room temperature. Which statement correctly describes the rates of these two reactions at the same, higher temperature?

(A) The reaction with the larger activation energy will be faster.

(B) The reaction with the smaller activation energy will be faster.

(C) The two reactions will continue to occur at the same rates.

(D) A prediction cannot be made without additional information.

Conceptual **PQ-23.** A catalyst increases the rate of a reaction by

(A) changing the mechanism of the reaction.

(B) increasing the activation energy of the reaction.

(C) increasing the concentration of one or more reactants.

(D) decreasing the difference in relative energy of the reactants and products.

Conceptual **PQ-24.** Which statement most accurately describes the behavior of a catalyst?

(A) A catalyst increases ΔG of a reaction and hence the forward rate.

(B) A catalyst reduces the ΔH of a reaction and hence the temperature needed to produce products.

(C) A catalyst reduces the activation energy for a reaction and increases the rate of a reaction.

(D) A catalyst increases the equilibrium constant and final product concentrations.

PQ-25. The rate constants for a specific reaction at two different temperatures are given in the table. What is the activation energy for the reaction?

Temperature	Rate constant
127 °C	3.0×10^{-4} s^{-1}
227 °C	6.0×10^{-2} s^{-1}

(A) –88.1 kJ·mol^{-1} **(B)** –12.7 kJ·mol^{-1} **(C)** 12.7 kJ·mol^{-1} **(D)** 88.1 kJ·mol^{-1}

PQ-26. Activation energy can be experimentally determined from the slope of the plot of

(A) k versus $1/T$. **(B)** k versus $\ln T$. **(C)** $\ln k$ versus T. **(D)** $\ln k$ versus $1/T$.

Conceptual **PQ-27.** Which energy diagram best matches the proposed mechanism for this exothermic reaction?

$$E + S \rightleftharpoons ES \quad \text{(fast)}$$
$$ES \rightarrow E + P \quad \text{(slow)}$$

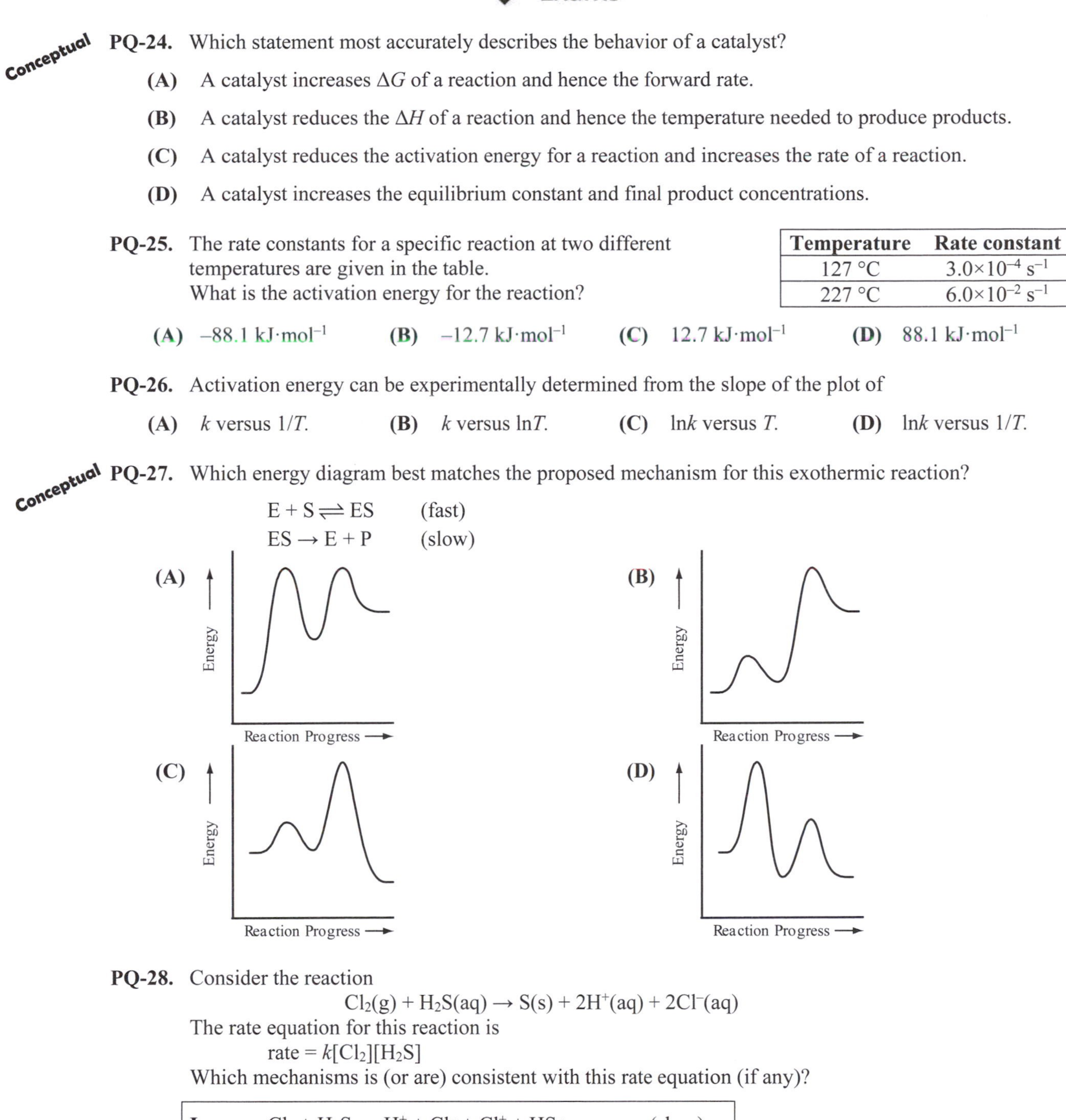

PQ-28. Consider the reaction

$$Cl_2(g) + H_2S(aq) \rightarrow S(s) + 2H^+(aq) + 2Cl^-(aq)$$

The rate equation for this reaction is

rate = $k[Cl_2][H_2S]$

Which mechanisms is (or are) consistent with this rate equation (if any)?

I	$Cl_2 + H_2S \rightarrow H^+ + Cl^- + Cl^+ + HS^-$ $Cl^+ + HS^- \rightarrow H^+ + Cl^- + S$	(slow) (fast)
II	$H_2S \rightarrow H^+ + HS^-$ $Cl_2 + HS^- \rightarrow 2Cl^- + H^+ + S$	(fast) (slow)

(A) **I** only **(B)** **II** only **(C)** Both **I** and **II** **(D)** Neither **I** or **II**

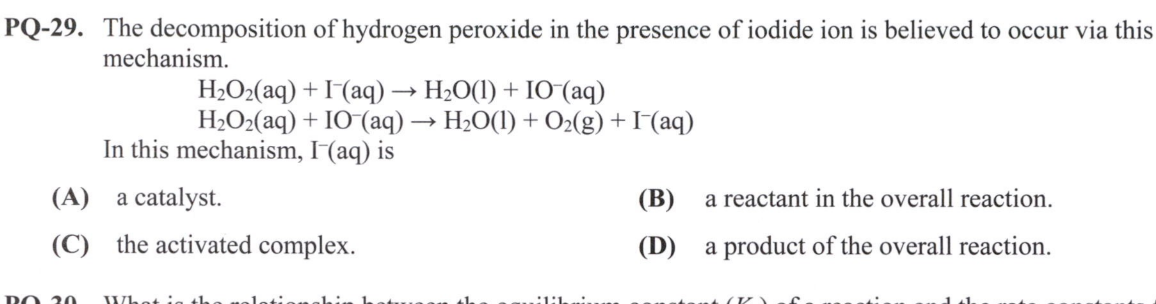

PQ-29. The decomposition of hydrogen peroxide in the presence of iodide ion is believed to occur via this mechanism.

$$H_2O_2(aq) + I^-(aq) \rightarrow H_2O(l) + IO^-(aq)$$
$$H_2O_2(aq) + IO^-(aq) \rightarrow H_2O(l) + O_2(g) + I^-(aq)$$

In this mechanism, $I^-(aq)$ is

(A) a catalyst.

(B) a reactant in the overall reaction.

(C) the activated complex.

(D) a product of the overall reaction.

PQ-30. What is the relationship between the equilibrium constant (K_c) of a reaction and the rate constants for the forward (k_f) and reverse (k_r) reactions in a single step reaction?

(A) $K_c = k_f k_r$

(B) $K_c = k_f/k_r$

(C) $K_c = 1/(k_f k_r)$

(D) $K_c = k_f - k_r$

Answers to Study Questions

1. B
2. B
3. C
4. A
5. D
6. A
7. C
8. B
9. C
10. A

Answers to Practice Questions

1. B
2. B
3. B
4. D
5. C
6. C
7. C
8. C
9. B
10. D
11. B
12. A
13. D
14. D
15. A
16. D
17. B
18. A
19. A
20. A
21. B
22. A
23. A
24. C
25. D
26. D
27. C
28. A
29. A
30. B

A C S
Exams

Chapter 11 – Equilibrium

Chapter Summary:

This chapter will focus on equilibrium including equilibrium constants and determining the amount of reactants and products at equilibrium.

Specific topics covered in this chapter are:

- Definition of equilibrium
- Law of Mass Action
- K_p and K_c
- Equilibrium calculations
- Reaction quotient (Q) of a reaction
- Le Chatlier's Principle

Previous material that is relevant to your understanding of questions in this chapter include:

- Stoichiometry (***Chapter 4***)
- Solutions and Aqueous Reactions, Part 1 (***Chapter 5***)
- Solutions and Aqueous Reactions, Part 2 (***Chapter 9***)

Where to find this in your textbook:

The material in this chapter typically aligns to "Fundamentals of Equilibrium Concepts" in your textbook. The name of your chapter(s) may vary.

Practice exam:

There are practice exam questions aligned to the material in this chapter. Because there are a limited number of questions on the practice exam, a review of the breadth of the material in this chapter is advised in preparation for your exam.

How this fits into the big picture:

The material in this chapter aligns to the Big Idea of Equilibrium (8) as listed on page 12 of this study guide.

Study Questions (SQ)

SQ-1. What is the equilibrium expression for this reaction?

$$Ni(CO)_4(g) \rightleftharpoons Ni(s) + 4CO(g)$$

(A) $K_c = \frac{[CO]^4}{[Ni(CO)_4]}$ **(B)** $K_c = \frac{[Ni][CO]^4}{[Ni(CO)_4]}$ **(C)** $K_c = \frac{[Ni(CO)_4]}{[CO]^4}$ **(D)** $K_c = \frac{[Ni(CO)_4]}{[Ni][CO]^4}$

Knowledge Required: (1) How to determine an equilibrium expression from a balanced equation.

Thinking it Through: The question asks you determine the equilibrium expression for a given reaction. The answer options are in K_c; therefore, you will consider the concentrations of the materials when constructing the equilibrium expression. An equilibrium constant is defined as the concentrations of products raised to a value equal to their stoichiometric coefficient in a balanced chemical equation divided by the concentrations of reactants raised to a value equal to their stoichiometric coefficients in a balanced chemical equation. Because the concentrations of solids and liquids do not change measurably, both are not included in the equilibrium expression. Therefore, when the concentration of gaseous or solution products are raised to their power you have a numerator of $[CO]^4$; likewise for reactants, when the concentration of gaseous or solution reactants are raised to their power you have a denominator of $[Ni(CO)_4]$. Combined, the numerator and denominator are Choice **(A)**.

Choice **(B)** is not correct because it includes a solid in the expression.

Choice **(C)** is not correct because it has the reactants in the numerator and the products in the denominator, which is reversed.

Choice **(D)** is not correct because it includes a solid in the expression and is the inverse of the balanced chemical equation as written.

Practice Questions Related to This: **PQ-1, PQ-2, PQ-3, PQ-4, PQ-5, PQ-6, PQ-7, PQ-8, PQ-9, PQ-10,** and **PQ-11**

SQ-2. Given the reaction and equilibrium constant:

$$2SO_3(g) \rightleftharpoons 2SO_2(g) + O_2(g) \qquad K_c = 2.3\times10^{-7}$$

What is the equilibrium constant for this reaction at the same temperature?

$$SO_3(g) \rightleftharpoons SO_2(g) + \tfrac{1}{2}O_2(g) \qquad K_c = ?$$

(A) 1.2×10^{-7} **(B)** 4.6×10^{-7} **(C)** 4.8×10^{-4} **(D)** 4.3×10^{6}

Knowledge Required: (1) The relationship between a balanced equation and an equilibrium constant.

Thinking it Through: The question asks you to determine an equilibrium constant for a given balanced chemical equation given a constant for a related balanced chemical equation. You note that the chemical equation for which you are to determine an equilibrium constant is halved compared to the chemical equation for which the equilibrium constant is known. You can write the equilibrium expressions then for both equations and compare:

$$2SO_3(g) \rightleftharpoons 2SO_2(g) + O_2(g) \qquad K = \frac{[SO_2]^2[O_2]}{[SO_3]^2}$$

$$SO_3(g) \rightleftharpoons SO_2(g) + \frac{1}{2}O_2(g) \qquad K' = \frac{[SO_2][O_2]^{1/2}}{[SO_3]} = K^{1/2} = \sqrt{K}$$

When the reaction is halved, the equilibrium constant is raised to the 0.5 power (or the square root). Thus, the equilibrium constant needed is $(2.3\times10^{-7})^{0.5}$ or 4.8×10^{-4}, Choice **(C)**.

Choice **(A)** is not correct because it is the known equilibrium constant divided by 2.
Choice **(B)** is not correct because it is the known equilibrium constant multiplied by 2.
Choice **(D)** is not correct because it is the inverse of the known equilibrium constant.

Practice Questions Related to This: **PQ-12**

SQ-3. Conceptual

Consider the equilibrium reactions:

$$2SO_2(g) + O_2(g) \rightleftharpoons 2SO_3(g) \qquad K_1$$

$$2CO(g) + O_2(g) \rightleftharpoons 2CO_2(g) \qquad K_2$$

What is the equilibrium constant, K, for this reaction?

$$2SO_2(g) + 2CO_2(g) \rightleftharpoons 2SO_3(g) + 2CO(g)$$

(A) $K = \left(\frac{K_1}{K_2}\right)$ **(B)** $K = \left(\frac{K_2}{K_1}\right)$ **(C)** $K = K_1 \times K_2$ **(D)** $K = K_1 + \frac{1}{K_2}$

Knowledge Required: (1) How to combine multiple chemical reactions and their respective equilibrium constants to form a new reaction and constant.

Thinking it Through: The question asks you to determine the equilibrium constant for a chemical reaction that results from the combination of two reactions with given equilibrium constants. First, you note that in order to combine the given reactions, the second reaction must be reversed; therefore, the value of the equilibrium constant for the reverse reaction is the inverse of the given K or K_2^{-1}.

$$2SO_2(g) + O_2(g) \rightleftharpoons 2SO_3(g) \qquad K_1$$

$$2CO_2(g) \rightleftharpoons 2CO(g) + O_2(g) \qquad K_2^{-1}$$

$$2SO_2(g) + 2CO_2(g) \rightleftharpoons 2SO_3(g) + 2CO(g) \qquad K_1 \times K_2^{-1}$$

The desired reaction is now the sum of the first reaction and the inverse of the second reaction; when reactions are summed, the new equilibrium constant is the product of the two *known* equilibrium constants or $K_1\times K_2^{-1}$ for the reaction given, Choice **(A)**.

Choice **(B)** is not correct because it represents the inverse of the desired chemical reaction.

Choice **(C)** is not correct because it does not account for the reverse of the second chemical reaction.

Choice **(D)** is not correct because it is the sum of the two constants; however, when reactions are summed, the respective equilibrium constants are multiplied.

Practice Questions Related to This: **PQ-13** and **PQ-14**

SQ-4. Consider the reaction: $X_2(g) + 2Y(g) \rightleftharpoons 2Z(g)$. 12.00 moles of Z are placed in an evacuated 2.00-liter flask. After the reactants and products reach equilibrium, the flask contains 6.00 moles of Y. What is the equilibrium constant, *K*, for the reaction?

(A) 0.333 **(B)** 0.667 **(C)** 1.50 **(D)** 3.00

Knowledge Required: (1) How to determine an equilibrium expression from a balanced equation. (2) How to calculate equilibrium amounts from an equilibrium expression.

Thinking it Through: The question asks you to determine the equilibrium constant given a chemical reaction, the amount of the single species present before equilibrium, and the amount of one species after equilibrium is reached. First, you are told that there are 12.0 moles of Z initially, and 6.00 moles of Y at equilibrium. You then also consider that you know how the amounts will change because you begin with only product (so the change for Z will be negative and the change for X_2 or Y will be positive) and the amounts of these changes from the stoichiometry of the balanced equation (an increase of **2**x because the coefficient for Y is **2**, etc.). This is best summarized in a table:

	X_2	+	2Y	$\rightleftharpoons$	2Z
Initial	0		0		12.00
Change	$+x$		$+2x$		$-2x$
Equilibrium	$+x$		6.00		$12.0 - 2x$

Now you can see that you have an equivalency of $2x = 6.00$ or $x = 3.00$. Using this value for x, you can complete all values in the table:

	X_2	+	2Y	$\rightleftharpoons$	2Z
Initial	0		0		12.00
Change	+3.00		+2(3.00)		–2(3.00)
Equilibrium	3.00		6.00		6.00

You now have all of the molar amounts of the substances at equilibrium. To obtain the molar concentrations, you divide each value by 2.00 L (given in the problem): $[X_2] = 1.50$ M $[Y] = 3.00$ M $[Z] = 3.00$ M

To determine the value of *K*, you write an equilibrium expression for this reaction using the concentration of the product raised to its coefficient divided by the concentrations of the reactants raised to their coefficients; you can then substitute the equilibrium concentrations and calculate *K*:

$$K = \frac{[Z]^2}{[X_2][Y]^2} = \frac{(3.00\text{ M})^2}{(1.50\text{ M})(3.00\text{ M})^2} = \frac{1}{1.50} = 0.667 \quad \text{Choice (B)}$$

Choice **(A)** is not correct because it is determined using moles and not molar concentrations.

Choice **(C)** is not correct because it is determined using the inverse of *K*.

Choice **(D)** is not correct because it is determined using moles and the inverse of *K*.

Practice Questions Related to This: **PQ-15** and **PQ-16**

SQ-5. Phosgene decomposes into carbon monoxide and elemental chlorine. If the initial concentration of $COCl_2(g)$ is 0.50 M, what is the equilibrium concentration of CO(g)?

$$COCl_2(g) \rightleftharpoons CO(g) + Cl_2(g) \qquad K_c = 6.6\times10^{-8}$$

(A) 1.8×10^{-4} M **(B)** 9.1×10^{-5} M **(C)** 1.7×10^{-8} M **(D)** 6.6×10^{-8} M

Knowledge Required: (1) How to write an equilibrium expression from a balanced equation. (2) How to use the equilibrium expression to find equilibrium amounts.

Thinking it Through: The question asks you to determine the equilibrium concentration of a product given that only the reactant is present before it is allowed to reach equilibrium. You can assume that the reaction is done at constant volume and therefore concentrations are proportional to amount.

Begin by setting up a table of initial-change-equilibrium concentrations. Initial concentrations are provided in the question. Mole ratios are used to determine the changes in concentration. Equilibrium values are determined by the addition/subtraction of the initial and change values:

	$COCl_2$	$\rightleftharpoons$	CO	Cl_2
Initial	0.50 M		0	0
Change	$-x$		$+x$	$+x$
Equilibrium	$0.50 - x$		x	x

The equilibrium expression for this reaction is the concentration of products raised to their respective coefficients divided by the concentration of the starting material raised to its respective coefficient:

$$K_c = \frac{[CO][Cl_2]}{[COCl_2]}$$

You then substitute the values obtained into this expression: $6.6\times10^{-8} = \dfrac{(x)(x)}{0.50 - x}$

And simplified further to: $6.6\times10^{-8} = \dfrac{x^2}{0.50 - x}$

At this point, you may be inclined to continue the algebra to set up the equation to use the quadratic equation. However, you also know you may be able to simplify the process. The value of the equilibrium constant is very small. You know this means that most of the species at equilibrium are reactants with only a very small quantity of products if you began with only reactants (which you did). Therefore, you then note that any change to the initial concentration of $COCl_2$ will be very small, leaving the initial concentration (0.50 M) essentially unchanged or:

	$COCl_2$	$\rightleftharpoons$	CO	Cl_2
Initial	0.50 M		0	0
Change	$-x$		$+x$	$+x$
Equilibrium	$0.50 - x \approx 0.50$		x	x

Which then makes the equilibrium expression: $6.6\times10^{-8} = \dfrac{(x)(x)}{0.50 - x} \approx \dfrac{(x)(x)}{0.50}$

Which can now be solved without the quadratic equation:

$$6.6\times10^{-8} = \frac{(x)(x)}{0.50} = \frac{x^2}{0.50} \qquad \sqrt{(6.6\times10^{-8})(0.50)} = x \qquad x = 1.8\times10^{-4}$$

Of course, you also know there is nothing wrong with continuing from before and setting up the equation and using the quadric equation:

$$6.6\times10^{-8}(0.50 - x) = x^2 \qquad x^2 + 6.6\times10^{-8}(x) - 3.3\times10^{-8} = 0$$

The quadratic equation can then be used to determine possible values of x: $\pm1.8\times10^{-4}$. Given that the concentrations of the products cannot be negative, x must be equal to $+1.8\times10^{-4}$ and thus the final concentration of CO is 1.8×10^{-4} M, choice **(A)**, which was the same answer you obtained using the approximation.

Note: *You can check your approximation by using the value of x you calculated and using this in the equilibrium expression to calculate K_c – the value will be the same as given if the approximation can be used. You can also approximate in the reverse if the value of K_c is very large and you only begin with product (so the change to the product concentration will be small).*

Choice **(B)** is not correct because it would be the answer if "$(2x)^2$" was the numerator of K_c.
Choice **(C)** is not correct because it would be the answer if "$2x$" was the numerator of K_c.
Choice **(D)** is not correct because the final concentration of CO is not the value of K_c.

Practice Questions Related to This: **PQ-17, PQ-18,** and **PQ-19**

SQ-6. BrCl(g) is in equilibrium with $Br_2(g)$ and $Cl_2(g)$ at 25.0 °C:

$$2\text{BrCl(g)} \rightleftharpoons \text{Br}_2(\text{g}) + \text{Cl}_2(\text{g}) \qquad K_p = 0.130$$

Initially, in a closed container at 25.0 °C, BrCl(g) has a partial pressure of 0.400 atm and $Br_2(g)$ and $Cl_2(g)$ each have partial pressures of 0.800 atm. What is the partial pressure of BrCl(g) once the system reaches equilibrium?

(A) 0.419 atm **(B)** 0.781 atm **(C)** 1.16 atm **(D)** 1.21 atm

Knowledge Required: (1) How to determine K_p from a balanced equation. (2) How to calculate equilibrium partial pressures from a K_p expression.

Thinking it Through: The question asks you to determine the partial pressure of a species given initial partial pressures of the starting material and products. Because this is a closed system, partial pressures are proportional to amount and therefore changes in partial pressures are proportional to changes in moles of a given species.

To decide which way the reaction proceeds, you calculate a value of Q using the initial pressures given in the question:

$$Q_p = \frac{(P_{\text{Br}_2})_0 (P_{\text{Cl}_2})_0}{(P_{\text{BrCl}})_0^2} = \frac{(0.800)(0.800)}{(0.400)^2} = 1.6 \quad (K_p = 0.130)$$

Because Q is greater than K, the reaction will go to the left.

You then continue by setting up a table of initial-change-equilibrium (ICE) partial pressures. Initial partial pressures are provided in the question. Your decision earlier that the reaction will proceed to the left is indicated in the ICE table by the negative (–) signs in the **change line** for the *products* and the positive (+) sign for the *reactant*.

Mole ratios are used to determine the changes in partial pressures (so +**2**x for **2**BrCl and –**1**x for **1**Br_2, etc.). Equilibrium values are determined by the addition/subtraction of the initial and change values.

	2BrCl(g)	⇌	$Br_2(g)$	$Cl_2(g)$
Initial	0.400 atm		0.800 atm	0.800 atm
Change	$+2x$		$-x$	$-x$
Equilibrium	$0.400 + 2x$		$0.800 - x$	$0.800 - x$

The K_p expression for this reaction is the partial pressures of products raised to their respective coefficients divided by the partial pressure of the reactant raised to its respective coefficient:

$$K_p = \frac{P_{\text{Br}_2} P_{\text{Cl}_2}}{P_{\text{BrCl}}^2}$$

You then substitute the values obtained into this expression:

$$0.130 = \frac{(0.800 - x)(0.800 - x)}{(0.400 + 2x)^2}$$

The expression is then further simplified: (The perfect square is used in the simplification.)

$$0.130 = \frac{(0.800 - x)^2}{(0.400 + 2x)^2} \qquad 0.130 = \left(\frac{0.800 - x}{0.400 + 2x}\right)^2$$

$$\sqrt{0.130} = \left(\frac{0.800 - x}{0.400 + 2x}\right) \qquad 0.361(0.400 + 2x) = 0.800 - x$$

$$0.144 + 0.722x = 0.800 - x \qquad 1.722x = 0.656 \qquad x = 0.381$$

Replacing this value of "x" into the table, we determine the partial pressures at equilibrium;

	2BrCl(g)	⇌	$Br_2(g)$	$Cl_2(g)$
Initial	0.400 atm		0.800 atm	0.800 atm
Change	+ 2(0.381 atm)		– (0.381 atm)	– (0.381 atm)
Equilibrium	1.16 atm		0.419 atm	0.419 atm

Thus, the equilibrium partial pressure of BrCl(g) is 1.16 atm, Choice **(C)**.

Choice **(A)** is not correct because it is the partial pressure of either product.

Choice **(B)** is not correct because the answer comes from adding one-half of x to the initial partial pressure of BrCl(g) instead of a full x; this is likely confused with subtracting one-half of x from either product partial pressure to determine their equilibrium partial pressures.

Choice **(D)** is not correct because K_p did not include the "2" exponent for the partial pressure of the reactant.

Practice Questions Related to This: **PQ-20** and **PQ-21**

SQ-7. Conceptual

Which will drive the equilibrium to form more Cu(s)?

$$Cu_2O(s) + CO(g) \rightleftharpoons 2Cu(s) + CO_2(g)$$

(A) remove CO(g)

(B) remove CO_2(g)

(C) add a catalyst

(D) increase the volume of the container

Knowledge Required: (1) Le Chatelier's principle.

Thinking it Through: The question asks you to determine what would influence the equilibrium such that more product is formed. Le Chatelier's principle would suggest that removing a reactant would lead to the formation of more reactants; therefore, choice **(A)** is incorrect. Adding a catalyst, choice **(C)**, would result in the rate of the reaction changing without impacting the position of equilibrium. Le Chatelier's principle would suggest that increasing the volume would favor the side of the reaction with more moles of gas; however, there is the same number of moles of gas for both the reactants and products of this reaction, and choice **(D)**, therefore, is incorrect. Finally, Le Chatelier's principle would suggest that removal of a product would lead to the formation of more product; therefore, choice **(B)** is correct, i.e., "remove CO_2(g)" would lead to the formation of more Cu(s).

Practice Questions Related to This: **PQ-22, PQ-23, PQ-24, PQ-25, PQ-26, PQ-27, PQ-28, PQ-29,** and **PQ-30**

Practice Questions (PQ)

Conceptual **PQ-1.** When the reversible reaction $N_2(g) + O_2(g) \rightleftharpoons 2NO(g)$ has reached a state of equilibrium,

(A) no further reactions occurs.

(B) the total moles of products must equal the remaining moles of reactant.

(C) the addition of a catalyst will cause formation of more NO.

(D) the concentration of each substance in the system will be constant.

PQ-2. Xenon tetrafluoride, XeF_4, can be prepared by heating Xe and F_2 together:

$$Xe(g) + 2F_2(g) \rightleftharpoons XeF_4(g)$$

What is the equilibrium constant expression for this reaction?

(A) $K_c = \frac{[XeF_4]}{[Xe][F_2]}$ **(B)** $K_c = \frac{[XeF_4]}{2[Xe][F_2]}$ **(C)** $K_c = \frac{[XeF_4]}{[Xe][F_2]^2}$ **(D)** $K_c = \frac{[Xe][F_2]}{[XeF_4]}$

Conceptual **PQ-3.** What is a proper description of chemical equilibrium?

(A) The reaction has stopped.

(B) The concentrations of products and reactants are equal.

(C) The rates of the forward and reverse reactions are zero.

(D) The rates of the forward and reverse reactions are the same.

PQ-4. What is the K_c for this reaction? $P_4(s) + 6Cl_2(g) \rightleftharpoons 4PCl_3(g)$

(A) $K_c = \dfrac{1}{[Cl_2]^6}$ (B) $K_c = \dfrac{[PCl_3]^4}{[P_4][Cl_2]^6}$ (C) $K_c = \dfrac{[PCl_3]^4}{[Cl_2]^6}$ (D) $K_c = \dfrac{[P_4][Cl_2]^6}{[PCl_3]^4}$

PQ-5. What is the K_p for this reaction? $2SO_3(g) \rightleftharpoons 2SO_2(g) + O_2(g)$

(A) $K_p = \dfrac{P_{SO_2}^2 P_{O_2}}{P_{SO_3}^2}$ (B) $K_p = \dfrac{P_{SO_2} P_{O_2}}{P_{SO_3}}$ (C) $K_p = \dfrac{(2P_{SO_2})^2 P_{O_2}}{(2P_{SO_3})^2}$ (D) $K_p = \dfrac{2P_{SO_2}^2 P_{O_2}}{2P_{SO_3}^2}$

PQ-6. What is the equilibrium expression for this reaction? $2C(s) + O_2(g) \rightleftharpoons 2CO(g)$

(A) $K_c = \dfrac{[CO]}{[C][O_2]}$ (B) $K_c = \dfrac{[CO]^2}{[C]^2[O_2]}$ (C) $K_c = \dfrac{[2CO]}{[2C][O_2]}$ (D) $K_c = \dfrac{[CO]^2}{[O_2]}$

Conceptual **PQ-7.** The value of an equilibrium constant can be used to predict each of these *except* the

(A) extent of a reaction.

(B) direction of a reaction.

(C) time required to reach equilibrium.

(D) quantity of reactant(s) remaining at equilibrium.

PQ-8. Chemical equilibrium is the result of

(A) a stoppage of further reaction.

(B) the unavailability of one of the reactants.

(C) opposing reactions attaining equal speeds.

(D) formation of products equal in mass to the mass of the reactants.

PQ-9. The photosynthetic conversion of CO_2 and O_2 can be represented by the reaction shown. What is the equilibrium expression for this reaction? $6CO_2(g) + 6H_2O(l) \rightleftharpoons C_6H_{12}O_6(s) + 6O_2(g)$

(A) $K_c = \dfrac{[CO_2]^6[C_6H_{12}O_6]}{[O_2]^6[H_2O]^6}$ (B) $K_c = \dfrac{[CO_2]^6}{[O_2]^6}$

(C) $K_c = \dfrac{[O_2]^6[H_2O]^6}{[CO_2]^6[C_6H_{12}O_6]}$ (D) $K_c = \dfrac{[O_2]^6}{[CO_2]^6}$

PQ-10. Which statement best describes general equilibrium?

(A) Equilibrium is reached when the reaction stops.

(B) There is only one set of equilibrium concentrations that equals the K_c value.

(C) At equilibrium, the rate of the forward reaction is the same as the rate of the reverse reaction.

(D) At equilibrium, the total concentration of products equals the total concentration of reactants.

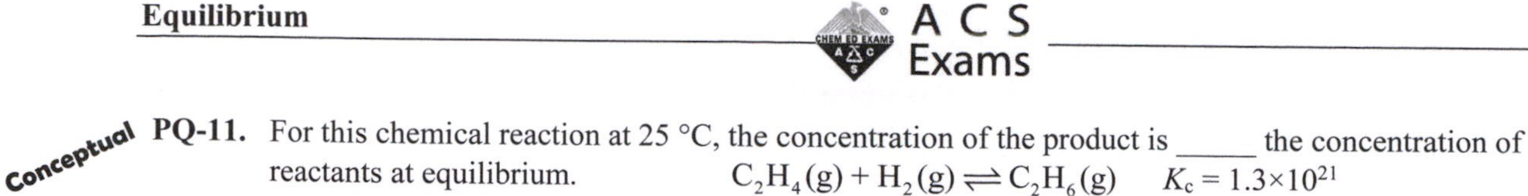

Conceptual **PQ-11.** For this chemical reaction at 25 °C, the concentration of the product is _____ the concentration of reactants at equilibrium. $C_2H_4(g) + H_2(g) \rightleftharpoons C_2H_6(g)$ $K_c = 1.3\times10^{21}$

(A) less than

(B) equal to

(C) greater than

(D) is not able to be compared to

PQ-12. In which case does $K_c = K_p$?

(A) $2A(g) + B(s) \rightleftharpoons 2C(s) + D(g)$

(B) $2A(g) + B(s) \rightleftharpoons C(s) + 2D(g)$

(C) $3A(g) + B(s) \rightleftharpoons 2C(s) + 2D(g)$

(D) K_c and K_p are equivalent in more than one of the above

Conceptual **PQ-13.** Given the equilibrium constants for these reactions:

$$4Cu(s) + O_2(g) \rightleftharpoons 2Cu_2O(s) \qquad K_{c_1}$$

$$2CuO(s) \rightleftharpoons Cu_2O(s) + \tfrac{1}{2}O_2(g) \qquad K_{c_2}$$

What is the value of K_c for this reaction?

$$2Cu(s) + O_2(g) \rightleftharpoons 2CuO(s)$$

(A) $\dfrac{\sqrt{K_{c_1}}}{K_{c_2}}$ **(B)** $\dfrac{K_{c_2}}{\sqrt{K_{c_1}}}$ **(C)** $\sqrt{K_{c_1}} \times K_{c_2}$ **(D)** $K_{c_1} \times K_{c_2}$

PQ-14. Given the equilibrium constants for these reactions:

$$2CH_4(g) \rightleftharpoons C_2H_6(g) + H_2(g) \qquad K_c = 9.5\times10^{-13}$$

$$CH_4(g) + H_2O(g) \rightleftharpoons CH_3OH(g) + H_2(g) \qquad K_c = 2.8\times10^{-21}$$

What is the value of K_c for this reaction?

$$2CH_3OH(g) + H_2(g) \rightleftharpoons C_2H_6(g) + 2H_2O(g)$$

(A) 9.5×10^{-13} **(B)** 2.9×10^{-9} **(C)** 3.4×10^{8} **(D)** 1.2×10^{29}

PQ-15. Carbon monoxide gas reacts with hydrogen gas at elevated temperatures to form methanol.

$$CO(g) + 2H_2(g) \rightleftharpoons CH_3OH(g)$$

When 0.40 mol of CO and 0.30 mol of H_2 are allowed to reach equilibrium in a 1.0 L container, 0.060 mol of CH_3OH are formed. What is the value of K_c?

(A) 0.50 **(B)** 0.98 **(C)** 1.7 **(D)** 5.4

PQ-16. At equilibrium in a 10 L vessel, there are 7.60×10^{-2} moles of SO_2, 8.60×10^{-2} moles of O_2, and 8.20×10^{-2} moles of SO_3. What is the equilibrium constant K_c under these conditions?

$$2SO_2(g) + O_2(g) \rightleftharpoons 2SO_3(g)$$

(A) 12.5 **(B)** 13.5 **(C)** 125 **(D)** 135

PQ-17. A mixture of 2.0 mol of CO(g) and 2.0 mol of H_2O(g) was allowed to come to equilibrium in a 10.0-L flask at a high temperature. If $K_c = 4.0$, what is the molar concentration of H_2(g) in the equilibrium mixture?

$$CO(g) + H_2O(g) \rightleftharpoons CO_2(g) + H_2(g)$$

(A) 0.67 M **(B)** 0.40 M **(C)** 0.20 M **(D)** 0.13 M

PQ-18. If a 1.0 L flask is filled with 0.22 mol of N_2 and 0.22 mol of O_2 at 2000°C, what is [NO] after the reaction establishes equilibrium? ($K_c = 0.10$ at 2000°C)

$$N_2(g) + O_2(g) \rightleftharpoons 2NO(g)$$

(A) 0.034 M **(B)** 0.060 M **(C)** 0.079 M **(D)** 0.12 M

PQ-19. For the reaction, $A(g) \rightleftharpoons B(g) + C(g)$,
5 moles of A are allowed to come to equilibrium in a closed rigid container. At equilibrium, how much of A and B are present if 2 moles of C are formed?

(A) 0 moles of A and 3 moles of B
(B) 1 mole of A and 2 moles of B
(C) 2 moles of A and 2 moles of B
(D) 3 moles of A and 2 moles of B

Conceptual **PQ-20.** The equilibrium constant, $K_c = 3.8\times10^{-5}$ for the reaction,

$$I_2(g) \rightleftharpoons 2I(g)$$

What is the state of the system if $[I_2(g)] = 1.0$ M and $[I(g)] = 1.0\times10^{-3}$ M?

(A) The system is at equilibrium.
(B) The system is shifting towards products.
(C) The system is shifting towards reactants.
(D) There is insufficient information to describe the status of the system.

Conceptual **PQ-21.** For the reaction: $\frac{1}{2}F_2(g) \rightleftharpoons F(g)$, a reaction mixture initially contains equal amounts of F_2(g) and F(g) in their standard states. If $K_p = 7.55\times10^{-2}$ at this temperature, which statement is true?

(A) $Q < K$, and the reaction proceeds towards the reactants.
(B) $Q < K$, and the reaction proceeds towards the products.
(C) $Q = K$, and the reaction is at equilibrium.
(D) $Q > K$, and the reaction proceeds towards the reactants.

Conceptual **PQ-22.** Consider this reaction at equilibrium.

$$2SO_2(g) + O_2(g) \rightleftharpoons 2SO_3(g) \qquad \Delta H = -198 \text{ kJ}$$

Which change would cause an increase in the SO_3 / SO_2 mole ratio?

(A) adding a catalyst
(B) removing O_2(g)
(C) decreasing the temperature
(D) decreasing the pressure

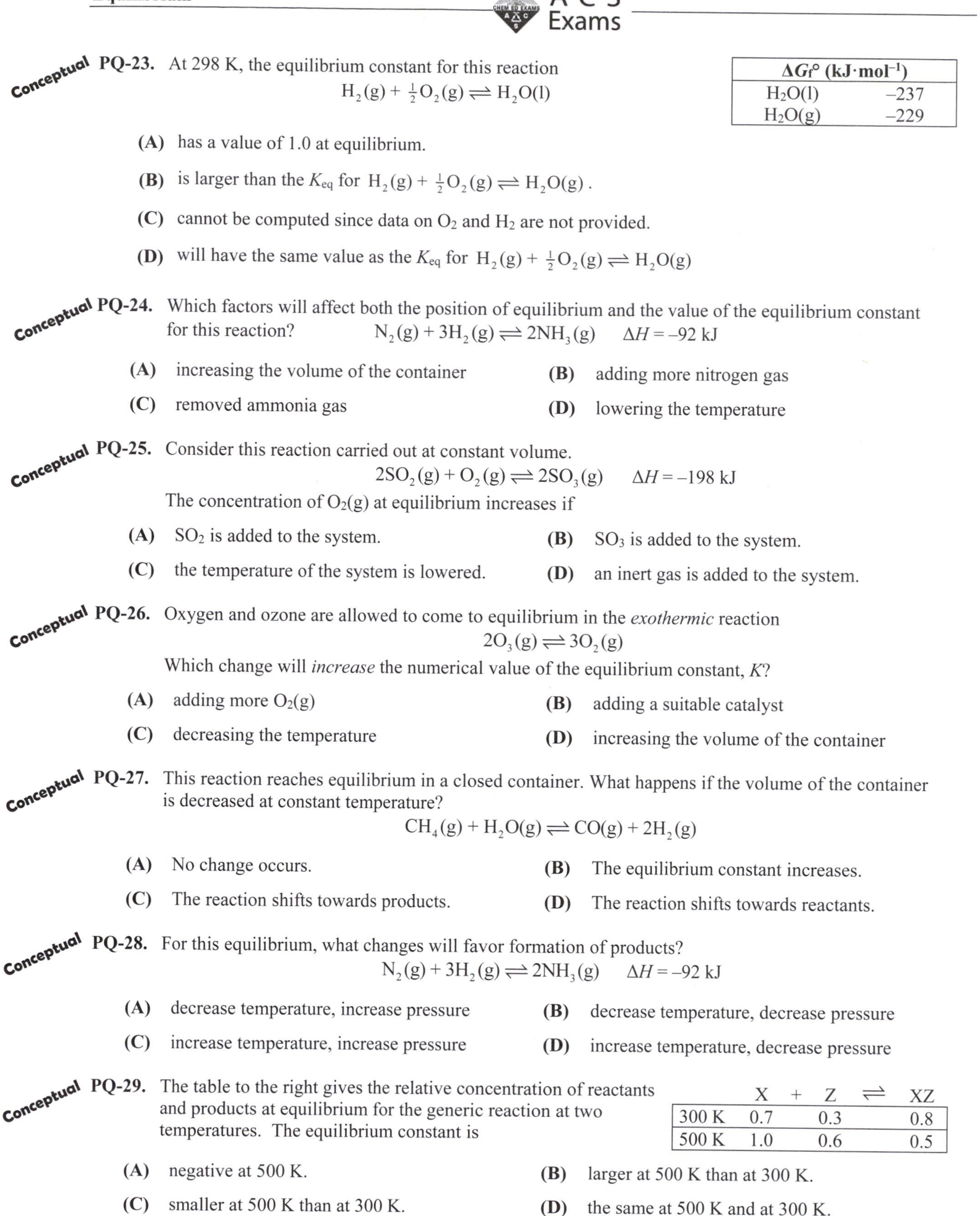

Conceptual **PQ-23.** At 298 K, the equilibrium constant for this reaction

$$H_2(g) + \tfrac{1}{2}O_2(g) \rightleftharpoons H_2O(l)$$

ΔG_f° (kJ·mol^{-1})	
$H_2O(l)$	–237
$H_2O(g)$	–229

(A) has a value of 1.0 at equilibrium.

(B) is larger than the K_{eq} for $H_2(g) + \tfrac{1}{2}O_2(g) \rightleftharpoons H_2O(g)$.

(C) cannot be computed since data on O_2 and H_2 are not provided.

(D) will have the same value as the K_{eq} for $H_2(g) + \tfrac{1}{2}O_2(g) \rightleftharpoons H_2O(g)$

Conceptual **PQ-24.** Which factors will affect both the position of equilibrium and the value of the equilibrium constant for this reaction? $N_2(g) + 3H_2(g) \rightleftharpoons 2NH_3(g)$ $\Delta H = -92$ kJ

(A) increasing the volume of the container

(B) adding more nitrogen gas

(C) removed ammonia gas

(D) lowering the temperature

Conceptual **PQ-25.** Consider this reaction carried out at constant volume.

$$2SO_2(g) + O_2(g) \rightleftharpoons 2SO_3(g) \qquad \Delta H = -198 \text{ kJ}$$

The concentration of $O_2(g)$ at equilibrium increases if

(A) SO_2 is added to the system.

(B) SO_3 is added to the system.

(C) the temperature of the system is lowered.

(D) an inert gas is added to the system.

Conceptual **PQ-26.** Oxygen and ozone are allowed to come to equilibrium in the *exothermic* reaction

$$2O_3(g) \rightleftharpoons 3O_2(g)$$

Which change will *increase* the numerical value of the equilibrium constant, *K*?

(A) adding more $O_2(g)$

(B) adding a suitable catalyst

(C) decreasing the temperature

(D) increasing the volume of the container

Conceptual **PQ-27.** This reaction reaches equilibrium in a closed container. What happens if the volume of the container is decreased at constant temperature?

$$CH_4(g) + H_2O(g) \rightleftharpoons CO(g) + 2H_2(g)$$

(A) No change occurs.

(B) The equilibrium constant increases.

(C) The reaction shifts towards products.

(D) The reaction shifts towards reactants.

Conceptual **PQ-28.** For this equilibrium, what changes will favor formation of products?

$$N_2(g) + 3H_2(g) \rightleftharpoons 2NH_3(g) \qquad \Delta H = -92 \text{ kJ}$$

(A) decrease temperature, increase pressure

(B) decrease temperature, decrease pressure

(C) increase temperature, increase pressure

(D) increase temperature, decrease pressure

Conceptual **PQ-29.** The table to the right gives the relative concentration of reactants and products at equilibrium for the generic reaction at two temperatures. The equilibrium constant is

	X	+ Z	⇌ XZ
300 K	0.7	0.3	0.8
500 K	1.0	0.6	0.5

(A) negative at 500 K.

(B) larger at 500 K than at 300 K.

(C) smaller at 500 K than at 300 K.

(D) the same at 500 K and at 300 K.

Conceptual **PQ-30.** A catalyst is added to a system at equilibrium. Which statement is TRUE?

(A) The temperature will decrease.

(B) The equilibrium constant will increase.

(C) The concentration of products will decrease.

(D) If the system is disturbed, it will return to equilibrium faster.

Answers to Study Questions

1. A
2. C
3. A
4. B
5. A
6. C
7. B

Answers to Practice Questions

1. D
2. C
3. D
4. C
5. A
6. D
7. C
8. C
9. D
10. C
11. C
12. B
13. A
14. D
15. D
16. D
17. D
18. B
19. D
20. B
21. D
22. C
23. B
24. D
25. B
26. C
27. D
28. A
29. C
30. D

Chapter 12 – Acids and Bases

Chapter Summary:

This chapter will focus on acids and bases including acid-base models, determining an acid dissociation constant, and calculating pH and pOH for a given acid dissociation.

Specific topics covered in this chapter are:

- Acid-base models (Arrhenius, Brønsted-Lowry, Lewis)
- pH and pOH
- Conjugate acid-base pairs
- Titrations and neutralizations
- Buffers

Previous material that is relevant to your understanding of questions in this chapter include:

- Solutions and Aqueous Reactions, Part 1 (***Chapter 5***)
- Solutions and Aqueous Reactions, Part 2 (***Chapter 9***)
- Equilibrium (***Chapter 11***)

Common representations used in questions related to this material:

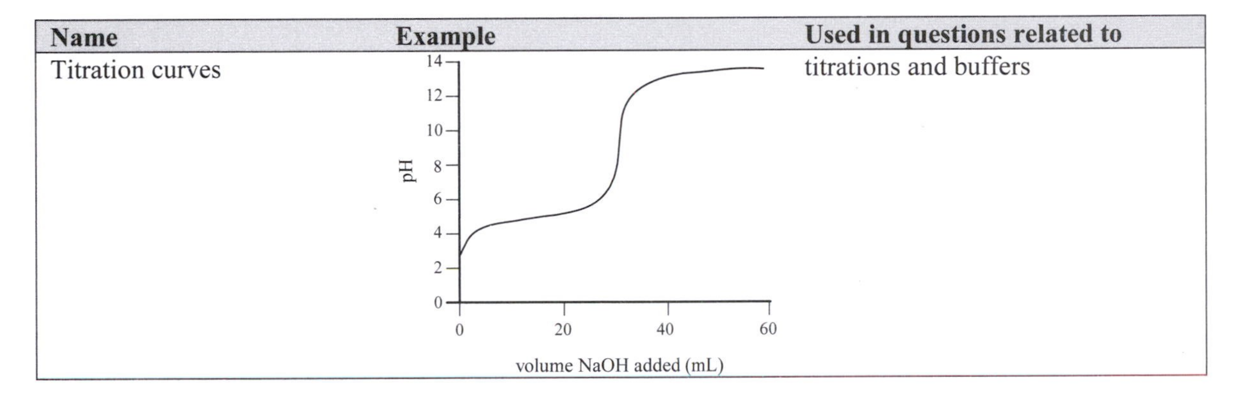

Name	Example	Used in questions related to
Titration curves	(titration curve)	titrations and buffers

Where to find this in your textbook:

The material in this chapter typically aligns to "Acid-Base Equilibria" in your textbook. The name of your chapter(s) may vary.

Practice exam:

There are practice exam questions aligned to the material in this chapter. Because there are a limited number of questions on the practice exam, a review of the breadth of the material in this chapter is advised in preparation for your exam.

How this fits into the big picture:

The material in this chapter aligns to the Big Idea of Equilibrium (8) as listed on page 12 of this study guide.

Study Questions (SQ)

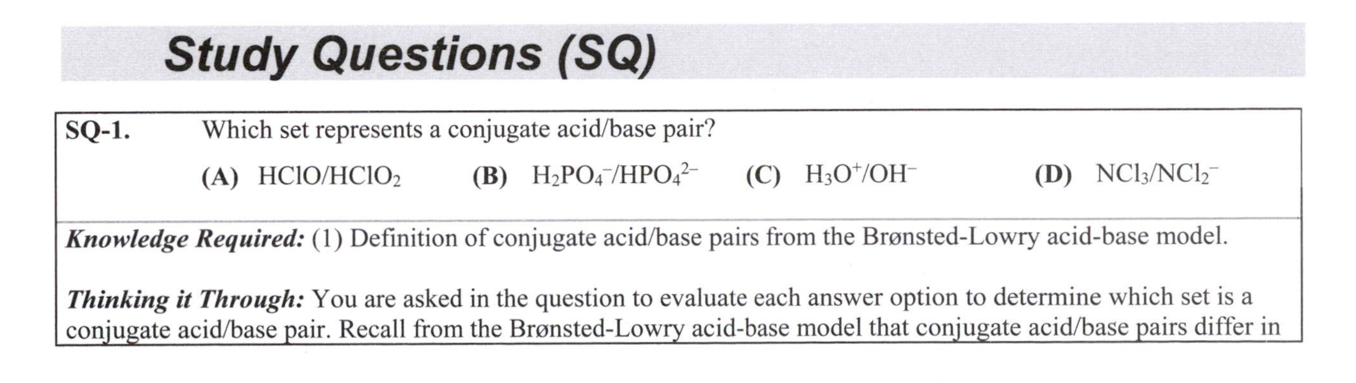

SQ-1. Which set represents a conjugate acid/base pair?

(A) $HClO/HClO_2$ **(B)** $H_2PO_4^-/HPO_4^{2-}$ **(C)** H_3O^+/OH^- **(D)** NCl_3/NCl_2^-

Knowledge Required: (1) Definition of conjugate acid/base pairs from the Brønsted-Lowry acid-base model.

Thinking it Through: You are asked in the question to evaluate each answer option to determine which set is a conjugate acid/base pair. Recall from the Brønsted-Lowry acid-base model that conjugate acid/base pairs differ in

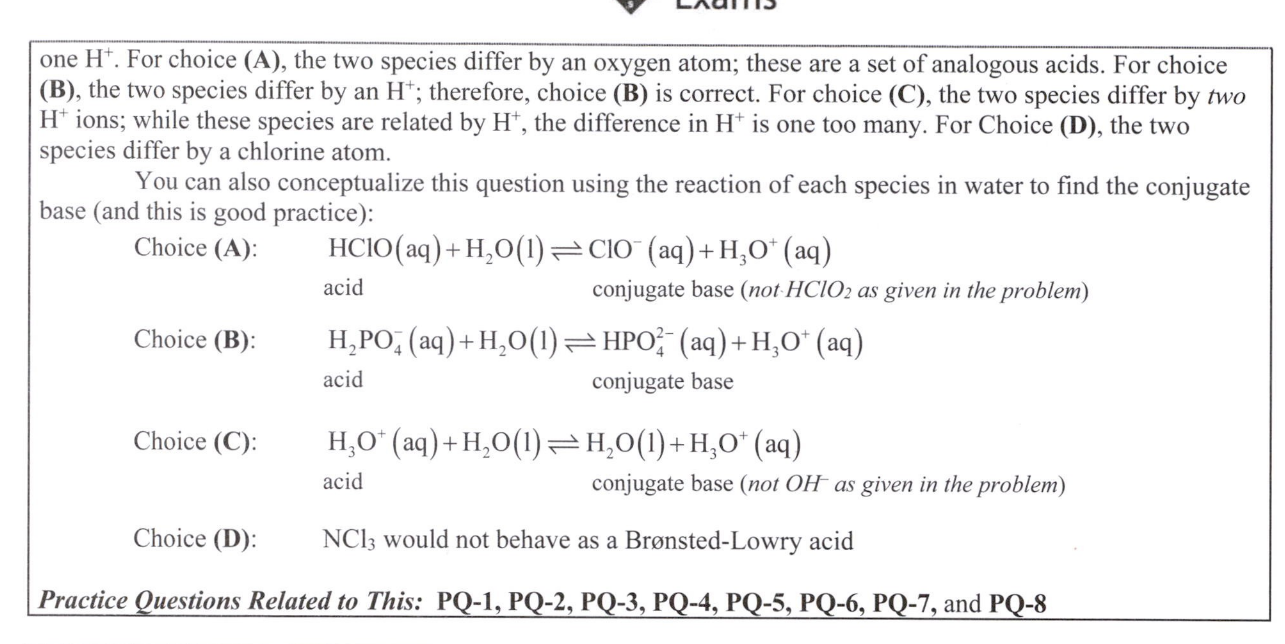

one H^+. For choice **(A)**, the two species differ by an oxygen atom; these are a set of analogous acids. For choice **(B)**, the two species differ by an H^+; therefore, choice **(B)** is correct. For choice **(C)**, the two species differ by *two* H^+ ions; while these species are related by H^+, the difference in H^+ is one too many. For Choice **(D)**, the two species differ by a chlorine atom.

You can also conceptualize this question using the reaction of each species in water to find the conjugate base (and this is good practice):

Choice **(A)**: $HClO(aq) + H_2O(l) \rightleftharpoons ClO^-(aq) + H_3O^+(aq)$
acid — conjugate base (*not $HClO_2$ as given in the problem*)

Choice **(B)**: $H_2PO_4^-(aq) + H_2O(l) \rightleftharpoons HPO_4^{2-}(aq) + H_3O^+(aq)$
acid — conjugate base

Choice **(C)**: $H_3O^+(aq) + H_2O(l) \rightleftharpoons H_2O(l) + H_3O^+(aq)$
acid — conjugate base (*not OH^- as given in the problem*)

Choice **(D)**: NCl_3 would not behave as a Brønsted-Lowry acid

Practice Questions Related to This: **PQ-1, PQ-2, PQ-3, PQ-4, PQ-5, PQ-6, PQ-7,** and **PQ-8**

SQ-2. What is the pH of a 0.820 M aqueous NH_3 solution? $K_b(NH_3) = 1.8\times10^{-5}$

(A) 2.42 **(B)** 9.25 **(C)** 11.58 **(D)** 13.91

Knowledge Required: (1) How to write a balanced equation for an aqueous acid-base reaction. (2) How to write the base dissociation constant expression (K_b) for a balanced chemical equation. (3) How to use the K_b expression to determine equilibrium concentrations. (4) How to convert between pH and pOH.

Thinking it Through: You are asked to determine the pH of a given aqueous solution for a known base and a known K_b. This problem has several steps: First, you write a balanced equation for the acid-base reaction; recall that you are told that the reaction is in water, and thus water will act as an acid with the given base:

$$NH_3(aq) + H_2O(l) \rightleftharpoons NH_4^+(aq) + OH^-(aq)$$

Next, you use the balanced acid-base equation to determine the K_b expression: concentrations of products raised to power of their stoichiometric coefficients divided by the concentrations of reactants raised to power of their stoichiometric coefficients. Note: pure liquids and solids are not included in the K_b expression (omitting water).

$$K_b = \frac{[NH_4^+][OH^-]}{[NH_3]}$$

Next, you construct an initial-change-equilibrium (ICE) table. Initial concentrations are provided in the question. Stoichiometric coefficients from the balanced chemical equation are used to determine the changes in concentrations. Equilibrium values are determined by the addition/subtraction of the initial and change values.

	$NH_3(aq)$	$\rightleftharpoons$	$NH_4^+(aq)$	+	$OH^-(aq)$
Initial	0.820		0		0
Change	$-x$		$+x$		$+x$
Equilibrium	$0.820 - x$		x		x

Because acid and base ionization constants are only about 95% accurate, you can compare the ratio of the equilibrium constant to the initial concentration of the acid or base. If the ratio is less than 2.5×10^{-3}, you can approximate or: $\frac{K_a}{[\text{acid}]_0} < 2.5\times10^{-3}$ $\frac{K_b}{[\text{base}]_0} < 2.5\times10^{-3}$.

For this question: $\frac{K_b(NH_3)}{[NH_3]_0} = \frac{1.8\times10^{-5}}{0.820\text{ M}} = 2.2\times10^{-5}$ which is less than 2.5×10^{-3} (so you can approximate)

Therefore, you can assume that x will be sufficiently small compared to the initial concentration of NH_3; therefore, the change to $[NH_3]_0$ is small and the approximation is $[NH_3]_0 = [NH_3]_{\text{equilibrium}}$ or $0.820 - x \approx 0.820$.

You now substitute the equilibrium concentrations into the K_b expression:

$$1.8\times10^{-5}=\frac{x\cdot x}{0.820}$$

The expression is then further simplified and solved:

$$1.5\times10^{-5}=x^2 \qquad x=\pm3.84\times10^{-3}$$

Because x is the final concentration of each of the products and because a concentration cannot be negative, then x is equal to $+3.84\times10^{-3}$ M which is $[NH_4^+]_{equilibrium}$ and $[OH^-]_{equilibrium}$.

Next, you determine the pOH from the concentration of $[OH^-]$:

$$pOH=-\log([OH^-])=-\log(3.84\times10^{-3})=2.42$$

Finally, you convert from pOH to pH knowing that the sum of pH and pOH is equal to 14 for an aqueous solution.

$$pOH+pH=14 \quad pH=14-pOH \qquad pH=14-2.42=11.58$$ Which is choice **(C)**

Choice **(A)** is not correct because it is the pOH of the solution.

Choice **(B)** is not correct because it was determined using the K_b value as the concentration of OH^- when determining pOH.

Choice **(D)** is not correct because it was determined using the initial concentration of NH_3 as the concentration of OH^- when determining pOH.

Practice Questions Related to This: **PQ-9, PQ-10, PQ-11, PQ-12,** and **PQ-13**

SQ-3. Conceptual

Which acid is the weakest in aqueous solution?

(A) acetic acid ($K_a = 1.8\times10^{-5}$)
(B) formic acid ($K_a = 1.8\times10^{-4}$)
(C) hydrocyanic acid ($K_a = 6.2\times10^{-10}$)
(D) nitrous acid ($K_a = 4.5\times10^{-4}$)

Knowledge Required: (1) How to write an acid dissociation equilibrium expression. (2) Relationship between acid dissociation constants (K_a) values and acid strength.

Thinking it Through: You are asked to evaluate the given acids and determine which acid is the weakest. You recall that K_a is expressed as the concentration of product(s) raised to the power of their stoichiometric coefficients (from the balanced chemical equation) divided by the concentration of reactant(s) raised to power of their stoichiometric coefficients. Thus, larger values of K_a are associated with a larger product to reactant ratio. Conversely, smaller values of K_a are associated with a smaller product to reactant ratio. For all of the choices, the values of K_a are less than one (so for all of these, starting with only reactants, at equilibrium, [reactants] > [products]). However, you know you can still determine relative acid strength when all K_a values are less than one by determining which value is the largest and how that corresponds to acid strength. Generically:

$$HA(aq)+H_2O(l)\rightleftharpoons A^-(aq)+H_3O^+(aq) \quad K_a=\frac{[A^-][H_3O^+]}{[HA]}$$

The larger the value of K_a, the higher $[H_3O^+]$, the stronger the acid

Thus, small values of K_a are associated with weak acids with the smallest K_a value associated with the weakest acid. You consider the answer options and note that hydrocyanic acid has the smallest K_a (6.2×10^{-10}) and conclude that it is the weakest of all the acids given as answer options; therefore, choice **(C)** is correct.

Practice Questions Related to This: **PQ-14** and **PQ-15**

SQ-4. Conceptual

Which anion is the most basic?

(A) ClO^-
(B) ClO_2^-
(C) ClO_3^-
(D) ClO_4^-

Knowledge Required: (1) How to determine acidity and basicity from an analogous series of oxoacids.

Thinking it Through: You are asked in the question to evaluate a series of bases to determine which is the most basic. The bases are analogous, considering the answer options from **(A)** to **(D)**, each has one more oxygen than the previous; thus, this a series of oxoacids. The general trend for a series of oxoacids is that as the number of oxygens increases (or more oxygen atoms in the acid) the acid gets stronger. This is due to the increase in electron withdrawing power of the added oxygens, which causes the O–H bond to weaken and the acid to get stronger.

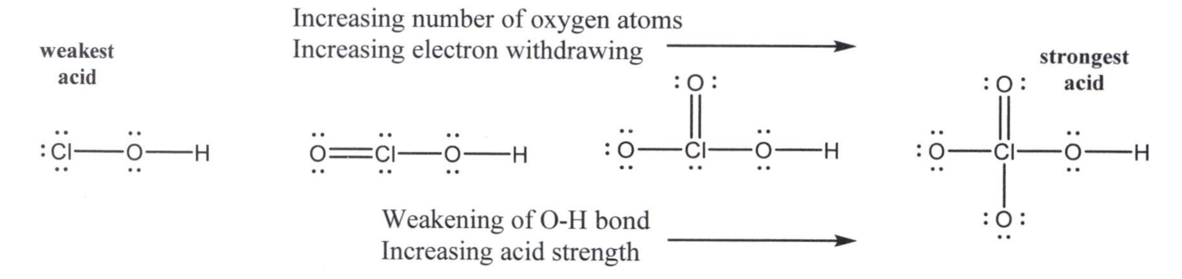

You can use the relative strength of the acid to determine the relative strengths of the conjugate bases as you know the value of K_b for the conjugate base is related to the K_a value of the acid through: $K_a \times K_b = K_w$.

Because K_w is a constant, at a given temperature, the larger the K_a value (stronger the acid) the smaller the K_b value (weaker the conjugate base). Therefore, the conjugate base of the weakest acid (fewest oxygen atoms) will be the strongest base – choice **(A)**. The next strongest base would be the conjugate base of the next weakest acid – choice **(B)**. The pattern would continue. the ranking of the base strength is:

strongest base → weakest base

$$ClO^- > ClO_2^- > ClO_3^- > ClO_4^-$$

Although not part of this question, when considering the structure of acids and relative strength, you can also be provided with the same number of oxygen atoms, but differing halogens, for example comparing $HClO_4$ and $HBrO_4$. Here, the difference in acid strength is due to the higher electronegativity of chlorine compared to bromine – with $HClO_4$ as the stronger acid.

Practice Questions Related to This: **PQ-16** and **PQ-17**

SQ-5. What is K_b of F^-? (K_a of HF is 6.8×10^{-4})

(A) 6.8×10^{10} **(B)** 1.5×10^{3} **(C)** 6.8×10^{-4} **(D)** 1.5×10^{-11}

Knowledge Required: (1) Relationship between K_a and K_b for a conjugate acid-base pair.

Thinking it Through: You are asked in the question to determine the K_b for a species given a known K_a for the conjugate acid of the species. To solve this problem, it is necessary to recall the relationship between K_a and K_b for an acid and its conjugate base dissociated in water:

$$HF(aq) \rightleftharpoons H^+(aq) + F^-(aq) \qquad K_a$$
$$F^-(aq) + H_2O(l) \rightleftharpoons HF(aq) + OH^-(aq) \qquad K_b$$
$$H_2O(l) \rightleftharpoons H^+(aq) + OH^-(aq) \qquad K_w = K_a \times K_b$$

Therefore: $K_b = \dfrac{K_w}{K_a} = \dfrac{1\times10^{-14}}{6.8\times10^{-4}} = 1.5\times10^{-11}$ which is choice **(D)**.

Choice **(A)** is not correct because it is determined by dividing K_a by K_w, which is the inverse of K_b.
Choice **(B)** is not correct because it is determined by taking the inverse of K_a.
Choice **(C)** is not correct because it is the value of K_a.

Practice Questions Related to This: **PQ-18**

SQ-6. *Conceptual* Which substance will dissolve in water to produce an acidic solution?

(A) $FeCl_3$ **(B)** Na_2O **(C)** $NaC_2H_3O_2$ **(D)** NH_3

Knowledge Required: (1) Dissociation of ions in aqueous solution. (2) Relative acidity/basicity of common species and ions.

Thinking it Through: You are asked in the question to evaluate a series of compounds to determine which will produce an acidic solution upon dissolution in water. The method you will use to do this is (in turn for each compound):

1. Consider the ions formed when the compound dissolves in water
2. Consider the reaction of each ion with water (hydrolysis)
3. Consider the acidic/basic ions

Choice **(A)** – $FeCl_3$

Step 1	$FeCl_3(s) \rightarrow Fe^{3+}(aq) + 3Cl^-(aq)$
Step 2	$Fe^{3+}(aq) + 6H_2O(l) \rightleftharpoons Fe(H_2O)_6(aq) \rightleftharpoons Fe(H_2O)_5OH^-(aq) + H^+(aq)$ $Cl^-(aq) + H_2O(l) \rightleftharpoons$ no reaction (HCl is a strong acid)
Step 3	Fe^{3+} is an acidic cation and Cl^- is a neutral anion

$FeCl_3$ produces an acidic solution

Choice **(B)** – Na_2O

Step 1	$Na_2O(s) \rightarrow 2Na^+(aq) + O^{2-}(aq)$
Step 2	$Na^+(aq) + H_2O(l) \rightleftharpoons$ no reaction (NaOH is a strong base) $O^{2-}(aq) + H_2O(l) \rightleftharpoons 2OH^-$
Step 3	Na^+ is neutral cation and O^{2-} is a basic anion

Na_2O produces a basic solution

Choice **(C)** – $NaC_2H_3O_2$

Step 1	$NaC_2H_3O_2(s) \rightarrow Na^+(aq) + C_2H_3O_2^-(aq)$
Step 2	$Na^+(aq) + H_2O(l) \rightleftharpoons$ no reaction (NaOH is a strong base) $C_2H_3O_2^-(aq) + H_2O(l) \rightleftharpoons HC_2H_3O_2(aq) + OH^-(aq)$
Step 3	Na^+ is neutral cation and $C_2H_3O_2^-$ is a basic anion

$NaC_2H_3O_2$ produces a basic solution

Choice **(D)** – NH_3 is a molecular compound and not a salt. Instead for this compound, you will consider the reaction of NH_3 is water directly (omitting the dissociation into ions):

$NH_3(aq) + H_2O(aq) \rightleftharpoons NH_4^+(aq) + OH^-(aq)$ which produces a basic solution.

Choices **(B)**, **(C)**, and **(D)** all lead to the formation of hydroxide anions and basic solutions. Therefore, Choice **(A)** is correct because it leads to the formation of hydronium ion (H_3O^+).

Practice Questions Related to This: **PQ-19,** and **PQ-20**

SQ-7. An acetate buffer contains equal volumes of 0.35 M $HC_2H_3O_2$ ($pK_a = 4.74$) and 0.55 M $NaC_2H_3O_2$. What is the pH of the buffer?

(A) 4.54 **(B)** 4.74 **(C)** 4.94 **(D)** 7.00

Knowledge Required: (1) How to use the Henderson-Hasselbalch equation. (2) How a buffer works.

Thinking it Through: You are asked in the question to determine the pH of a buffer solution given the pK_a of the acid, concentrations of the acid and its conjugate base. The Henderson-Hasselbalch equation is used to relate pH, pK_a, and concentrations of conjugate acid-base pairs.

$$pH = pK_a + \log\left(\frac{[A^-]}{[HA]}\right)$$

You are given the concentration of the acid ($HC_2H_3O_2$, 0.35 M) and the conjugate base ($NaC_2H_3O_2$, 0.55 M) in the question. In addition, you are given the pK_a of the acid ($HC_2H_3O_2$, 4.74). These values are entered into the Henderson-Hasselbalch equation and pH is determined:

$$pH = 4.74 + \log\left(\frac{0.55}{0.35}\right) \qquad pH = 4.74 + \log(1.57)$$

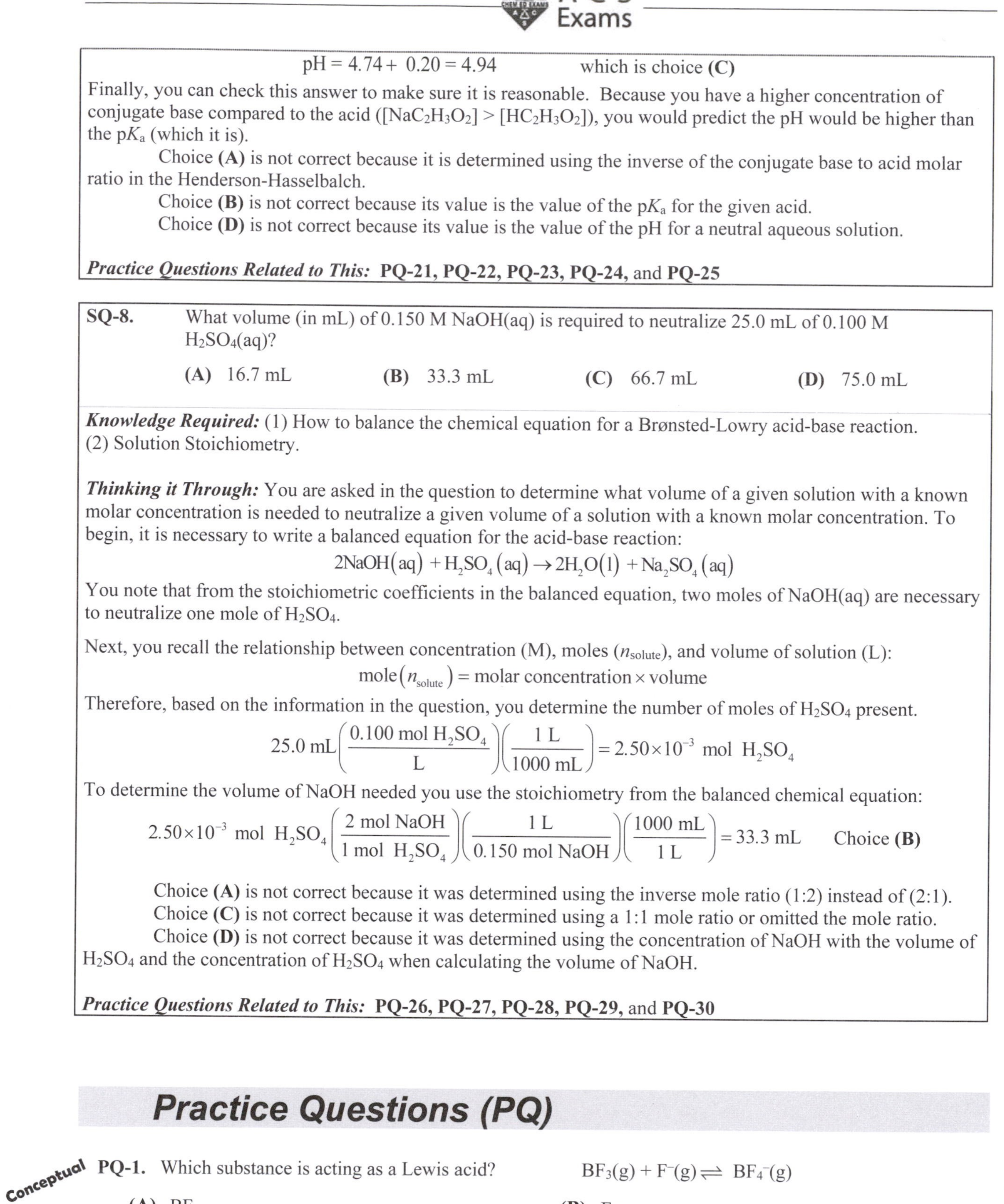

$$pH = 4.74 + 0.20 = 4.94 \qquad \text{which is choice (C)}$$

Finally, you can check this answer to make sure it is reasonable. Because you have a higher concentration of conjugate base compared to the acid ($[NaC_2H_3O_2] > [HC_2H_3O_2]$), you would predict the pH would be higher than the pK_a (which it is).

Choice **(A)** is not correct because it is determined using the inverse of the conjugate base to acid molar ratio in the Henderson-Hasselbalch.

Choice **(B)** is not correct because its value is the value of the pK_a for the given acid.

Choice **(D)** is not correct because its value is the value of the pH for a neutral aqueous solution.

Practice Questions Related to This: **PQ-21, PQ-22, PQ-23, PQ-24,** and **PQ-25**

SQ-8. What volume (in mL) of 0.150 M NaOH(aq) is required to neutralize 25.0 mL of 0.100 M H_2SO_4(aq)?

(A) 16.7 mL **(B)** 33.3 mL **(C)** 66.7 mL **(D)** 75.0 mL

Knowledge Required: (1) How to balance the chemical equation for a Brønsted-Lowry acid-base reaction. (2) Solution Stoichiometry.

Thinking it Through: You are asked in the question to determine what volume of a given solution with a known molar concentration is needed to neutralize a given volume of a solution with a known molar concentration. To begin, it is necessary to write a balanced equation for the acid-base reaction:

$$2NaOH(aq) + H_2SO_4(aq) \rightarrow 2H_2O(l) + Na_2SO_4(aq)$$

You note that from the stoichiometric coefficients in the balanced equation, two moles of NaOH(aq) are necessary to neutralize one mole of H_2SO_4.

Next, you recall the relationship between concentration (M), moles (n_{solute}), and volume of solution (L):

$$\text{mole}(n_{solute}) = \text{molar concentration} \times \text{volume}$$

Therefore, based on the information in the question, you determine the number of moles of H_2SO_4 present.

$$25.0 \text{ mL}\left(\frac{0.100 \text{ mol } H_2SO_4}{L}\right)\left(\frac{1 \text{ L}}{1000 \text{ mL}}\right) = 2.50\times10^{-3} \text{ mol } H_2SO_4$$

To determine the volume of NaOH needed you use the stoichiometry from the balanced chemical equation:

$$2.50\times10^{-3} \text{ mol } H_2SO_4\left(\frac{2 \text{ mol NaOH}}{1 \text{ mol } H_2SO_4}\right)\left(\frac{1 \text{ L}}{0.150 \text{ mol NaOH}}\right)\left(\frac{1000 \text{ mL}}{1 \text{ L}}\right) = 33.3 \text{ mL} \qquad \text{Choice (B)}$$

Choice **(A)** is not correct because it was determined using the inverse mole ratio (1:2) instead of (2:1).

Choice **(C)** is not correct because it was determined using a 1:1 mole ratio or omitted the mole ratio.

Choice **(D)** is not correct because it was determined using the concentration of NaOH with the volume of H_2SO_4 and the concentration of H_2SO_4 when calculating the volume of NaOH.

Practice Questions Related to This: **PQ-26, PQ-27, PQ-28, PQ-29,** and **PQ-30**

Practice Questions (PQ)

Conceptual

PQ-1. Which substance is acting as a Lewis acid? $BF_3(g) + F^-(g) \rightleftharpoons BF_4^-(g)$

(A) BF_3

(B) F^-

(C) BF_4^-

(D) This is not a Lewis acid-base reaction.

Conceptual **PQ-2.** What is a true statement about Lewis acids and bases?

(A) A Lewis acid must have a nonbonding (lone) pair of electrons.

(B) A Lewis base must have a nonbonding (lone) pair of electrons.

(C) In a Lewis acid-base reaction, a pair of electrons is donated from the acid to the base.

(D) In a Lewis acid-base reaction, one species always goes from having a charge to being electronically neutral.

Conceptual **PQ-3.** What are the Brønsted-Lowry bases in this reaction?

$$NH_3(aq) + H_2O(aq) \rightleftharpoons NH_4^+(aq) + OH^-(aq)$$

(A) NH_3 and OH^- **(B)** H_2O and NH_4^+ **(C)** NH_3 and H_3O **(D)** NH_4^+ and OH^-

PQ-4. What is the hydroxide ion concentration in an aqueous solution with a pH of 7.0 at 25 °C?

(A) 0.0 M **(B)** 1.0×10^{-14} M **(C)** 1.0×10^{-7} M **(D)** 7.0 M

PQ-5. What is the pH of a 0.0050 M solution of $Ba(OH)_2(aq)$ at 25 °C?

(A) 2.00 **(B)** 2.30 **(C)** 11.70 **(D)** 12.00

Conceptual **PQ-6.** The formation of a complex ion such as $Cu(NH_3)_4^{2+}(aq)$ can best be categorized as a(n) ___ reaction.

$$Cu^{2+}(aq) + 4NH_3(aq) \rightarrow Cu(NH_3)_4^{2+}(aq)$$

(A) Arrhenius acid-base **(B)** Brønsted-Lowry acid-base

(C) Lewis acid-base **(D)** oxidation-reduction

PQ-7. A 10.0 mL portion of 0.010 M HCl is added to 100.0 mL of water. What is the pH of the resulting solution?

(A) between 3.02 and 3.10 **(B)** between 2.90 and 3.01

(C) between 2.02 and 2.10 **(D)** between 1.90 and 2.01

PQ-8. The pOH of pure water at 40 °C is 6.8. What is the hydronium concentration, $[H_3O^+]$, in pure water at this temperature?

(A) 1.0×10^{-14} M **(B)** 6.3×10^{-8} M **(C)** 1.0×10^{-7} M **(D)** 1.6×10^{-7} M

PQ-9. What is the pH of a 0.053 M solution of KOH at 25 °C?

(A) 0.89 **(B)** 1.28 **(C)** 12.72 **(D)** 13.11

PQ-10. What is the equilibrium concentration of nitrous acid, HNO_2 ($K_a = 4.5\times10^{-4}$), in a solution that has a pH of 1.65?

(A) 0.0032 M **(B)** 0.022 M **(C)** 0.49 M **(D)** 1.1 M

PQ-11. The pain reliever codeine is a weak base with a K_b equal to 1.6×10^{-6}. What is the pH of a 0.05 M aqueous codeine solution?

(A) 7.1 **(B)** 10.5 **(C)** 11.1 **(D)** 12.7

Conceptual **PQ-12.** Besides water molecules, what species is/are present at the greatest concentration when $NH_3(g)$ is bubbled into water? (K_b for $NH_3(aq)$ is 1.8×10^{-5})

(A) $NH_3(aq)$ **(B)** $NH_4OH(aq)$

(C) $NH_4^+(aq)$ and $OH^-(aq)$ **(D)** $NH_2^-(aq)$ and $H_3O^+(aq)$

Conceptual **PQ-13.** Which 0.10 M solution will have the largest concentration of hydroxide ion?

(A) NH_3 (K_b of $NH_3 = 1.8\times10^{-5}$) **(B)** NaCN (K_a of HCN $= 4.9\times10^{-10}$)

(C) $NaClO_2$ (K_a of $HClO_2 = 1.7\times10^{-4}$) **(D)** $NaHCO_3$ (K_b of $HCO_3^- = 2.3\times10^{-8}$)

Conceptual **PQ-14.** Which is the strongest acid in aqueous solution?

(A) acetic acid ($K_a = 1.8\times10^{-5}$) **(B)** benzoic acid ($K_a = 6.3\times10^{-5}$)

(C) formic acid ($K_a = 1.7\times10^{-4}$) **(D)** hydrofluoric acid ($K_a = 7.1\times10^{-4}$)

Conceptual **PQ-15.** Which aqueous acid has the largest K_a value?

(A) HBrO **(B)** $HBrO_2$ **(C)** HClO **(D)** $HClO_2$

Conceptual **PQ-16.** In non-aqueous solution, the trend in acid strength is observed to be: K_a (HI) > K_a (HBr) > K_a (HCl)
Which periodic trend best explains this observed pattern?

(A) Atomic radius: as the bond between H and the halogen becomes shorter, acid strength decreases

(B) Electron affinity: as the halogen becomes more attracted to electrons, acid strength increases

(C) Electronegativity: as the bond between the two atoms becomes less polar, acid strength decreases

(D) Ionization energy: as it becomes harder to remove an electron from the halogen, acid strength increases

Conceptual **PQ-17.** The reaction is observed to have a $K_{eq} > 1$.

$$HSO_3^-(aq) + HPO_4^{2-}(aq) \rightleftharpoons SO_3^{2-}(aq) + H_2PO_4^-(aq)$$

What is the strongest acid present in this equilibrium?

(A) $H_2PO_4^-$ **(B)** HPO_4^{2-} **(C)** HSO_3^- **(D)** SO_3^{2-}

PQ-18. What is the pH of a 0.400 M sodium formate ($NaCHO_2$) solution? K_a ($HCHO_2$) $= 1.8\times10^{-4}$

(A) 2.07 **(B)** 5.33 **(C)** 8.67 **(D)** 11.93

Conceptual **PQ-19.** Which salt will form a basic aqueous solution?

(A) NaF **(B)** KBr **(C)** LiCl **(D)** NH_4NO_3

PQ-20. The pH of a 0.050 M aqueous solution of ammonium chloride (NH_4Cl) falls within what range?

(A) 0 to 2 **(B)** 2 to 7 **(C)** 7 to 12 **(D)** 12 to 14

PQ-21. What is the pH of a buffer solution containing equal volumes of 0.11 M $NaCH_3COO$ and 0.090 M. CH_3COOH? K_a (CH_3COOH) $= 1.8\times10^{-5}$

(A) 2.42 **(B)** 4.83 **(C)** 11.58 **(D)** 13.91

Conceptual **PQ-22.** Which pair of compounds would be the best choice to make a buffer solution with a pH around 5?

(A) HF and NaF $K_a = 6.9\times10^{-4}$

(B) $C_2H_5CO_2H$ and $C_2H_5CO_2Na$ $K_a = 1.3\times10^{-5}$

(C) HClO and NaClO $K_a = 2.8\times10^{-8}$

(D) NH_4Cl and NH_3 $K_a = 5.6\times10^{-10}$

Conceptual **PQ-23.** Which pair could be used to make a buffer solution?

(A) acetic acid and sodium chloride

(B) acetic acid and sodium hydroxide

(C) hydrochloric acid and potassium chloride

(D) hydrochloric acid and sodium hydroxide

PQ-24. What is the pH of a buffer solution made by adding 0.010 mole of solid NaF to 50. mL of 0.40 M HF? Assume no change in volume. K_a (HF) = 6.9×10^{-4}

(A) 1.6 **(B)** 1.9 **(C)** 2.9 **(D)** 3.2

Conceptual **PQ-25.** An aqueous buffer solution contains only HCN ($pK_a = 9.31$) and KCN and has a pH of 8.50. What can be concluded about the relative concentrations of HCN and KCN in the buffer?

(A) [HCN] > [KCN]

(B) [HCN] < [KCN]

(C) [HCN] = [KCN]

(D) nothing can be concluded about the relative concentrations

Conceptual **PQ-26.** Methyl orange is an indicator that changes color from red to yellow-orange over the pH range from 2.9 to 4.5. Methyl orange would be the most appropriate indicator for which type of acid -base titration?

(A) A weak acid titrated with a strong base

(B) A weak base titrated with a strong acid

(C) A strong acid titrated with a strong base

(D) Methyl orange would not be an appropriate indicator for any acid-base titration.

Conceptual **PQ-27.** Which graph best represents the titration of ammonia with hydrochloric acid?

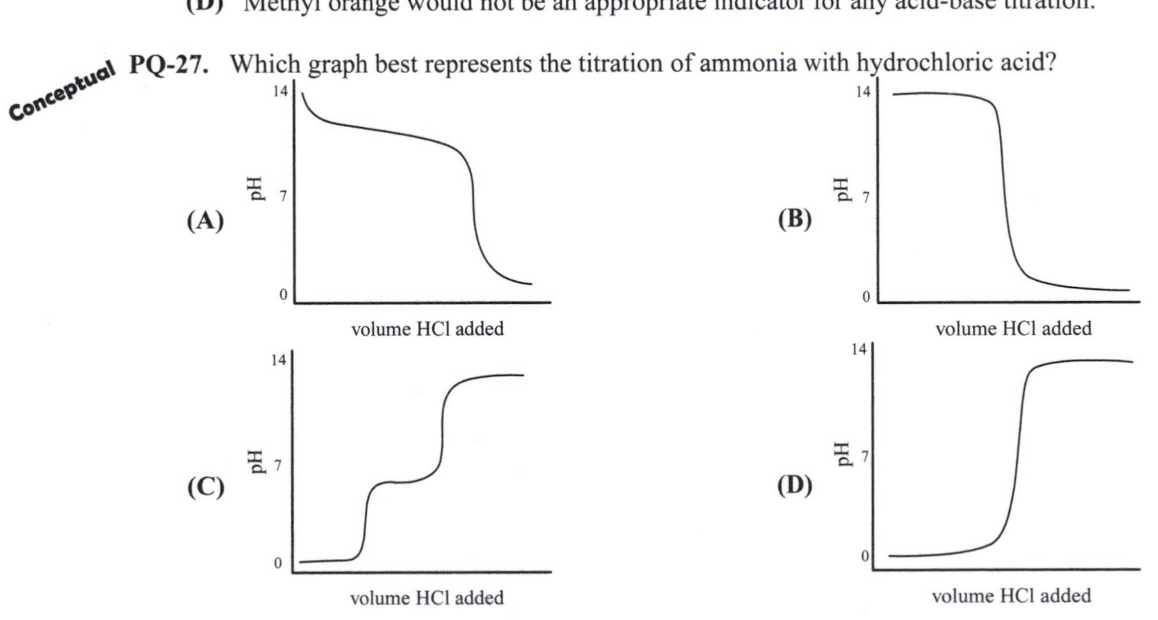

Conceptual **PQ-28.** In the laboratory, 50 mL of 0.1 M HCl is mixed with 50 mL of 0.1 M NaOH and the solution is stirred gently. At equilibrium, what ionic species, if any, will be present in large amounts in the reaction mixture?

(A) Na^+ and Cl^- only

(B) H_3O^+ and OH^- only

(C) H_3O^+, OH^-, Na^+, and Cl^-

(D) No ions will be present

PQ-29. What is the approximate pK_a of the weak acid in the titration curve?

(A) 2.0

(B) 4.8

(C) 8.8

(D) 11.0

PQ-30. What is the pH of a solution that results from mixing 25.0 mL of 0.200 M HA with 12.5 mL of 0.400 M NaOH? ($K_a = 1.0\times10^{-5}$)

(A) 2.94

(B) 4.94

(C) 9.06

(D) 11.06

Answers to Study Questions

1. B
2. C
3. C
4. A
5. D
6. A
7. C
8. B

Answers to Practice Questions

1. A
2. B
3. A
4. C
5. D
6. C
7. A
8. D
9. C
10. D
11. B
12. A
13. B
14. D
15. D
16. A
17. C
18. C
19. A
20. B
21. B
22. B
23. B
24. C
25. A
26. B
27. A
28. A
29. B
30. C

Chapter 13 – Solubility Equilibria

Chapter Summary:

This chapter will focus on the solubility of solids in aqueous media and other solvents.

Specific topics covered in this chapter are:

- Molar solubility
- Equilibrium expressions for solubility reactions
- Solubility product constants (K_{sp}) and related calculations
- Common ion effect
- Formation reactions, expressions and constants (K_f)

Previous material that is relevant to your understanding of questions in this chapter includes:

- Solutions and Aqueous Reactions, Part 1 (***Chapter 5***)
- Solutions and Aqueous Reactions, Part 2 (***Chapter 9***)
- Equilibrium (***Chapter 11***)

Common representations used in questions related to this material:

Name	Example	Used in questions related to
Compound units	$mol \cdot L^{-1}$	molar solubility
Macroscopic diagrams		common ion effect

Where to find this in your textbook:

The material in this chapter typically aligns to "Aqueous Equilibria" in your textbook. The name of your chapter(s) may vary.

Practice exam:

There are practice exam questions aligned to the material in this chapter. Because there are a limited number of questions on the practice exam, a review of the breadth of the material in this chapter is advised in preparation for your exam.

How this fits into the big picture:

The material in this chapter aligns to the Big Idea of Equilibrium (8) as listed on page 12 of this study guide.

Study Questions (SQ)

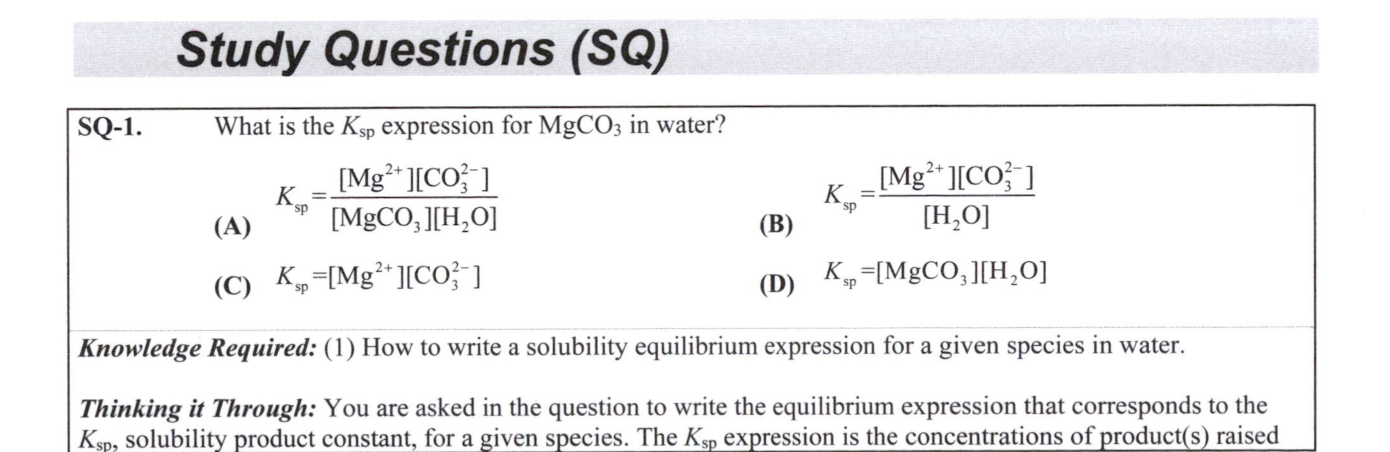

SQ-1. What is the K_{sp} expression for $MgCO_3$ in water?

(A) $K_{sp} = \dfrac{[Mg^{2+}][CO_3^{2-}]}{[MgCO_3][H_2O]}$

(B) $K_{sp} = \dfrac{[Mg^{2+}][CO_3^{2-}]}{[H_2O]}$

(C) $K_{sp} = [Mg^{2+}][CO_3^{2-}]$

(D) $K_{sp} = [MgCO_3][H_2O]$

Knowledge Required: (1) How to write a solubility equilibrium expression for a given species in water.

Thinking it Through: You are asked in the question to write the equilibrium expression that corresponds to the K_{sp}, solubility product constant, for a given species. The K_{sp} expression is the concentrations of product(s) raised to the power of their respective stoichiometric coefficients from a balanced chemical equation divided by the concentrations of reactant(s) raised to the power of their respective stoichiometric coefficients. Solids and liquids are not included in the expression because their concentrations are assumed to be constant. You, therefore, begin by constructing the balanced equation for the dissociation of $MgCO_3(s)$ in water:

$$MgCO_3(s) \rightleftharpoons Mg^{2+}(aq) + CO_3^{2-}(aq)$$

From the balanced equation, you can construct the K_{sp} expression:

$$K_{sp} = [Mg^{2+}][CO_3^{2-}] \quad \text{Choice (C)}$$

Choice **(A)** is not correct because it includes the concentrations of the solid and the liquid.
Choice **(B)** is not correct because it includes the concentration of the liquid.
Choice **(D)** is not correct because it includes the concentrations of the solid and the liquid, and does not include the concentrations of the dissolved ions.

Practice Questions Related to This: **PQ-1** and **PQ-2**

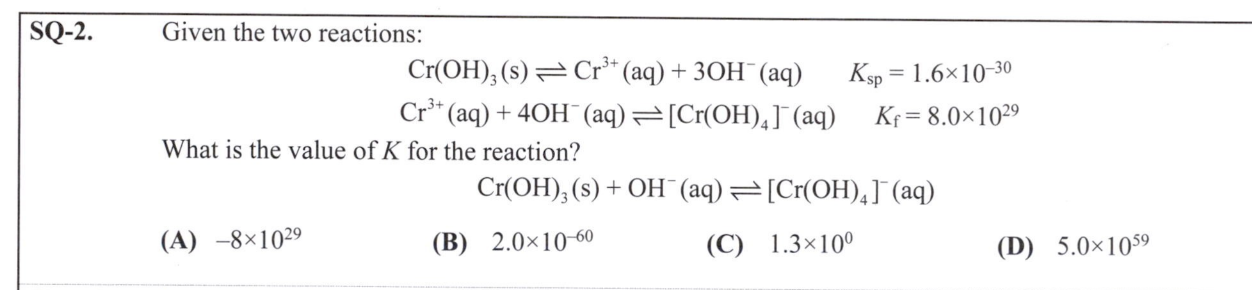

SQ-2. Given the two reactions:

$$Cr(OH)_3(s) \rightleftharpoons Cr^{3+}(aq) + 3OH^-(aq) \qquad K_{sp} = 1.6\times10^{-30}$$

$$Cr^{3+}(aq) + 4OH^-(aq) \rightleftharpoons [Cr(OH)_4]^-(aq) \qquad K_f = 8.0\times10^{29}$$

What is the value of K for the reaction?

$$Cr(OH)_3(s) + OH^-(aq) \rightleftharpoons [Cr(OH)_4]^-(aq)$$

(A) -8×10^{29} **(B)** 2.0×10^{-60} **(C)** 1.3×10^{0} **(D)** 5.0×10^{59}

Knowledge Required: (1) How to write an equilibrium expression. (2) How to combine multiple chemical reactions and their respective equilibrium constants to form a new reaction and equilibrium constant.

Thinking it Through: You are asked to determine the value of the equilibrium constant for a chemical reaction using two reactions with known constants, K_{sp} and K_f. You begin by writing the two equations and corresponding equilibrium constants. You observe that the target equation is the sum of the two given equations. You also then know that when two equations are summed, the corresponding equilibrium constant (K) is the product of the two equilibrium constants (K_{sp} and K_f):

$$Cr(OH)_3(s) \rightleftharpoons Cr^{3+}(aq) + 3OH^-(aq) \qquad K_{sp} = [Cr^{3+}][OH^-]^3$$

$$Cr^{3+}(aq) + 4OH^-(aq) \rightleftharpoons [Cr(OH)_4]^-(aq) \qquad K_f = \frac{[[Cr(OH)_4]^-]}{[Cr^{3+}][OH^-]^4}$$

$$Cr(OH)_3(s) + OH^-(aq) \rightleftharpoons [Cr(OH)_4]^-(aq) \qquad K = \frac{[[Cr(OH)_4]^-]}{[OH^-]}$$

$$K = [Cr^{3+}][OH^-]^3 \times \frac{[[Cr(OH)_4]^-]}{[Cr^{3+}][OH^-]^4} = K_{sp} \times K_f$$

When reactions are summed, the new $K = K_{sp} \times K_f = 1.3$ or 1.3×10^0, Choice **(C)**.

Choice **(A)** is not correct because it is the value obtained when K_f is substracted from K_{sp}.
Choice **(B)** is not correct because it is the value obtained when $K_{sp} \div K_f$.
Choice **(D)** is not correct because it is the value obtained when $K_f \div K_{sp}$.

Practice Questions Related to This: **PQ-3**

SQ-3. What is the solubility product constant, K_{sp}, of $Mg(OH)_2$ if its molar solubility in water is 1.6×10^{-4} $mol\cdot L^{-1}$?

(A) 2.6×10^{-8} **(B)** 1.6×10^{-11} **(C)** 1.0×10^{-12} **(D)** 4.1×10^{-12}

Knowledge Required: (1) Definition of molar solubility. (2) How to write a solubility equilibrium expression. (3) How to determine a solubility product constant (K_{sp}) given molar solubility.

Thinking it Through: You are asked in the question to determine the solubility product constant (K_{sp}) for a given species with a known solubility in water. To begin, you construct a balanced chemical equation for the dissociation and a corresponding K_{sp} expression:

$$Mg(OH)_2(s) \rightleftharpoons Mg^{2+}(aq) + 2OH^-(aq)$$

$$K_{sp} = [Mg^{2+}][OH^-]^2$$

You see that you are given the "molar solubility" of $Mg(OH)_2$ in the question. To understand what this is, you proceed to set up the initial-change-equilibrium (ICE) table for this process where you have purposely included the states to remind yourself that $Mg(OH)_2(s)$ is not included in the K_{sp} expression:

	$Mg(OH)_2(s)$	$\rightleftharpoons$	$Mg^{2+}(aq)$	+	$2OH^-(aq)$
Initial	--		0		0
Change	--		$+x$		$+2x$
Equilibrium	--		x		$2x$

From here you see that the molar solubility (the value of "x" in the table) corresponds to the number of moles solute in 1.0 L of a saturated solution or $[Mg(OH)_2]$.

> *Note:* You know this is used to standardize the comparison of solubility of various ionic compounds where the ion ratios are not equivalent (such that you can compare $[Al(OH)_3]$ or the molar solubility of $Al(OH)_3$ to $[AgOH]$ or the molar solubility of AgOH but cannot compare the respective concentrations of the hydroxide ion to predict which compound is more soluble).

Based on the coefficients from the balanced chemical equation, the solubility concentration has a 1:1 ratio with the concentration of Mg^{2+} and a 1:2 ratio with the concentration of OH^-. You can thus enter these values into the K_{sp} expression and solve:

$$K_{sp} = \left[Mg^{2+}\right]\left[OH^-\right]^2 = (x)(2x)^2 = 4x^3$$

$$K_{sp} = \left[1.6\times10^{-4}\right]\left[2\times\left(1.6\times10^{-4}\right)\right]^2 = 1.6\times10^{-11}$$

Choice **(B)**

Choice **(A)** is not correct because the value is obtained using a K_{sp} expression that does not account for the coefficient of hydroxide ion in the balanced chemical equation.

Choice **(C)** is not correct because the value is obtained using the inverse of the 1:2 mole ratio when calculating the concentration of hydroxide ion.

Choice **(D)** is not correct because the value is obtained without using the 1:2 mole ratio when calculating the concentration of hydroxide ion.

Practice Questions Related to This: **PQ-3, PQ-4, PQ-5, PQ-6** and **PQ-7**

SQ-4. Which compound has the highest molar solubility?

Conceptual

(A) AgI, $K_{sp} = 8.52\times10^{-17}$

(B) $BaCO_3$, $K_{sp} = 2.58\times10^{-9}$

(C) $Fe(OH)_3$, $K_{sp} = 2.79\times10^{-39}$

(D) ZnS, $K_{sp} = 3.00\times10^{-23}$

Knowledge Required: (1) Definition of molar solubility. (2) How to write a solubility equilibrium expression. (3) How to determine molar solubility given a solubility product constant (K_{sp}).

Thinking it Through: You are asked in the question to evaluate a series of species and their provided K_{sp} values to determine which species has the highest molar solubility. To make this evaluation, you consider each species. To begin, you write the balanced equation and the corresponding K_{sp} expression for each.

Choice **(A)**: $AgI(s) \rightleftharpoons Ag^+(aq) + I^-(aq)$ $\quad K_{sp} = \left[Ag^+\right]\left[I^-\right]$

Choice **(B)**: $BaCO_3(s) \rightleftharpoons Ba^{2+}(aq) + CO_3^{2-}(aq)$ $\quad K_{sp} = \left[Ba^{2+}\right]\left[CO_3^{2-}\right]$

Choice **(C)**: $Fe(OH)_3(s) \rightleftharpoons Fe^{3+}(aq) + 3OH^-(aq)$ $\quad K_{sp} = \left[Fe^{3+}\right]\left[OH^-\right]^3$

Choice **(D)**: $ZnS(s) \rightleftharpoons Zn^{2+}(aq) + S^{2-}(aq)$ $K_{sp} = [Zn^{2+}][S^{2-}]$

Next, you assume that equilibrium is reached and the solution is saturated. Additionally, you assume there is not a starting concentration of any species. The variable x will be used to represent one mole equivalent of the solid (for example, AgI) when expressing the concentrations of each species, so then x represents the molar solubility. Therefore, when x values and the K_{sp} values are substituted for each, you obtain these expressions:

Choice **(A)**: $8.52 \times 10^{-17} = x \cdot x$

Choice **(B)**: $2.58 \times 10^{-9} = x \cdot x$

Choice **(C)**: $2.79 \times 10^{-39} = x \cdot (3x)^3$

Choice **(D)**: $3.00 \times 10^{-23} = x \cdot x$

For each answer option, you solve for x; because x cannot be negative, only positive values of x are retained.

Choice **(A)**: $x = 9.2 \times 10^{-9}$

Choice **(B)**: $x = 5.1 \times 10^{-5}$

Choice **(C)**: $x = 1.0 \times 10^{-10}$

Choice **(D)**: $x = 5.5 \times 10^{-12}$

Therefore, the species with the highest molar solubility (the highest value of x) is $BaCO_3$, Choice **(B)**.

Practice Questions Related to This: **PQ-8, PQ-9, PQ-10, PQ-11, PQ-12, PQ-13, PQ-14, PQ-15, PQ-16,** and **PQ-17**

SQ-5. Conceptual

A solution contains 0.002 M Pb^{2+} and 0.002 M Ag^+. What happens when NaCl(s) is added to bring the chloride ion concentration to 0.01 M?

	K_{sp}
$PbCl_2$	1.2×10^{-5}
AgCl	1.8×10^{-10}

(A) Neither $PbCl_2$ nor AgCl will precipitate.

(B) Both $PbCl_2$ and AgCl precipitate.

(C) Only $PbCl_2$ precipitates.

(D) Only AgCl precipitates.

Knowledge Required: (1) How to write a solubility equilibrium expression. (2) How to determine a reaction quotient, Q. (3) How to predict precipitation by comparing Q with K_{sp}.

Thinking it Through: You are asked in the question to determine what occurs when a concentration of a given ion is brought to a certain level. To do this you will calculate the value of Q for each set of conditions and compare to the value of K_{sp}. The criteria for the comparison are:

If $Q_{sp} < K_{sp}$ the reaction will proceed to form more products: **no precipitate forms.**
Using AgCl as the example, the concentration of the ions is sufficiently small that the solution is **unsaturated** and the reaction will shift to the **right**: $AgCl(s) \rightleftharpoons Ag^+(aq) + Cl^-(aq)$

If $Q_{sp} > K_{sp}$ the reaction will proceed to form more reactants: **a precipitate forms.**
Using AgCl as the example, the concentration of the ions is sufficiently large that the solution is **saturated** and the reaction will shift to the **left**: $AgCl(s) \rightleftharpoons Ag^+(aq) + Cl^-(aq)$

If $Q_{sp} > K_{sp}$ the reaction is at equilibrium: **the solution is just saturated.**
Using AgCl as the example, the concentration of the ions is at the saturation point such that the solution is **saturated** and the reaction is at equilibrium: $AgCl(s) \rightleftharpoons Ag^+(aq) + Cl^-(aq)$

Therefore, for this question, the relevant Q expressions are:

$AgCl(s) \rightleftharpoons Ag^+(aq) + Cl^-(aq)$ $Q = [Ag^+]_0[Cl^-]_0$

$PbCl_2(s) \rightleftharpoons Pb^{2+}(aq) + 2Cl^-(aq)$ $Q = [Pb^{2+}]_0[Cl^-]_0^2$

For $PbCl_2$: $Q = (0.002)(0.01)^2 = 2 \times 10^{-7}$ For AgCl: $Q = (0.002)(0.01) = 2 \times 10^{-5}$

The Q for $PbCl_2$ is less than K_{sp} for $PbCl_2$, so $PbCl_2$ will **not** precipitate.

The Q for AgCl is greater than K_{sp} for AgCl, so AgCl will precipitate.

Choice **(D)** is the correct answer.

Practice Questions Related to This: **PQ-18** and **PQ-19**

SQ-6. A sample of "hard" water contains about 2.0×10^{-2} mol of Ca^{2+} ions per liter. What is the maximum concentration of fluoride ion that could be present in hard water? Assume fluoride is the only anion present that will precipitate calcium ion. (K_{sp} of CaF_2(s) at 25 °C = 4.0×10^{-11})

(A) 2.0×10^{-9} M **(B)** 2.2×10^{-5} M **(C)** 4.4×10^{-5} M **(D)** 2.0×10^{-2} M

Knowledge Required: (1) How to write a solubility equilibrium expression. (2) How to determine the concentration of one ion given the solubility product constant (K_{sp}) and concentration of a common ion.

Thinking it Through: You are asked to determine the maximum amount of a species that can be dissolved into a solution that contains a common ion with the species being dissolved. You begin by constructing a balanced chemical equation and K_{sp} expression for the dissolution:

$$CaF_2(s) \rightleftharpoons Ca^{2+}(aq) + 2F^-(aq)$$

$$K_{sp} = [Ca^{2+}][F^-]^2$$

Next, you consider what this looks like if CaF_2 was first in pure water. By slowly adding solid CaF_2 to water, initially (when $[Ca^{2+}]$ and $[F^-]$ are very small), the solution would remain homogeneous. After some point, the concentration of the ions reaches a concentration such that the solution is saturated and no more solid CaF_2 will dissolve, and a heterogeneous solution has formed. You know this is called the "molar solubility" and determine this using the process outlined in **SQ-3**. Using this, you find the molar solubility for CaF_2 in pure water is equal to 2.2×10^{-4} M or in pure water, $[Ca^{2+}] = 2.2\times10^{-4}$ M and $[F^-] = 4.4\times10^{-4}$ M.

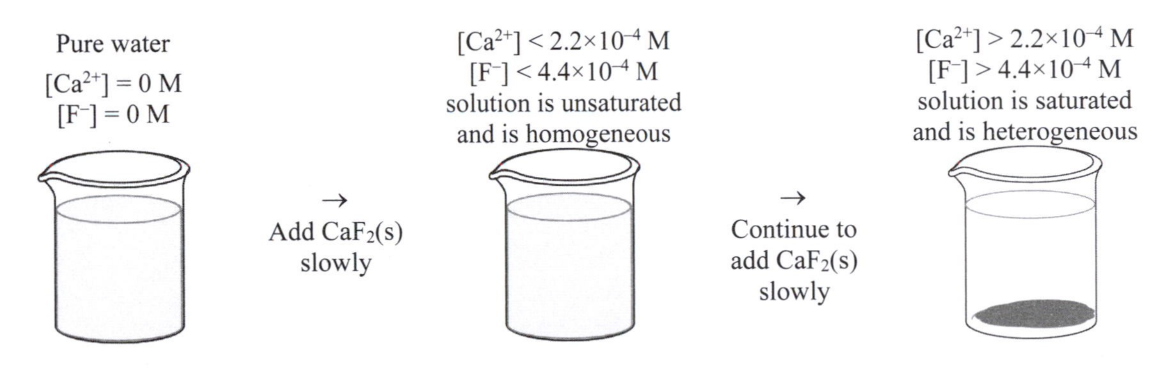

Now, you consider what this looks like for CaF_2 in tap water containing $[Ca^{2+}] = 2.0\times10^{-2}$ M. By slowly adding solid CaF_2 to tap water, initially (when $[F^-]$ is still very small), the solution would remain homogeneous. After some point, the concentration of the fluoride reaches a concentration such that the solution is saturated and no more solid CaF_2 will dissolve, and a heterogeneous solution has formed. However, this is different because you started with $[Ca^{2+}]_0$. So to determine the concentration of the fluoride ion, [F–], you will include the initial concentration of the calcium ion:

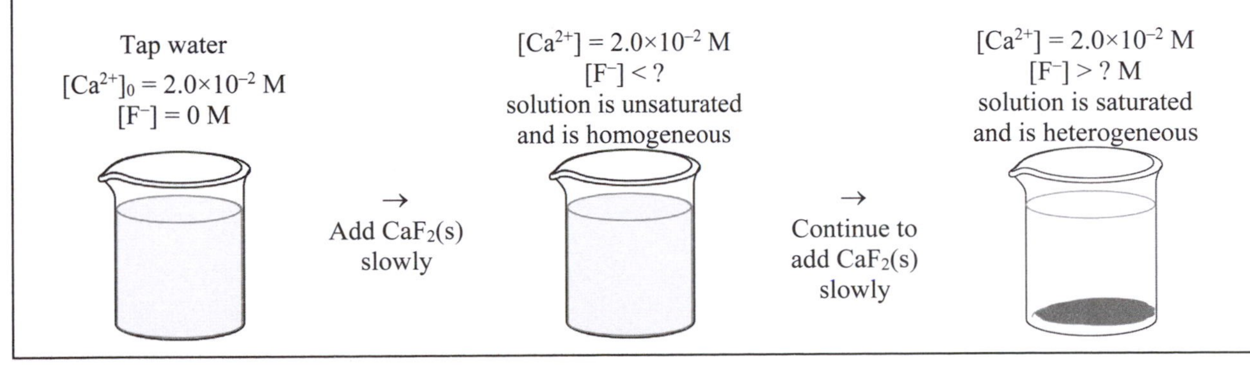

	$CaF_2(s)$	$\rightleftharpoons$	$Ca^{2+}(aq)$	+	$2F^-(aq)$
Initial	--		2.0×10^{-2}		0
Change	--		$+x$		$+2x$
Equilibrium	--		$2.0\times10^{-2}+x$		$2x$

Because the concentration of the calcium ion is higher, you can assume the change in the concentration of the calcium ion (from the added CaF_2) will be sufficiently small (or assume $2.0\times10^{-2}+x\approx2.0\times10^{-2}$) to be unchanged from the initial concentration:

$$K_{sp}=[Ca^{2+}][F^-]^2=[Ca^{2+}](2x)^2=[Ca^{2+}]4x^2 \qquad 4x^2=\frac{K_{sp}}{[Ca^{2+}]} \qquad x=\sqrt{\frac{K_{sp}}{4[Ca^{2+}]}}=2.2\times10^{-5}$$

$$[F^-]=2x=2(2.2\times10^{-5})=4.4\times10^{-5}\text{ M}$$

Corresponding to choice **(C)**.

Choice **(A)** is not correct because it is the value of x if the coefficient of "2" for F^- was not considered in the K_{sp} expression and initial-change-equilibrium table.

Choice **(B)** is not correct because it is the value of x; however, the final concentration of F^- is $2\cdot x$.

Choice **(D)** is not correct because it is the initial concentration of Ca^{2+} and the approximate equilibrium concentration of Ca^{2+}.

Practice Questions Related to This: **PQ-20**

SQ-7. What will change the value of K_{sp} for silver azide (AgN_3)?

Conceptual

(A) Adding water to the solution.

(B) Adding silver ions to the solution.

(C) Removing azide ions from the solution.

(D) Increasing the temperature of the solution.

Knowledge Required: (1) What influences the value of K_{sp}

Thinking it Through: You are asked in the question to evaluate what changes to a system at equilibrium would result in a change in the value of K_{sp}. Choice **(A)** would result in decreasing concentrations and thus increases in the amount of silver azide dissolved. Choice **(B)** would result in a common ion effect whereby the addition of silver ions would lead to more silver azide precipitation. Choice **(C)** would have the opposite effect whereby the removal of azide ions would result in more silver azide being dissolved. Choices **(A)**, **(B)**, and **(C)** impact Q for the dissolution, but do not impact the value of K_{sp}. Choice **(D)**, though, does have a direct impact on the value of K_{sp} whereby an increase in temperature results in a different K_{sp} value.

Practice Questions Related to This: **PQ-21, PQ-22, PQ-23, PQ-24, PQ-25, PQ-26, PQ-27, PQ-28** and **PQ-29**

SQ-8. When comparing Q and K_{sp} of an unsaturated solution, ________.

Conceptual

(A) $Q < K_{sp}$

(B) $Q = K_{sp}$

(C) $Q > K_{sp}$

(D) $Q = K_{sp} = 0$

Knowledge Required: (1) The definition of Q. (2) How Q and K_{sp} are related.

Thinking it Through: You are asked in the question to determine the relationship between Q and K_{sp} for an unsaturated solution. Recall that for an unsaturated solution, more starting material can be dissolved into solution. Therefore, the molar solubility has not been reached. The solubility product constant, K_{sp}, is defined as the product of the maximum concentration of dissolved species, i.e. the molar solubility. The value of Q is defined as the product of the actual concentrations of dissolved species. Therefore, if Q is greater than K_{sp}, there are concentrations of the dissolved species greater than the maximum allowable concentrations and thus species will precipitate out until the concentrations of dissolved species matches K_{sp}, Choice **(C)**. If Q is equal to K_{sp}, there are exactly the concentration of dissolved species that match K_{sp}, Choice **(B)**. If Q is less than K_{sp}, there are concentrations of the dissolved species that are less than the maximum concentrations (or molar solubility values) and thus more species could be dissolved into solution, Choice **(A)**. Choice **(D)** would only be true if the species is entirely insoluble.

An unsaturated solution is a solution whereby more species could be dissolved; therefore, Choice **(A)** where Q is less than K_{sp} is correct for an unsaturated solution.

Practice Questions Related to This: **PQ-30**

Practice Questions (PQ)

PQ-1. What is the K_{sp} expression for $CoCO_3(s)$ in water?

(A) $K_{sp} = \dfrac{[Co^{2+}][CO_3^{2-}]}{[CoCO_3][H_2O]}$ **(B)** $K_{sp} = \dfrac{[Co^{2+}][CO_3^{2-}]}{[H_2O]}$

(C) $K_{sp} = [Co^{2+}][CO_3^{2-}]$ **(D)** $K_{sp} = [CoCO_3][H_2O]$

PQ-2. What is the K_{sp} expression for $Ni_3(PO_4)_2(s)$ in water?

(A) $K_{sp} = [Ni^{2+}]^3[PO_4^{3-}]^2$ **(B)** $K_{sp} = [Ni^{2+}]^2[PO_4^{3-}]^3$

(C) $K_{sp} = \left(3\times[Ni^{2+}]\right)\left(2\times[PO_4^{3-}]\right)$ **(D)** $K_{sp} = \left(3\times[Ni^{2+}]\right)^3\left(2\times[PO_4^{3-}]\right)^2$

Conceptual

PQ-3. Given the two reactions:

$$CuCl(s) + Cl^-(aq) \rightleftharpoons [CuCl_2]^-(aq) \qquad K = 5.2\times10^{-2}$$

$$Cu^+(aq) + 2Cl^-(aq) \rightleftharpoons [CuCl_2]^-(aq) \qquad K_f = 3.0\times10^{5}$$

What is the value of K_{sp} for CuCl?

(A) $K \times K_f$ **(B)** $\dfrac{K}{K_f}$ **(C)** $\dfrac{K_f}{K}$ **(D)** $\dfrac{1}{K\times K_f}$

PQ-4. A saturated solution of MgF_2 contains 1.6×10^{-3} mol of MgF_2 per liter at a certain temperature. What is the K_{sp} of MgF_2 at this temperature?

(A) 2.7×10^{-6} **(B)** 1.6×10^{-8} **(C)** 3.1×10^{-9} **(D)** 6.2×10^{-9}

PQ-5. A saturated solution of $Al(OH)_3$ has a molar solubility of 2.9×10^{-9} M at a certain temperature. What is the solubility product constant, K_{sp}, of $Al(OH)_3$ at this temperature?

(A) 2.1×10^{-34} **(B)** 1.9×10^{-33} **(C)** 8.4×10^{-18} **(D)** 2.5×10^{-17}

Conceptual

PQ-6. Silver chromate (Ag_2CrO_4; $K_{sp} = 1.1\times10^{-12}$) has a molar solubility of 6.5×10^{-5} M. What calculation provides the concentration of silver ions in a saturated solution of Ag_2CrO_4?

(A) $2 \times (6.5\times10^{-5})$ M **(B)** $1 \times (6.5\times10^{-5})$ M **(C)** $[2 \times (6.5\times10^{-5})]^2$ M **(D)** $\sqrt[3]{1\times(1.1\times10^{-12})}$ M

PQ-7. The solubility of lead(II) iodide in water is 0.62 grams per 1.0 L of solution. What is the solubility product constant, K_{sp}, for the dissolution? $PbI_2(s) \rightleftharpoons Pb^{2+}(aq) + 2I^-(aq)$

(A) 2.4×10^{-9} **(B)** 9.7×10^{-9} **(C)** 1.8×10^{-6} **(D)** 4.0×10^{-3}

PQ-8. For Ag_2S the solubility product constant, K_{sp}, is 6×10^{-30}. What is the molar solubility of Ag_2S?

(A) 1×10^{-10} M **(B)** 2×10^{-10} M **(C)** 2×10^{-15} M **(D)** 2×10^{-30} M

PQ-9. For BaF_2 $K_{sp} = 1.0\times10^{-6}$. What is the molar solubility of BaF_2 in a solution containing 0.10 M NaF?

(A) 6.3×10^{-3} M **(B)** 1.0×10^{-4} M **(C)** 2.5×10^{-5} M **(D)** 1.0×10^{-5} M

Conceptual **PQ-10.** Which saturated solution has the highest concentration of F^- ions at 25 °C?

(A) BaF_2, $K_{sp} = 1.8\times10^{-7}$ **(B)** CaF_2, $K_{sp} = 3.5\times10^{-11}$

(C) MgF_2, $K_{sp} = 7.4\times10^{-11}$ **(D)** SrF_2, $K_{sp} = 2.5\times10^{-9}$

PQ-11. What is the molar solubility of Ag_3PO_4 in water? K_{sp} (Ag_3PO_4) = 1.4×10^{-16}

(A) 1.1×10^{-4} M **(B)** 4.8×10^{-5} M **(C)** 5.2×10^{-6} M **(D)** 6.8×10^{-9} M

PQ-12. Given K_{sp} of CaF_2 is 3.5×10^{-11}. The molar solubility of CaF_2 in 0.200 M NaF is ________.

(A) 2.1×10^{-4} M **(B)** 1.8×10^{-10} M **(C)** 8.8×10^{-10} M **(D)** 3.5×10^{-11} M

PQ-13. What is the maximum mass of solid barium sulfate (233 g·mol^{-1}) that can be dissolved in 1.00 L of 0.100 M Na_2SO_4 solution? K_{sp} ($BaSO_4$) = 1.5×10^{-9}

(A) 1.5×10^{-8} g **(B)** 2.1×10^{-6} g **(C)** 3.5×10^{-6} g **(D)** 9.0×10^{-3} g

Conceptual **PQ-14.** Rank the following species from most to least soluble in water.

I. $Ba(NO_3)_2$, $K_{sp} = 4.64\times10^{-3}$
II. $Cd(OH)_2$, $K_{sp} = 7.20\times10^{-15}$
III. $FeCO_3$, $K_{sp} = 3.13\times10^{-11}$

(A) I > II > III **(B)** I > III > II **(C)** II > III > I **(D)** II > I > III

Conceptual **PQ-15.** Rank the following species from least to most soluble in water.

I. $PbBr_2$, $K_{sp} = 6.60\times10^{-6}$
II. $PbCl_2$, $K_{sp} = 1.70\times10^{-5}$
III. PbI_2, $K_{sp} = 9.80\times10^{-9}$

(A) I < III < II **(B)** II < III < I **(C)** III < I < II **(D)** III < II < I

PQ-16. For LiF $K_{sp} = 1.84\times10^{-3}$. What is the molar solubility of LiF?

(A) 6.78×10^{-1} M **(B)** 4.29×10^{-2} M **(C)** 8.78×10^{-3} M **(D)** 3.00×10^{-6} M

Conceptual **PQ-17.** Which saturated solution has the highest concentration of OH^- ions at 25 °C?

(A) $Al(OH)_3$, $K_{sp} = 3.0\times10^{-34}$ **(B)** $Ca(OH)_2$, $K_{sp} = 5.02\times10^{-6}$

(C) $Cd(OH)_2$, $K_{sp} = 7.2\times10^{-15}$ **(D)** $Pr(OH)_3$, $K_{sp} = 3.4\times10^{-24}$

Conceptual **PQ-18.** The solubility of solid silver chromate, Ag_2CrO_4, which has a K_{sp} equal to 9.0×10^{-12} at 25 °C, is determined in water and in two different aqueous solutions. Predict the relative solubility of silver chromate in the three solutions.

I	pure water
II	0.1 M $AgNO_3$
III	0.1 M Na_2CrO_4

(A) I = II = III **(B)** I < II < III **(C)** II = III < I **(D)** II < III < I

Conceptual **PQ-19.** What will be the result if 100 mL of 0.06 M $Mg(NO_3)_2$ is added to 50 mL of 0.06 M $Na_2C_2O_4$? Assume the reaction is taking place at 25 °C. (K_{sp} of MgC_2O_4(s) at 25 °C = 8.6×10^{-5})

(A) No precipitate will form.

(B) A precipitate will form and an excess of Mg^{2+} ions will remain in solution.

(C) A precipitate will form and an excess of $C_2O_4^{2-}$ ions will remain in solution.

(D) A precipitate will form and both Mg^{2+} and $C_2O_4^{2-}$ ions are present in excess.

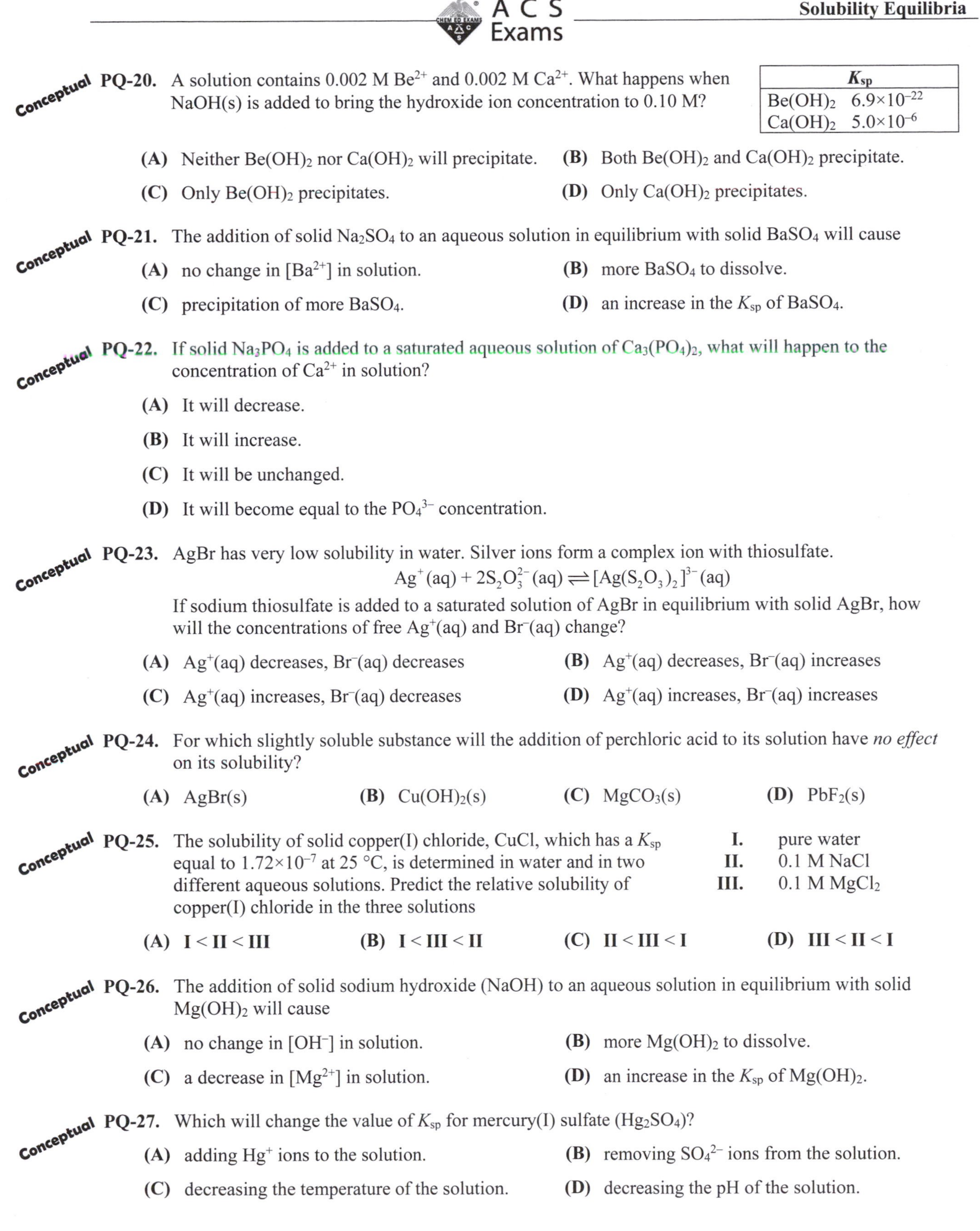

Conceptual **PQ-20.** A solution contains 0.002 M Be^{2+} and 0.002 M Ca^{2+}. What happens when NaOH(s) is added to bring the hydroxide ion concentration to 0.10 M?

	K_{sp}
$Be(OH)_2$	6.9×10^{-22}
$Ca(OH)_2$	5.0×10^{-6}

(A) Neither $Be(OH)_2$ nor $Ca(OH)_2$ will precipitate. **(B)** Both $Be(OH)_2$ and $Ca(OH)_2$ precipitate.

(C) Only $Be(OH)_2$ precipitates. **(D)** Only $Ca(OH)_2$ precipitates.

Conceptual **PQ-21.** The addition of solid Na_2SO_4 to an aqueous solution in equilibrium with solid $BaSO_4$ will cause

(A) no change in $[Ba^{2+}]$ in solution. **(B)** more $BaSO_4$ to dissolve.

(C) precipitation of more $BaSO_4$. **(D)** an increase in the K_{sp} of $BaSO_4$.

Conceptual **PQ-22.** If solid Na_3PO_4 is added to a saturated aqueous solution of $Ca_3(PO_4)_2$, what will happen to the concentration of Ca^{2+} in solution?

(A) It will decrease.

(B) It will increase.

(C) It will be unchanged.

(D) It will become equal to the PO_4^{3-} concentration.

Conceptual **PQ-23.** AgBr has very low solubility in water. Silver ions form a complex ion with thiosulfate.

$$Ag^+(aq) + 2S_2O_3^{2-}(aq) \rightleftharpoons [Ag(S_2O_3)_2]^{3-}(aq)$$

If sodium thiosulfate is added to a saturated solution of AgBr in equilibrium with solid AgBr, how will the concentrations of free $Ag^+(aq)$ and $Br^-(aq)$ change?

(A) $Ag^+(aq)$ decreases, $Br^-(aq)$ decreases **(B)** $Ag^+(aq)$ decreases, $Br^-(aq)$ increases

(C) $Ag^+(aq)$ increases, $Br^-(aq)$ decreases **(D)** $Ag^+(aq)$ increases, $Br^-(aq)$ increases

Conceptual **PQ-24.** For which slightly soluble substance will the addition of perchloric acid to its solution have *no effect* on its solubility?

(A) AgBr(s) **(B)** $Cu(OH)_2$(s) **(C)** $MgCO_3$(s) **(D)** PbF_2(s)

Conceptual **PQ-25.** The solubility of solid copper(I) chloride, CuCl, which has a K_{sp} equal to 1.72×10^{-7} at 25 °C, is determined in water and in two different aqueous solutions. Predict the relative solubility of copper(I) chloride in the three solutions

I. pure water
II. 0.1 M NaCl
III. 0.1 M $MgCl_2$

(A) **I < II < III** **(B)** **I < III < II** **(C)** **II < III < I** **(D)** **III < II < I**

Conceptual **PQ-26.** The addition of solid sodium hydroxide (NaOH) to an aqueous solution in equilibrium with solid $Mg(OH)_2$ will cause

(A) no change in $[OH^-]$ in solution. **(B)** more $Mg(OH)_2$ to dissolve.

(C) a decrease in $[Mg^{2+}]$ in solution. **(D)** an increase in the K_{sp} of $Mg(OH)_2$.

Conceptual **PQ-27.** Which will change the value of K_{sp} for mercury(I) sulfate (Hg_2SO_4)?

(A) adding Hg^+ ions to the solution. **(B)** removing SO_4^{2-} ions from the solution.

(C) decreasing the temperature of the solution. **(D)** decreasing the pH of the solution.

Conceptual **PQ-28.** For AgI $K_{sp} = 8.5\times10^{-17}$ and for $[Ag(CN)_2]^-$ $K_f = 1.0\times10^{21}$. Which statement is true?

(A) AgI is completely insoluble in a solution that contains CN^-.

(B) AgI is less soluble in water than in a solution that contains CN^-.

(C) AgI is more soluble in water than in a solution that contains CN^-.

(D) AgI is equally soluble in water and in a solution that contains CN^-.

Conceptual **PQ-29.** Ammonia (NH_3) is slowly added to a saturated solution of Ag_2CO_3 ($K_{sp} = 8.46\times10^{-12}$). Which statement is true if K_f for $[Ag(NH_3)_2]^+$ is 1.6×10^7?

(A) Ag_2CO_3 is completely miscible in a solution that contains NH_3.

(B) Ag_2CO_3 is more soluble in water than in a solution that contains NH_3.

(C) Ag_2CO_3 is less soluble in water than in a solution that contains NH_3.

(D) Ag_2CO_3 is equally soluble in water and in a solution that contains NH_3.

Conceptual **PQ-30.** When comparing Q and K_{sp} of a saturated solution, ________.

(A) $Q < K_{sp}$ **(B)** $Q = K_{sp}$ **(C)** $Q \geq K_{sp}$ **(D)** $Q = K_{sp} = 0$

Answers to Study Questions

1. C
2. C
3. B
4. B
5. D
6. C
7. D
8. A

Answers to Practice Questions

1. C
2. A
3. B
4. B
5. B
6. A
7. B
8. A
9. B
10. A
11. B
12. C
13. C
14. A
15. C
16. B
17. B
18. D
19. D
20. B
21. C
22. A
23. B
24. A
25. D
26. C
27. C
28. B
29. C
30. B

A C S
Exams

Chapter 14 – Thermodynamics

Chapter Summary:

This chapter will focus on entropy and free energy, energy, and enthalpy. The transfer of energy during chemical and physical changes will be calculated. The usefulness of entropy and free energy being state functions will be used to calculate the values of entropy and free energy changes using Hess's law.

Specific topics covered in this chapter are:

- Entropy and trends in molar entropy values
- Predicting the sign of the entropy change for a process
- Calculating the value of ΔS for a chemical reaction
- Definition of a spontaneous process
- Calculating ΔG for a reaction using Hess's law
- Predicting the temperature dependence of spontaneity
- Relating values of $\Delta G°$ and equilibrium coefficients

Previous material that is relevant to your understanding of questions in this chapter include:

- Significant figures (***Toolbox***)
- Scientific notation (***Toolbox***)
- Balancing chemical reactions ***(Chapter 3)***
- Heat and Enthalpy ***(Chapter 6)***
- Equilibrium ***(Chapter 11)***

Common representations used in questions related to this material:

Name	Example	Used in questions related to
compound units	$kJ \cdot mol^{-1}$ and $J \cdot mol^{-1} \cdot K^{-1}$	density, enthalpy changes for reactions (ΔH), Gibbs free energy changes for reactions (ΔG), entropy changes for reactions (ΔS)
Particulate diagrams	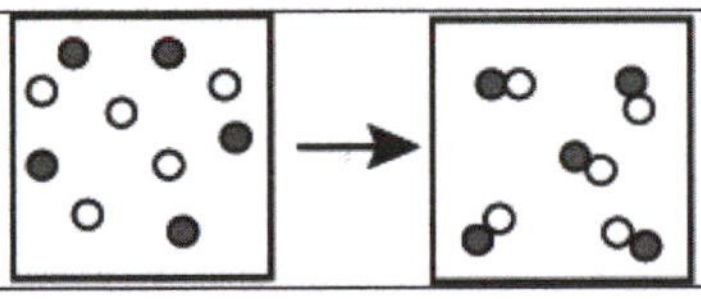	entropy changes

Where to find this in your textbook:

The material in this chapter typically aligns to "Thermodynamics". The name of your chapter may vary.

Practice exam:

There are practice exam questions aligned to the material in this chapter. Because there are a limited number of questions on the practice exam, a review of the breadth of the material in this chapter is advised in preparation for your exam.

How this fits into the big picture:

The material in this chapter aligns to the Big Ideas of Energy and Thermodynamics (6) and Equilibrium (8) as listed on page 12 of this study guide.

Study Questions (SQ)

SQ-1.

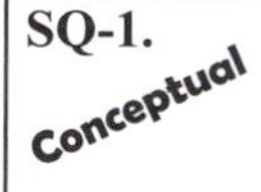

Of the four listed compounds, three have correct values for molar entropy, $S°$, associated with them. For which species is the entropy value *incorrect* at 298 K?

	Substance	$S°$ **(J·mol^{-1}·K^{-1})**
(A)	Ca(s)	202
(B)	$Br_2(l)$	152
(C)	$CO_2(g)$	214
(D)	$AlCl_3(s)$	109

Knowledge Required: (1) Trends in molar entropy values.

Thinking it Through: You need to identify the incorrect pairing of species and molar entropy values. You remember the general trends for molar entropy values:

(1) The more restricted the motion of the physical state, the lower the entropy value, $S°(s) < S°(l) < S°(g)$
(2) The more complex the species, more atoms/ions present, the larger the value of $S°$.
(3) As temperature increases, the value of $S°$ increases.

You would expect the values of the solids to be less than the values of the liquids and these would be less than the values of the gases. Using this you predict that the $S°$ values for Ca(s) and $AlCl_3(s)$ should be smaller than the value of $Br_2(l)$ and the $S°$ value for CO_2 should be the largest. You also know that the value of $S°$ for Ca(s) should be smaller than the value for $S°$ for $AlCl_3(s)$. You predict the values should go in the order

$$S°(Ca(s)) < S°(AlCl_3(s)) < S°(Br_2(l)) < S°(CO_2(g))$$

When you substitute the given values into the prediction you get:

$$202 < 109 < 152 < 214$$

It is clear that the $S°$ value for Ca(s) is the incorrect one. Choice **(A)** is the incorrect pair, and therefore the correct answer. The other species and molar entropy values are correctly paired up.

Practice Questions Related to This: **PQ-1** and **PQ-2**

SQ-2. Conceptual

Which change is likely to be accompanied by an increase in entropy?

(A) $N_2(g) + 3H_2(g) \rightarrow 2NH_3(g)$

(B) $Ag^+(aq) + Cl^-(aq) \rightarrow AgCl(s)$

(C) $CO_2(s) \rightarrow CO_2(g)$

(D) $H_2O(g) \rightarrow H_2O(l)$

Knowledge Required: (1) Definition of entropy. (2) Trends in molar entropy values.

Thinking it Through: You recall that the entropy change for a process is equal to:

$$\Delta S = S°(\text{products}) - S°(\text{reactants})$$

You also know that for an increase in entropy you are looking for a process with a positive value of ΔS ($\Delta S > 0$), this occurs when the entropy of the products is greater than the entropy of the reactants:

$$\Delta S > 0 \qquad S^\circ(\text{products}) - S^\circ(\text{reactants}) > 0 \qquad S^\circ(\text{products}) > S^\circ(\text{reactants})$$

You consider each choice and predict the sign of ΔS for each.

In choice **(A)** you have: Reactants (4 moles gas) $\rightarrow$ Products (2 moles gas)
You predict the sign of ΔS for this process to be (–) because the $\Delta n_{gas} = -2$ or a decrease in entropy.

In choice **(B)** you have: Reactants (2 moles solution) → Products (1 mole solid)
You predict the sign of ΔS for this process to be (–) because the transition from solution to solid results in a decrease in entropy.

In choice **(C)** you have: Reactants (1 mole solid) → Products (1 mole gas)
You predict the sign of ΔS for this process to be (+) because $\Delta n_{gas} = +1$ or an increase in entropy.

In choice **(D)** you have: Reactants (1 mole gas) → Products (1 mole liquid)
You predict the sign of ΔS for this process to be (–) because $\Delta n_{gas} = -1$ or a decrease in entropy.

The correct choice is **(C)**.

Practice Questions Related to This: **PQ-3, PQ-4, P-5, PQ-6, PQ-7,** and **PQ-8**

SQ-3. For this reaction at 25 °C, $\Delta H° = -1854\ \text{kJ·mol}^{-1}$ and $\Delta S° = -236\ \text{J·mol}^{-1}\text{·K}^{-1}$.

$$CH_3COCH_3(g) + 4O_2(g) \rightarrow 3CO_2(g) + 3H_2O(l)$$

What is the value of $\Delta G°$ for this reaction?

(A) $-1784\ \text{kJ·mol}^{-1}$ **(B)** $-1848\ \text{kJ·mol}^{-1}$ **(C)** $-1924\ \text{kJ·mol}^{-1}$ **(D)** $68500\ \text{kJ·mol}^{-1}$

Knowledge Required: (1) The calculation of free energy changes from enthalpy and entropy data. (2) The correct use of units.

Thinking it Through: You recognize that you are given a temperature and values for $\Delta H°$ and $\Delta S°$ and are being asked to calculate $\Delta G°$. You know to use the relationship:

$$\Delta G° = \Delta H° - T\Delta S°.$$

You also recall that the units of $\Delta H°$ are in kJ and the units of $\Delta S°$ are in J, so you must convert. The temperature you are given is in °C and you must use the value expressed in kelvin. You begin by doing the necessary conversions:

$T = 25 + 273 = 298\ \text{K}$ $\Delta S° = -236\ \text{J·mol}^{-1}\text{·K}^{-1} = -0.236\ \text{kJ·mol}^{-1}\text{·K}^{-1}$

Substituting these values into the expression for $\Delta G°$:

$\Delta G° = -1854\ \text{kJ} \cdot \text{mol}^{-1} - (298\ \text{K})(-0.236\ \text{kJ} \cdot \text{mol}^{-1} \cdot \text{K}^{-1}) = -1784\ \text{kJ} \cdot \text{mol}^{-1}$ which is choice **(A).**

Choice **(B)** is incorrect because the calculation was done without changing the temperature from °C to K. Choice **(C)** is incorrect because the calculation was done making a sign error in the second part of the expression. Choice **(D)** is incorrect because the calculation was done without converting the value of the $\Delta S°$ to kJ.

Practice Questions Related to This: **PQ-9** and **PQ-10**

SQ-4. *Conceptual*

When NH_4NO_3 dissolves in water, the temperature of the solution decreases. What describes the enthalpy and entropy changes of the system and which change drives the process?

(A) $\Delta H = (-)$ and $\Delta S = (-)$ and the process is driven by the enthalpy change.

(B) $\Delta H = (-)$ and $\Delta S = (+)$ and the process is driven by the enthalpy change.

(C) $\Delta H = (+)$ and $\Delta S = (-)$ and the process is driven by the entropy change.

(D) $\Delta H = (+)$ and $\Delta S = (+)$ and the process is driven by the entropy change.

Knowledge Required: (1) Definition of endothermic and exothermic processes. (2) Prediction of entropy change for a process. (3) The sign of ΔG for a spontaneous process.

Thinking it Through: You are asked to use an observation to determine the signs of the enthalpy and entropy change for the dissolving of NH_4NO_3 in water. The first thing you notice is the salt dissolves so the process is spontaneous. The sign of ΔG for a spontaneous process is negative.

You recall that endothermic processes take energy in from the surroundings, this results in a (+) sign for ΔH. Exothermic processes release energy into the surroundings and the sign of ΔH would be negative (–). The question tells you that the solution gets cold when the NH_4NO_3 dissolves in water. You recognize that this means the process is endothermic, $\Delta H > 0$ because the dissolution process is absorbing energy from the solution (part of the surroundings).

You also know that when the solid NH_4NO_3 dissolves in water the ions become hydrated:

$NH_4NO_3(s) \rightarrow NH_4^+(aq) + NO_3^-(aq)$ This results in an increase of entropy, $\Delta S > 0$

You recall that for a spontaneous process, the value of $\Delta G < 0$ and the relationship for ΔG is:

$$\Delta G = \Delta H - T\Delta S.$$

Putting these all together: $\Delta G < 0$, so $\Delta H - T\Delta S < 0$ and therefore, $\Delta H < T\Delta S$. Recall that, $\Delta H > 0$ and $\Delta S > 0$ and the process is driven by the entropy change.

The correct answer is choice **(D)**.

Choice **(A)** is incorrect because the signs of ΔH and ΔS are incorrect and the process is being driven by the entropy change. Choice **(B)** is incorrect because the sign of ΔH is incorrect and the process is being driven by the entropy change. Choice **(C)** is incorrect because the sign of ΔS is incorrect.

Practice Questions Related to This: **PQ- 11, PQ-12,** and **PQ-13**

SQ-5. What is the free energy change for the formation of one mole of ammonia from its elements under standard conditions?

$2NH_3(g) \rightarrow N_2(g) + 3H_2(g)$ $\Delta G° = 33.0\ kJ \cdot mol^{-1}$

(A) $-66.0\ kJ \cdot mol^{-1}$ **(B)** $-33.0\ kJ \cdot mol^{-1}$

(C) $-16.5\ kJ \cdot mol^{-1}$ **(D)** $66.0\ kJ \cdot mol^{-1}$

Knowledge Required: (1) How to manipulate thermochemical equations.

Thinking it Through: You are being asked to find the free energy of formation for a substance from a given reaction. You recall that the definition of a formation reaction: A reaction that produces one mole of the substance from its elements in their standard states. You write the formation reaction for ammonia:

$\frac{1}{2} N_2(g) + \frac{3}{2}H_2(g) \rightarrow NH_3(g)$

You now go back to the equation you were given in the question:

$2NH_3(g) \rightarrow N_2(g) + 3H_2(g)$ $\Delta G° = 33.0\ kJ \cdot mol^{-1}$

You reverse the reaction, remembering to change the sign of $\Delta G°$:

$N_2(g) + 3H_2(g) \rightarrow 2NH_3(g)$ $\Delta G° = -33.0\ kJ \cdot mol^{-1}$

You recognize that this reaction produces ammonia from its elements, but it produces 2 moles of ammonia. To get the formation reaction, you need to multiply the reaction by ½:

$\frac{1}{2} N_2(g) + \frac{3}{2}H_2(g) \rightarrow NH_3(g)$ $\Delta G° = \frac{1}{2}(-33.0\ kJ \cdot mol^{-1}) = -16.5\ kJ \cdot mol^{-1}$

The correct choice is **(C)**.

Choice **(A)** is incorrect because to get this answer the reaction was multiplied by 2 instead of by ½. Choice **(B)** is incorrect because to get this answer the reaction has not been multiplied by ½ . Choice **(D)** is incorrect because to get this answer the sign on ΔG was not inverted and was multiplied by 2 instead of ½.

Practice Questions Related to This: **PQ-14** and **PQ-15**

SQ-6. Calculate the standard Gibbs free energy change at 298 K for the reaction:

$Fe_3O_4(s) + 4CO(g) \rightleftharpoons 3Fe(s) + 4CO_2(g)$

ΔG°_f / kJ·mol^{-1}	
CO(g)	–137.2
CO_2(g)	–394.4
Fe_3O_4(s)	–1014

(A) –757 kJ·mol^{-1} **(B)** –14.8 kJ·mol^{-1}

(C) 14.8 kJ·mol^{-1} **(D)** 757 kJ·mol^{-1}

Knowledge Required: (1) Hess's law. (2) Rules for manipulating free energy values.

Thinking it Through: You recall that Hess's law allows you to calculate the Δ*G* of a reaction using tabulated free energies of formation values. This works because we can imagine any reaction as consisting of first converting the reactants back into their elements and then forming the products from these elements. This is, in effect reversing the formation reactions of the reactants and then using the formation reactions of the products. You recall that this process is represented in the relationship:

$$\Delta G^\circ_{rxn} = \sum n\Delta G^\circ_f(\text{products}) - \sum m\Delta G^\circ_f(\text{reactants})$$

You recall that the symbol ∑ means to sum and *n* and *m* are the stoichiometric coefficients of the reactants and products. You also notice that there is no standard free energy of formation for Fe(s) given, this is because you know the standard free energy of formation of an element in its standard state is zero. The value of ΔG°_{rxn} can be calculated as:

$$\Delta G^\circ_{rxn} = \left[4\Delta G^\circ_f(CO_2(g)) + 3\Delta G^\circ_f(Fe(s))\right] - \left[4\Delta G^\circ_f(CO(g)) + \Delta G^\circ_f(Fe_3O_4(s))\right]$$

$$\Delta G^\circ_{rxn} = \left[4\text{ mol }CO_2\left(\frac{-394\text{ kJ}}{\text{mol }CO_2}\right) + 3\text{ mol Fe}\left(\frac{0\text{ kJ}}{\text{mol Fe}}\right)\right] - \left[4\text{ mol CO}\left(\frac{-137.2\text{ kJ}}{\text{mol CO}}\right) + 1\text{ mol }Fe_3O_4\left(\frac{-1014\text{ kJ}}{\text{mol }Fe_3O_4}\right)\right]$$

$$\Delta G^\circ_{rxn} = -14.8\text{ kJ}\cdot\text{mol}^{-1}$$

Choice **(B)** is the correct choice.

Choice **(A)** is incorrect because it did not use the coefficients in the balanced equation and used reactants minus products. Choice **(C)** is incorrect because it calculated the answer using reactants minus products. Choice **(D)** is incorrect because it did not use the coefficients in the balanced equation.

Practice Questions Related to This: **PQ-16, PQ-17, PQ-18,** and **PQ-19**

SQ-7. An equilibrium mixture of I_2(g), Cl_2(g), and ICl(g) at 298 K has partial pressures of 0.0100 atm, 0.0100 atm, and 0.0900 atm, respectively. What is Δ*G*° at 298 K for the reaction:

$I_2(g) + Cl_2(g) \rightleftharpoons 2ICl(g)$

(A) –16.9 kJ·mol^{-1} **(B)** –10.9 kJ·mol^{-1}

(C) 10.9 kJ·mol^{-1} **(D)** 16.9 kJ·mol^{-1}

Knowledge Required: (1) Relationship between equilibrium and free energy. (2) How to write an equilibrium expression for a reaction. (3) How to calculate the value of *K*.

Thinking it Through: You are given the equilibrium pressures of a system at equilibrium and need to calculate the value of Δ*G*°. You recall the relationship between free energy and equilibrium,

$$\Delta G = \Delta G^\circ + RT\ln Q$$

where *Q* is the reaction quotient. At equilibrium the value of Δ*G* is zero and *Q* = *K*

The relationship becomes: $\Delta G^\circ = -RT\ln K$

You know you can calculate the value of *K* from the given information using the equilibrium expression:

$$I_2(g) + Cl_2(g) \rightleftharpoons 2ICl(g) \qquad K = \frac{(P_{ICl})^2}{(P_{Cl_2})(P_{I_2})} = \frac{(0.0900)^2}{(0.0100)(0.0100)} = 81$$

$$\Delta G° = -RT\ln K = -(8.314\ J{\cdot}mol^{-1}{\cdot}K^{-1})(298K)\ \ln(81) = -10851\ J{\cdot}mol^{-1}$$

Converting to kJ this value becomes $-10.9\ kJ{\cdot}mol^{-1}$. This is choice **(B)**. You also know you can check your answer to see if it is reasonable because you have more products than reactants, so you predict a larger value of K ($K > 1$) and this would then be considered a spontaneous process, so you would expect a negative free energy ($\Delta G < 0$) as you calculated.

Choice **(A)** is not correct because the value of K was calculated without the exponent for ICl.

Choice **(C)** is not correct because the negative sign was dropped and the value of K was calculated without the exponent for ICl.

Choice **(D)** is not correct because the negative sign was dropped.

Practice Questions Related to This: **PQ-20, PQ-21, PQ-22, PQ-23,** and **PQ-24**

SQ-8. Conceptual

Which statement is correct for the reaction below under standard conditions?

$$2Cl_2(g) + 2NO(g) \rightarrow N_2(g) + 2Cl_2O(g) \qquad \Delta H° = -22.0\ kJ{\cdot}mol^{-1}$$

(A) The reaction is spontaneous at all temperatures.

(B) The reaction is spontaneous only at low temperatures.

(C) The reaction is spontaneous only at high temperatures.

(D) The reaction is not spontaneous at any temperature.

Knowledge Required: (1) Relationship between ΔG, ΔH, ΔS, and T. (2) Definition of spontaneous.

Thinking it Through: You are being asked to predict the spontaneity of a chemical reaction. You notice that you are only given a value of $\Delta H°$ and not for $\Delta S°$. You also remember that a reaction will be spontaneous under standard conditions when $\Delta G°$ is negative. You also recall the relationship for $\Delta G°$, $\Delta H°$, $\Delta S°$, and T.

$$\Delta G° = \Delta H° - T\Delta S°.$$

The sign of $\Delta G°$ depends on the signs and magnitudes of $\Delta H°$ and $\Delta S°$ and the value of the temperature. These relationships are summarized below:

$\Delta H°$	$\Delta S°$	$-T\Delta S°$	$\Delta G°$	Result
–	+	–	–	Spontaneous at all temperatures
+	–	+	+	Nonspontaneous at all temperatures
–	–	+	+ or –	Spontaneous at low temperature/nonspontaneous at high temperature
+	+	–	+ or –	Spontaneous at high temperature/nonspontaneous at low temperature

You know the sign of $\Delta H°$ is negative. You need to determine the sign of $\Delta S°$. Recalling what you know about molar entropy values, you notice that the reaction produces 3 moles of gas from 4 moles of gas. You predict that the reaction will have a negative $\Delta S°$.

The table tells you that this type of reaction will be spontaneous at low temperature. This is choice **(B)** and the other choices are incorrect.

Practice Questions Related to This: **PQ-25, PQ-26,** and **PQ-27**

SQ-9. Conceptual

For a reaction where $K > 1$ at all temperatures, which statement(s) must be true?

I.	$\Delta H°$ is negative
II.	$\Delta S°$ is positive
III.	$\Delta G°$ is positive

(A) Only **I** **(B)** Only **III** **(C)** Both **I** and **II** **(D)** Both **I** and **III**

Knowledge Required: (1) Relationship between ΔG, ΔH, ΔS, and T. (2) Relationship between ΔG and K.

Thinking it Through: You are told that the reaction has a value of K greater than 1 at all temperatures. The relationship between $\Delta G°$ and K is:

$$\Delta G^\circ = -RT \ln K \quad \text{or} \quad K = e^{\left(\frac{-\Delta G^\circ}{RT}\right)}$$

When $\Delta G°$ is a negative number the value of K is always greater than 1, and when $\Delta G°$ is positive, the value K is less than one. Because the value of $K > 1$ at all temperatures, the value of $\Delta G°$ must be negative. This can happen at all temperatures when $\Delta H°$ is negative AND $\Delta S°$ is positive. See the table in **SQ-8** for all the possible combinations. The correct choice is **(C).**

Choice **(A)** is incorrect because a negative $\Delta H°$ is not sufficient to make a reaction spontaneous at all temperatures. Choice **(B)** and **(D)** are not correct because if $\Delta G°$ were positive, K would be less than one at all temperatures.

Practice Questions Related to This: **PQ-28, PQ-29,** and **PQ-30**

Practice Questions (PQ)

Conceptual **PQ-1.** Which gas has the largest molar entropy at 298 K and 1 atm?

(A) Ar **(B)** C_3H_8 **(C)** CO_2 **(D)** HCl

Conceptual **PQ-2.** Which best explains why the entropy of a sample of water increases as the temperature changes from 25 °C to 75 °C?

(A) There is more hydrogen bonding at the higher temperature.

(B) More possible energy states are available at the higher temperature.

(C) There are fewer possible ways to distribute energy at the higher temperature.

(D) More water molecules can be broken into hydrogen and oxygen at the higher temperature.

Conceptual **PQ-3.** In which reaction will the entropy of the system increase?

(A) $N_2(g) + 3H_2(g) \rightarrow 2NH_3(g)$

(B) $2Mg(s) + O_2(g) \rightarrow 2MgO(s)$

(C) $CaCO_3(s) \rightarrow CaO(s) + CO_2(g)$

(D) $H_2O(l) + CO_2(g) \rightarrow H_2CO_3(aq)$

Conceptual **PQ-4.** Which process is associated with a negative entropy change ($\Delta S°_{rxn} < 0$)?

(A) $NaCl(s) \rightarrow Na^+(aq) + Cl^-(aq)$

(B) $N_2(g) + 3H_2(g) \rightarrow 2NH_3(g)$

(C) $2BrCl(l) \rightarrow Br_2(l) + Cl_2(g)$

(D) $H_2O(l) \rightarrow H_2O(g)$

Conceptual **PQ-5.** For which of these processes is the value of ΔS expected to be negative?

I. Sugar is dissolved in water.
II. Steam is condensed.
III. $CaCO_3$ is decomposed in to CaO and CO_2.

(A) **I** only **(B)** **I** and **III** only **(C)** **II** only **(D)** **II** and **III** only

Conceptual **PQ-6.** For which process is the entropy change per mole the largest at constant temperature?

(A) $H_2O(l) \rightarrow H_2O(g)$

(B) $H_2O(s) \rightarrow H_2O(g)$

(C) $H_2O(s) \rightarrow H_2O(l)$

(D) $H_2O(l) \rightarrow H_2O(s)$

Conceptual **PQ-7.** In which process is entropy *decreased*?

(A) dissolving sugar in water **(B)** expanding a gas

(C) evaporating a liquid **(D)** freezing water

Conceptual **PQ-8.** Which figure represents a process with a positive entropy change?

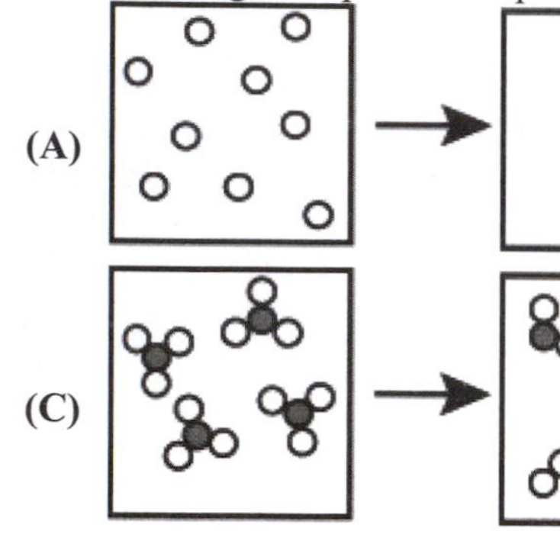

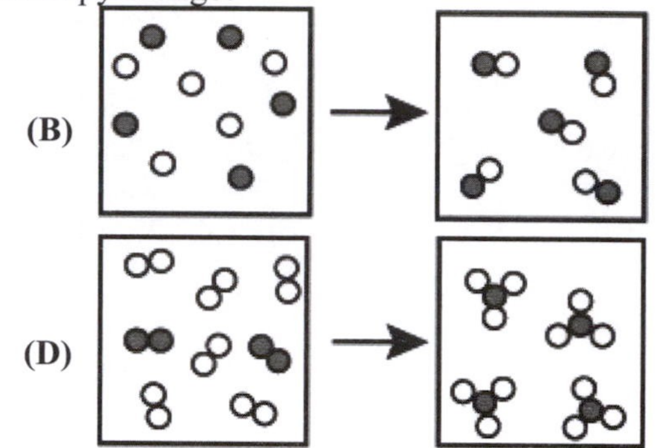

PQ-9. For the reaction: $2PCl_3(l) + O_2(g) \rightarrow 2OPCl_3(g)$

$\Delta H° = -508\ kJ{\cdot}mol^{-1}$ and $\Delta S° = -178\ J{\cdot}mol^{-1}{\cdot}K^{-1}$ at 298 K.

What is $\Delta G°$ for this reaction?

(A) $-561\ kJ{\cdot}mol^{-1}$ **(B)** $-504\ kJ{\cdot}mol^{-1}$ **(C)** $-455\ kJ{\cdot}mol^{-1}$ **(D)** $52{,}500\ kJ{\cdot}mol^{-1}$

PQ-10. What is $\Delta G°_{rxn}$ for $2CO(g) + O_2(g) \rightarrow 2CO_2(g)$ if $\Delta H°_{rxn} = -566\ kJ{\cdot}mol^{-1}$ and $\Delta S°_{rxn} = -173\ J{\cdot}mol^{-1}{\cdot}K^{-1}$ at 298 K.

(A) $-319\ kJ{\cdot}mol^{-1}$ **(B)** $-514\ kJ{\cdot}mol^{-1}$ **(C)** $-739\ kJ{\cdot}mol^{-1}$ **(D)** $-5099\ kJ{\cdot}mol^{-1}$

Conceptual **PQ-11.** What are the signs of the entropy change and enthalpy change for the combustion reaction of magnesium? $2Mg(s) + O_2(g) \rightarrow 2MgO(s)$

(A) $\Delta H > 0, \Delta S > 0$ **(B)** $\Delta H > 0, \Delta S < 0$ **(C)** $\Delta H < 0, \Delta S > 0$ **(D)** $\Delta H < 0, \Delta S < 0$

Conceptual **PQ-12.** Which of the relationships are true about water boiling in a container that is open to the atmosphere?

(A) $\Delta H > 0, \Delta S > 0$ **(B)** $\Delta H > 0, \Delta S < 0$ **(C)** $\Delta H < 0, \Delta S > 0$ **(D)** $\Delta H < 0, \Delta S < 0$

Conceptual **PQ-13.** What are the signs of ΔH and ΔS for the deposition of Zn?

	ΔH	ΔS
(A)	positive	positive
(B)	positive	negative
(C)	negative	positive
(D)	negative	negative

PQ-14. Given:

$C_3H_8(g) + 5O_2(g) \rightarrow 3CO_2(g) + 4H_2O(g)$ $\Delta G°_{rxn} = -2074\ kJ{\cdot}mol^{-1}$

What is the value of $\Delta G°_{rxn}$ for the reaction:

$6CO_2(g) + 8H_2O(g) \rightarrow 2C_3H_8(g) + 10O_2(g)$

(A) $-1037\ kJ{\cdot}mol^{-1}$ **(B)** $1037\ kJ{\cdot}mol^{-1}$ **(C)** $2074\ kJ{\cdot}mol^{-1}$ **(D)** $4148\ kJ{\cdot}mol^{-1}$

PQ-15. What is $\Delta G°$ for the reaction $2N_2O(g) + 3O_2(g) \rightarrow 4NO_2(g)$?

$4NO(g) \rightarrow 2N_2O(g) + O_2(g)$	$\Delta G° = -139.56\ kJ \cdot mol^{-1}$
$2NO(g) + O_2(g) \rightarrow 2NO_2(g)$	$\Delta G° = -69.70\ kJ \cdot mol^{-1}$

(A) $0.16\ kJ \cdot mol^{-1}$ **(B)** $2.09\ kJ \cdot mol^{-1}$ **(C)** $69.86\ kJ \cdot mol^{-1}$ **(D)** $-209.26\ kJ \cdot mol^{-1}$

PQ-16. Which has a value of zero at 25 °C?

(A) $\Delta G°_f$ of $O(g)$ **(B)** $\Delta G°_f$ of $O_2(g)$ **(C)** $\Delta G°_f$ of $O_3(g)$ **(D)** all of these

PQ-17. What is the value of $\Delta G°_{rxn}$ for the reaction below?

$$4PCl_3(g) \rightarrow 6Cl_2(g) + P_4(g)$$

	$\Delta G°_f$ / $kJ \cdot mol^{-1}$
$PCl_3(g)$	–269.6
$P_4(g)$	24.1

(A) $-1103\ kJ \cdot mol^{-1}$ **(B)** $-1054\ kJ \cdot mol^{-1}$ **(C)** $-321\ kJ \cdot mol^{-1}$ **(D)** $1103\ kJ \cdot mol^{-1}$

PQ-18. What is the value $\Delta G°$ for this reaction?

$$CO_2(g) + 2H_2O(g) \rightarrow CH_4(g) + 2O_2(g)$$

	$\Delta G°_f$ / $kJ \cdot mol^{-1}$
$CO_2(g)$	–394
$H_2O(g)$	–229
$CH_4(g)$	–50.8

(A) $-801\ kJ \cdot mol^{-1}$ **(B)** $-118\ kJ \cdot mol^{-1}$ **(C)** $572\ kJ \cdot mol^{-1}$ **(D)** $801\ kJ \cdot mol^{-1}$

Conceptual **PQ-19.** What is always true for a spontaneous process at constant pressure?

(A) $\Delta G_{system} < 0$ **(B)** $\Delta S_{system} > 0$

(C) $\Delta S_{system} > 0$ and $\Delta G_{system} > 0$ **(D)** $\Delta S_{system} > 0$ and $\Delta G_{system} = 0$

PQ-20. Consider the ionization of nitrous acid at 25 °C:

$$HNO_2(aq) + H_2O(l) \rightleftharpoons H_3O^+(aq) + NO_2^-(aq) \qquad \Delta G° = 19.1\ kJ \cdot mol^{-1}$$

What is the value of K for this equilibrium?

(A) 2.0×10^{-7} **(B)** 4.5×10^{-4} **(C)** 2.2 **(D)** 2.2×10^{3}

PQ-21. The ionization of aqueous hydrofluoric acid has an equilibrium constant 6.31×10^{-4} at 25 °C. What is the value of $\Delta G°_{rxn}$ for this reaction?

(A) $0.665\ kJ \cdot mol^{-1}$ [1] **(B)** $1.53\ kJ \cdot mol^{-1}$ **(C)** $7.93\ kJ \cdot mol^{-1}$ **(D)** $18.3\ kJ \cdot mol^{-1}$

Conceptual **PQ-22.** For the generic reaction shown, $K = 0.355$ (at 950 K). What can you conclude about ΔG and $\Delta G°$ in a reaction mixture for which $Q = 0.205$ at 950 K?

$$2AZ_2(g) + Z_2(g) \rightleftharpoons 2AZ_3(g)$$

(A) ΔG and $\Delta G°$ are both negative **(B)** ΔG and $\Delta G°$ are both positive

(C) ΔG is positive and $\Delta G°$ is negative **(D)** ΔG is negative and $\Delta G°$ is positive

PQ-23. What is the value of the equilibrium constant K for a reaction for which $\Delta G°$ is equal to $5.20\ kJ \cdot mol^{-1}$ at 50 °C?

(A) 0.144 **(B)** 0.287 **(C)** 6.93 **(D)** 86.4

PQ-24. For the reaction

$$NH_4Cl(s) \rightleftharpoons NH_3(g) + HCl(g)$$

$\Delta H° = 176\ kJ \cdot mol^{-1}$ and $\Delta G° = 91.2\ kJ \cdot mol^{-1}$ at 298 K. What is the value of ΔG at 1000 K?

(A) $-109\ kJ \cdot mol^{-1}$ **(B)** $-64\ kJ \cdot mol^{-1}$ **(C)** $64\ kJ \cdot mol^{-1}$ **(D)** $109\ kJ \cdot mol^{-1}$

Conceptual **PQ-25.** The reaction is ______

$Fe_2O_3(s) + 3C(s) \rightarrow 2Fe(s) + 3CO(g)$ $\Delta H° = +490.7\ kJ \cdot mol^{-1}$ $\Delta S° = +541\ J \cdot mol^{-1} \cdot K^{-1}$

(A) never spontaneous.

(B) always spontaneous.

(C) spontaneous only below a certain temperature.

(D) spontaneous only above a certain temperature.

Conceptual **PQ-26.** A spontaneous reaction has a negative ΔS. What must be true about the reaction?

(A) The reaction is exothermic and the temperature is sufficiently low.

(B) The reaction is exothermic and the temperature is sufficiently high.

(C) The reaction is endothermic and the temperature is sufficiently low.

(D) The reaction is endothermic and the temperature is sufficiently high.

Conceptual **PQ-27.** For the process $O_2(g) \rightarrow 2O(g)$, $\Delta H° = 498\ kJ \cdot mol^{-1}$. What would be the predicted sign of $\Delta S°_{rxn}$ and the conditions under which this reaction would be spontaneous?

	$\Delta S°_{rxn}$	Spontaneous
(A)	positive	at low temperatures only
(B)	positive	at high temperatures only
(C)	negative	at high temperatures only
(D)	negative	at low temperatures only

Conceptual **PQ-28.** Which acid has a more positive value of $\Delta G°$ for its ionization in water, and which acid is stronger?

Acid	K_a
HNO_2	3×10^{-4}
HCN	5×10^{-10}

	more positive $\Delta G°$	stronger acid
(A)	HNO_2	HCN
(B)	HNO_2	HNO_2
(C)	HCN	HCN
(D)	HCN	HNO_2

Conceptual **PQ-29.** Given that the $\Delta G°$ value for a particular reaction, $A \rightleftharpoons B$, is negative, what can be said about the relative portions of reactants and products when the reaction reaches equilibrium?

(A) There are fewer reactant molecules than product molecules.

(B) There are equal numbers of reactant and product molecules.

(C) There are more reactant molecules than product molecules.

(D) A relationship cannot be drawn without calculating K_c and $\Delta G°$.

Conceptual **PQ-30.** When the reaction $A \rightleftharpoons B$ reaches equilibrium, which quantity equals zero?

(A) The concentrations of A and B.

(B) The enthalpy change, ΔH.

(C) The equilibrium constant, K_{eq}.

(D) The free energy change ΔG.

Answers to Study Questions

1. A
2. C
3. A
4. D
5. C
6. B
7. B
8. B
9. C

Answers to Practice Questions

1. B
2. B
3. C
4. B
5. C
6. B
7. D
8. C
9. C
10. B
11. D
12. A
13. D
14. D
15. A
16. B
17. D
18. D
19. A
20. B
21. D
22. D
23. A
24. A
25. D
26. A
27. B
28. D
29. A
30. D

A C S
Exams

Chapter 15 – Electrochemistry

Chapter Summary:

This chapter will focus on concepts and reactions that involve the transfer of electrons.

Specific topics covered in this chapter are:

- Galvanic cells and cell diagrams
- Calculating cell potentials for galvanic reactions
- Relating values of E°_{cell} and ΔG°, and K_{eq}
- Using the Nernst equation
- Calculating quantities in electrolytic cells
- Calculations using concentration cells

Previous material that is relevant to your understanding of questions in this chapter include:

- Significant figures (***Toolbox***)
- Scientific notation (***Toolbox***)
- Balancing redox reactions ***(Chapter 5)***
- Definitions of anode, cathode, reduction, and oxidation ***(Chapter 5)***
- Equilibrium ***(Chapter 11)***
- Thermodynamics ***(Chapter 14)***

Common representations used in questions related to this material:

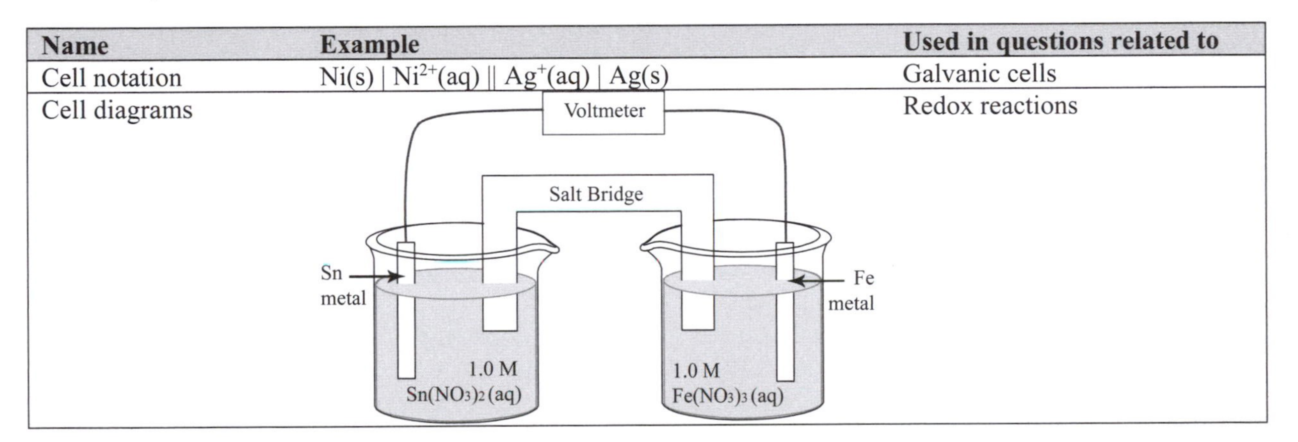

Name	Example	Used in questions related to
Cell notation	$Ni(s) \mid Ni^{2+}(aq) \parallel Ag^{+}(aq) \mid Ag(s)$	Galvanic cells
Cell diagrams		Redox reactions

Where to find this in your textbook:

The material in this chapter typically aligns to "Electrochemistry" in your textbook. The name of your chapter may vary.

Practice exam:

There are practice exam questions aligned to the material in this chapter. Because there are a limited number of questions on the practice exam, a review of the breadth of the material in this chapter is advised in preparation for your exam.

How this fits into the big picture:

The material in this chapter aligns to the Big Idea of Energy and Thermodynamics (6) and Equilibrium (8) as listed on page 12 of this study guide.

ACS Exams

Study Questions (SQ)

SQ-1. *Conceptual*

What is the coefficient for the nitrite ion (NO_2^-) when the redox reaction is balanced in basic solution? $MnO_4^-(aq) + NO_2^-(aq) \rightarrow MnO_2(s) + NO_3^-(aq)$

(A) 1 (B) 2 (C) 3 (D) 4

Knowledge Required: (1) Ability to write half-reactions. (2) Rules for balancing redox equations.

Thinking it Through: You are given a redox reaction and are asked to determine the coefficient of a species when the equation is balanced in basic solution. You recall that the rules for balancing in acidic and basic solution are the same until the second to last step (where you balance charge – using H^+ for acidic solutions and OH^- for basic solutions). The first step is to separate the reaction into two half-reactions and proceed as shown in the table for each half reaction:

Step	Rxn 1	Rxn 2
Assign oxidation states (shown above):	$\overset{+3}{N}O_2^-(aq) \rightarrow \overset{+5}{N}O_3^-(aq)$	$\overset{+7}{Mn}O_4^-(aq) \rightarrow \overset{+4}{Mn}O_2(s)$
Assign as ox/red	Oxidation *(going from +3 to +5 is a loss of electrons)*	reduction *(going from +7 to +4 is a gain of electrons)*
Balance element undergoing ox/red	$\overset{+3}{N}O_2^-(aq) \rightarrow \overset{+5}{N}O_3^-(aq)$ *N already balanced*	$\overset{+7}{Mn}O_4^-(aq) \rightarrow \overset{+4}{Mn}O_2(s)$ *Mn already balanced*
Add e⁻/balance ox state with electrons	$\overset{+3}{N}O_2^-(aq) \rightarrow \overset{+5}{N}O_3^-(aq) + 2e^-$	$\overset{+7}{Mn}O_4^-(aq) + 3e^- \rightarrow \overset{+4}{Mn}O_2(s)$
Balance charge with OH^- (charge below)*	$\underset{-2}{2OH^-}(aq) + \underset{-1}{NO_2^-}(aq) \rightarrow \underset{-1}{NO_3^-}(aq) + \underset{-2}{2e^-}$	$\underset{-1}{MnO_4^-}(aq) + \underset{-3}{3e^-} \rightarrow \underset{0}{MnO_2}(aq) + \underset{-4}{4OH^-}(aq)$
Balance H with H_2O	$2OH^-(aq) + NO_2^-(aq) \rightarrow NO_3^-(aq) + 2e^- + H_2O(l)$	$2H_2O(l) + MnO_4^-(aq) + 3e^- \rightarrow MnO_2(aq) + 4OH^-(aq)$

**As a reminder, you know you would use H^+ if this was in acidic solution rather than basic solution.*

The next step is to multiple the half-reactions so that the number of electrons produced equals the number of electrons used. The overall equation is then:

Ox: $[2OH^-(aq) + NO_2^-(aq) \rightarrow NO_3^-(aq) + 2e^- + H_2O(l)] \times 3$

Red: $[2H_2O(l) + MnO_4^-(aq) + 3e^- \rightarrow MnO_2(aq) + 4OH^-(aq)] \times 2$

Overall: $3NO_2^-(aq) + 2MnO_4^-(aq) + H_2O(l) \rightarrow 3NO_3^-(aq) + 2MnO_2(aq) + 2OH^-(aq)$

Where the electrons are balanced and canceled by multiplying each (or both as here) half reaction by a factor. Finally, you canceled the extra water molecules (from the right or products side) and the extra hydroxide ions (from the left or reactants side). You can now find the coefficient for the nitrite ion (listed first) which is 3.

Choice **(C)** is the correct answer.

Choice **(A)** is incorrect because it is the result of not multiplying the half-reactions to make the number of electrons cancel.

Choice **(B)** and **(D)** are incorrect because they are the result of incorrectly multiplying the half-reactions.

Practice Questions Related to This: **PQ-1, PQ-2, PQ-3**

SQ-2 A spontaneous electrochemical cell is set up as shown. Which statement is true?

$Sn^{2+} + 2e^- \rightarrow Sn$	$E° = –0.136$ V
$Fe^{3+} + 3e^- \rightarrow Fe$	$E° = –0.036$ V

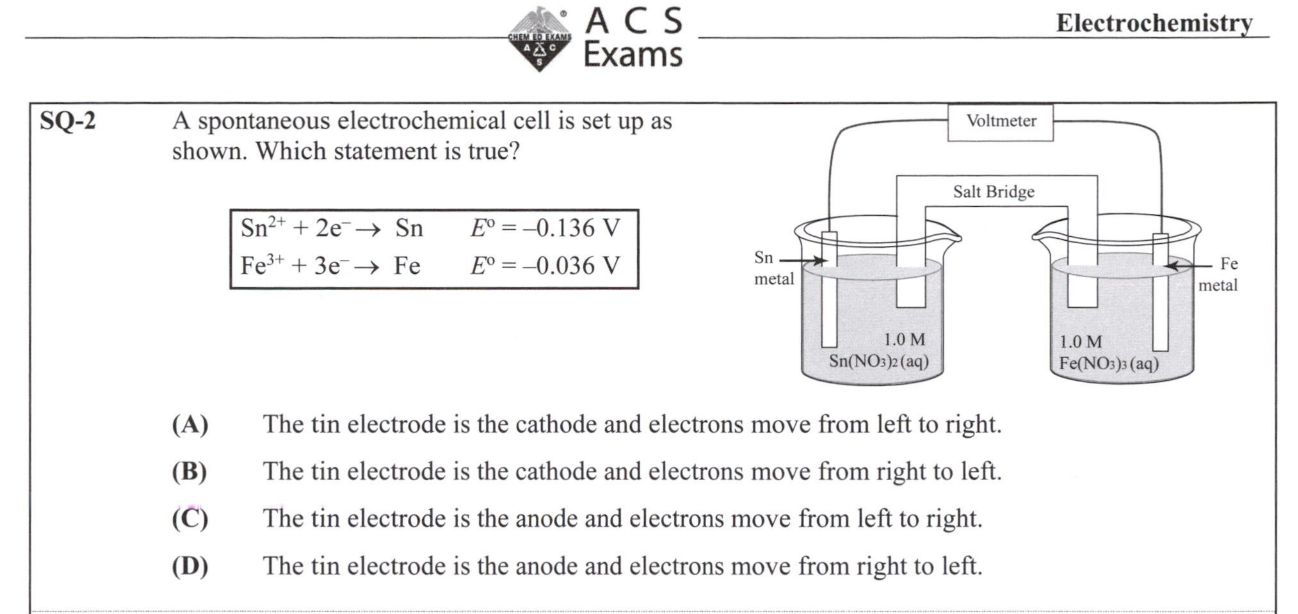

(A) The tin electrode is the cathode and electrons move from left to right.

(B) The tin electrode is the cathode and electrons move from right to left.

(C) The tin electrode is the anode and electrons move from left to right.

(D) The tin electrode is the anode and electrons move from right to left.

Knowledge Required: (1) Definition of spontaneous reaction. (2) Definition of anode and cathode.

Thinking it Through: You are being asked to evaluate some statements about a spontaneous electrochemical cell. You begin by using the given cell potentials to determine the spontaneous reaction. Recall that for a reaction to be spontaneous the value of $E°_{cell}$ is positive because reactions with positive cell potentials have negative free energy changes, and are spontaneous. To get the spontaneous reaction you need to reverse the reaction involving Sn. You also have to multiply the half-reactions so that electrons cancel (here 6 electrons are being transferred as written). Remember that the oxidation half-reaction provides the electrons for the reduction half-reaction. You also remember to **NOT** multiply the values of $E°$ since these cell potentials are *intensive* properties and so their value does not depend on the number of times the half-reaction occurs.

$Sn \rightarrow Sn^{2+} + 2e^-$	$E° = 0.136$ V	oxidation
$Fe^{3+} + 3e^- \rightarrow Fe$	$E° = –0.036$ V	reduction

$3(Sn \rightarrow Sn^{2+} + 2e^-)$	$E° = 0.136$ V
$2(Fe^{3+} + 3e^- \rightarrow Fe)$	$E° = –0.036$ V
$3Sn + 2Fe^{3+} \rightarrow 3Sn^{2+} + 2Fe$	$E° = 0.100$ V

You also know that the oxidation reaction occurs at the anode and the reduction occurs at the cathode. For the spontaneous reaction, the anode is the Sn and the Fe is the cathode. You know that electrons will move from where they are produced (the oxidation half-reaction or the reaction at anode) to where they are consumed (the reduction half-reaction or the reaction at the cathode). The source of electrons for the reduction is the anode. Therefore, the correct choice is **(C)**.

Choices **(A)** and **(B)** are not correct because tin is the anode.

Choice **(D)** is not correct because the electrons move from the anode to the cathode.

Practice Questions Related to This: **PQ-4, PQ-5, PQ-6**

SQ-3. Given the standard reduction potentials at 25 °C, what is the standard cell potential of a $Ni(s) \mid Ni^{2+}(aq) \parallel Ag^+(aq) \mid Ag(s)$ galvanic cell?

Half Reaction	$E°$
$Ag^+(aq) + e^- \rightarrow Ag(s)$	0.79 V
$Ni^{2+}(aq) + 2e^- \rightarrow Ni(s)$	–0.23 V

(A) 0.56 V **(B)** 1.02 V **(C)** 1.35 V **(D)** 1.81 V

Knowledge Required: (1) Ability to interpret electrochemical cell notation. (2) Definition of a galvanic cell.

Thinking it Through: You are told the cell is galvanic and you remember that galvanic cells are spontaneous reactions. For a reaction to be spontaneous under standard conditions the value of $\Delta G°$ must be negative. You also recall that when $\Delta G°$ is negative, then $E°_{cell}$ positive. You are given the reaction in electrochemical cell notation. You know that in this notation, the first pair represents the reaction at the anode (oxidation) and the second pair is the reaction at the cathode (reduction). The double lines ("||") represent the salt bridge.

oxidation half-reaction||reduction half-reaction
oxidation reactant | oxidation product || reduction reactant | reduction product

Converting the given notation $Ni(s) \mid Ni^{2+}(aq) \parallel Ag^{+}(aq) \mid Ag(s)$ to half-reactions you write:

$Ni(s) \rightarrow Ni^{2+}(aq) + 2e^-$ $E° = 0.23$ V (***Note*** you changed the sign of $E°$ from the reduction in the table)
$Ag^+(aq) + e^- \rightarrow Ag(s)$ $E° = 0.79$ V

Multiplying the Ag^+ half-reaction by 2 to make the electrons cancel, and remembering to **NOT** multiply the values of the cell potentials, since they are intensive properties, you get.

$Ni(s) + 2Ag^+(aq) \rightarrow 2Ag(s) + Ni^{2+}(aq)$ $E° = 1.02$ V which is choice **(B)**.

Choice **(A)** is not correct because the half-reaction for Ni was not reversed. Choice **(C)** is not correct because the half-reaction reaction for silver was multiplied by 2 and the Ni reaction was not reversed. Choice **(D)** is not correct because the half-reaction reaction for silver was multiplied by 2 and the Ni reaction was reversed.

***Practice Questions Related to This:* PQ-7, PQ-8, PQ-9, PQ-10, PQ-11**

SQ-4. An oxidation-reduction reaction in which 3 electrons are transferred has a $\Delta G° = 18.55\ kJ \cdot mol^{-1}$ at 25°C. What is the value of $E°$?

(A) 0.192 V **(B)** –0.064 V **(C)** –0.192 V **(D)** –0.577 V

Knowledge Required: (1) The relationship of free energy to cell potential.

Thinking it Through: You are given a value of $\Delta G°$ and are asked for the cell potential. You recall the relationship between $\Delta G°$ and $E°$: $\Delta G° = -nFE°$

[**Note**: It isn't needed in this problem but the value of $E°$ is also related to the equilibrium constant.]

$$\Delta G° = -RT\ln K \quad \text{so} \quad -RT\ln K = -nFE°$$

$$E° = (RT/nF)\ln K$$

where n is the number of electrons transferred and F is the Faraday constant. *Note:* The value of the Faraday constant will typically be given on the reference page of the exam with other constants.

Solving for $E°$ you get: $E° = -\Delta G°/nF$

Substituting values results in:

$$E^{o} = \frac{-(18.55\text{ kJ})\left(\frac{1000\text{ J}}{1\text{ kJ}}\right)}{(3\text{ mol electrons})\left(\frac{96485\text{ C}}{1\text{ mol electrons}}\right)} = -0.0642\text{ J}\cdot\text{C}^{-1} = -0.0642\text{ V}$$

Where you have used the definition $1\ V = 1\ J \cdot C^{-1}$. The correct choice is **(B)**.
Choice **(A)** is not correct because it omitted the number of electrons and dropped the negative sign.
Choice **(C)** is not correct because it omitted the number of electrons.
Choice **(D)** is not correct because it multiplied by the number of electrons rather than dividing.

***Practice Questions Related to This:* PQ-12, PQ-13**

SQ-5. Conceptual

Consider the reaction.

$Cu^{2+}(aq) + Fe(s) \rightarrow Cu(s) + Fe^{2+}(aq)$ $E° = 0.78$ V

What is the value of E when $[Cu^{2+}]$ is equal to 0.040 M and $[Fe^{2+}]$ is equal to 0.40 M?

(A) 0.72 V **(B)** 0.75 V **(C)** 0.81 V **(D)** 0.84 V

Knowledge Required: (1) Use the Nernst equation to calculate cell potential under nonstandard conditions.

Thinking it Through: You recall the relationship between E and $E°$ is given by the Nernst equation. [**Note**: the equation will be provided either in the problem or on the reference sheet.] There are two common versions of the Nernst equation:

$$E = E^{\circ} - \frac{RT}{nF}\ln Q \qquad E = E^{\circ} - \frac{0.0592}{n}\log Q$$

The 0.0592 comes from using the values for R, F, and 25 °C (298 K) and the conversion between the natural and base-10 log. You then use the reaction to write an expression for Q; recall from chapter 11 that Q is an expression written exactly like an equilibrium expression, but the values substituted into the expression are not necessarily equilibrium values. Once the expression is written and the appropriate values are used you calculate Q.

$$Q = \frac{\text{Products}}{\text{Reactants}} = \frac{\left[Fe^{2+}\right]}{\left[Cu^{2+}\right]} = \frac{0.40\text{ M}}{0.040\text{ M}} = 10$$

Now, using the value of $n = 2$ because 2 mole of electrons are transferred for every mole of reaction:

$$E = 0.78\text{ V} - \frac{0.0592}{2}\log(10) = 0.75\text{ V}$$ which is choice **(B)**.

Choice **(A)** is incorrect because it omits n. Choice **(C)** is incorrect because it added the nonstandard cell potential term, $(0.0592/2)\log Q$. Choice **(D)** is incorrect because it added the nonstandard cell potential term and didn't use n =2.

Practice Questions Related to This: **PQ-14, PQ-15, PQ-16, PQ-17, PQ-18, PQ-19, PQ-20**

SQ-6. What half-reaction occurs at the anode during the electrolysis of molten sodium iodide?

(A) $2I^- \rightarrow I_2 + 2e^-$ **(B)** $I_2 + 2e^- \rightarrow 2I^-$

(C) $Na \rightarrow Na^+ + e^-$ **(D)** $Na^+ + e^- \rightarrow Na$

Knowledge Required: (1) The definition of electrolysis. (2) Definitions of anode and cathode in electrochemical cells.

Thinking it Through: You are being asked to identify the reaction occurring at the anode of an electrolytic cell. You remember that electrolysis reactions are reactions that use electrical energy to drive a chemical process. Specifically, these reactions have a positive ΔG and can be driven by the application of electrical energy. This is different from the spontaneous reactions used in galvanic cells, which have negative ΔG values and the excess free energy can be used to do electrical work. You also remember that no matter if the cell is a galvanic or electrolytic cell, oxidation occurs at the anode and reduction occurs at the cathode.

Because the electrolysis involves molten sodium iodide, you do not have to consider the oxidation or reduction of water. Remember that in aqueous solutions, these reactions are often the ones that occur.

Recall that in molten salts, the species of interest are the ions. For molten sodium iodide the ions are: Na^+ and I^-. These are the only options for reactants for the half-reactions. Because the process at the anode will be oxidation, you eliminate Na^+ because it is in its highest oxidation state. Therefore, the only possible reaction is the oxidation of I^-: $2I^- \rightarrow I_2 + 2e^-$ which is choice **(A)**.

Choice **(B)** is not correct because this is the reduction of I_2. Choice **(C)** is not correct because it is the oxidation of Na, a species not present in the molten salt. Choice **(D)** is not correct because it is the reduction of Na^+.

Practice Questions Related to This: **PQ-21, PQ-22, PQ-23**

SQ-7. How many minutes are required to plate 2.08 g of copper at a constant current flow of 1.26 A?

$$Cu^{2+}(aq) + 2e^- \rightarrow Cu(s)$$

Constants	
Molar mass, Cu	63.5 g·mol^{-1}
1 faraday (F)	96485 C

(A) 41.8 min **(B)** 83.6 min

(C) 133 min **(D)** 5016 min

Knowledge Required: (1) The stoichiometry of electrolytic processes. (2) The relationships among amperes, faradays, time, and moles of electrons.

Thinking it Through: You are given a mass and current. You recall that an Ampere is defined as 1 C of charge per second: $1\text{ A} = \frac{1\text{ C}}{1\text{ s}}$. Therefore, the 1.26 A can be written as 1.26 C·s^{-1} or as a conversion factor; 1.26 C = 1 s.

The stoichiometric relationship involving electrons and copper is: 2 mol e$^-$ = 1 mol Cu
The Faraday is also a conversation between coulomb and moles of electrons: 1 mol e$^-$ = 96485 C
You can use these relationships to find the time required. You start with the mass of copper desired:

$$2.08\text{ g Cu}\left(\frac{1\text{ mol Cu}}{63.5\text{ g Cu}}\right)\left(\frac{2\text{ mol e}^-}{1\text{ mol Cu}}\right)\left(\frac{96485\text{ C}}{1\text{ mol e}^-}\right)\left(\frac{1\text{ s}}{1.26\text{ C}}\right)\left(\frac{1\text{ min}}{60\text{ s}}\right) = 83.6\text{ min}$$ which is choice **(B)**.

Choice **(A)** is incorrect because it didn't use the 2 mol of electrons per 1 mol Cu relationship. Choice **(C)** is incorrect because it incorrectly multiplied by 1.26. Choice **(D)** is incorrect because it is the number of seconds, and did not convert the time to minutes.

Practice Questions Related to This: **PQ-24, PQ-25, PQ-26, PQ-27, PQ-28**

SQ-8. For the concentration cell

$$Fe(s)\ |\ Fe^{3+}(aq),\ 0.100\text{ M}\ ||\ Fe^{3+}(aq),\ 1.00\text{ M}\ |\ Fe(s)$$

what is E_{cell} at 25 °C (in mV)?

(A) 9.87 mV **(B)** 19.7 mV **(C)** 29.6 mV **(D)** 59.2 mV

Knowledge Required: (1) Definition of a concentration cell. (2) Ability to interpret electrochemical cell notation.

Thinking it Through: You are being asked about a concentration cell. You recall that in a concentration cell the source of the electrical potential, and the current flow, is the differences in the concentrations of the species. The given cell notation can be converted to half-reactions:

$$Fe(s)\ |\ Fe^{3+}(aq),\ 0.100\text{ M}\ ||\ Fe^{3+}(aq),\ 1.00\text{ M}\ |\ Fe(s)$$

oxidation ↙ ↘ reduction

$Fe(s) \rightarrow Fe^{3+}(aq,\ 0.100\text{ M}) + 3e^-$ $Fe^{3+}(aq,\ 1.00\text{ M}) + 3e^- \rightarrow Fe(s)$

After adding the half-reactions, you get: $Fe(s) + Fe^{3+}(aq,\ 1.00\text{ M}) \rightarrow Fe^{3+}(aq,\ 0.100\text{ M}) + Fe(s)$

The Nernst equation takes into account the dependence of cell potential on concentrations:

$$E_{cell} = E^{\circ}_{cell} - \left(\frac{0.0592}{n}\right)\log Q$$ remembering that Q is the ratio of the concentrations of products to reactants.

You also recall that the E°_{cell} for a concentration cell is zero, because the two half-reactions are the same:

$$E_{cell} = 0 - \left(\frac{0.0592}{3}\right)\log\left(\frac{0.100\text{ M}}{1.00\text{ M}}\right) = 0.0197\text{ V}$$

which is choice **(B)**.

$$0.0197\text{ V}\left(\frac{1000\text{ mV}}{1\text{ V}}\right) = 19.7\text{ mV}$$

Choice **(A)** is not correct because it used 6 for the number of electrons in the Nernst equation. Choice **(C)** is not correct because it used 2 for the number of electrons in the Nernst equation. Choice **(D)** is not correct because it did not use the number of electrons in the Nernst equation.

***Practice Questions Related to This:* PQ-29, PQ-30**

Practice Questions (PQ)

Conceptual **PQ-1.** What is the coefficient of the hydroxide ion for the half reaction when the hypobromite ion reacts to form the bromide ion in basic solution? $BrO^-(aq) \rightarrow Br^-(aq)$ *unbalanced*

(A) 1 **(B)** 2 **(C)** 3 **(D)** 4

Conceptual **PQ-2.** What is the balanced half reaction of the nitrate ion reacting to form nitrogen monoxide in acidic solution?

(A) $NO_3^-(aq) + H_2O(l) \rightarrow NO(g) + 3e^- + 2H^+(aq)$

(B) $NO_3^-(aq) + 3e^- + 4H^+(aq) \rightarrow NO(g) + 2H_2O(l)$

(C) $NO_3^-(aq) + 3e^- + 2H_2O(l) \rightarrow NO(g) + 4OH^-(aq)$

(D) $NO_3^-(aq) + 4e^- + 5H^+(aq) \rightarrow NO(g) + 2H_2O(l)$

Conceptual **PQ-3.** What is the coefficient for water when the reaction is balanced in acidic solution?

$Cr_2O_7^{2-}(aq) + Mn^{2+}(aq) \rightarrow Cr^{3+}(aq) + MnO_2(s)$ *unbalanced*

(A) 1 **(B)** 2 **(C)** 4 **(D)** 5

PQ-4. Which reaction occurs at the cathode of a galvanic cell constructed from these two half-cells?

The standard reduction potential of gold(III) is +1.50 V.
The standard reduction potential of tin(II) is –0.14 V.

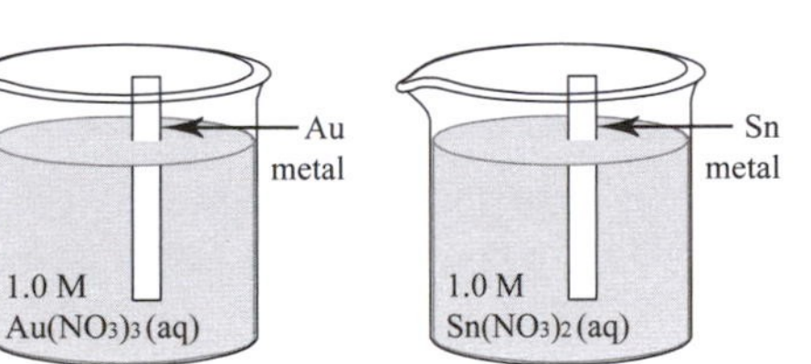

(A) $Au^{3+}(aq) + 3e^- \rightarrow Au(s)$

(B) $Sn^{2+}(aq) + 2e^- \rightarrow Sn(s)$

(C) $Au(s) \rightarrow Au^{3+}(aq) + 3e^-$

(D) $Sn(s) \rightarrow Sn^{2+}(aq) + 2e^-$

Conceptual **PQ-5.** Which reaction will occur if each substance is in its standard state?
Assume potentials are given in water at 25 °C.

Standard Reduction Potentials	*E*°
$Ni^{2+}(aq) + 2e^- \rightarrow Ni(s)$	–0.28 V
$Sn^{4+}(aq) + 2e^- \rightarrow Sn^{2+}(aq)$	0.15 V
$Br_2(l) + 2e^- \rightarrow 2Br^-(aq)$	1.06 V

(A) Ni^{2+} will oxidize Sn^{2+} to give Sn^{4+}

(B) Sn^{4+} will oxidize Br^- to give Br_2

(C) Br_2 will oxidize Ni(s) to give Ni^{2+}

(D) Ni^{2+} will oxidize Br_2 to give Br^-

PQ-6. Which represents a spontaneous reaction and what is the correct $E°_{cell}$?

Standard Reduction Potentials	$E°$
$Fe^{3+}(aq) + 3e^- \rightarrow Fe(s)$	–0.04 V
$Cl_2(g) + 2e^- \rightarrow 2Cl^-(aq)$	1.36 V

	Reaction	$E°_{cell}$
(A)	$Fe(s) + Cl_2(g) \rightarrow Fe^{3+}(aq) + 2Cl^-(aq)$	1.40 V
(B)	$Fe^{3+}(aq) + Cl_2(g) \rightarrow Fe(s) + 2Cl^-(aq)$	1.40 V
(C)	$2Fe(s) + 3Cl_2(g) \rightarrow 2Fe^{3+}(aq) + 6Cl^-(aq)$	1.40 V
(D)	$2Fe(s) + 3Cl_2(g) \rightarrow 2Fe^{3+}(aq) + 6Cl^-(aq)$	4.16 V

PQ-7. Which combination of reactants will produce the greatest voltage based on these standard electrode potentials?

Standard Reduction Potentials	$E°$
$Cu^+(aq) + e^- \rightarrow Cu(s)$	0.52 V
$Sn^{4+}(aq) + 2e^- \rightarrow Sn^{2+}(aq)$	0.15 V
$Cr^{3+}(aq) + e^- \rightarrow Cr^{2+}(aq)$	–0.41 V

(A) Cu^+ and Sn^{2+} **(B)** Cu^+ and Cr^{2+} **(C)** Cu and Sn^{2+} **(D)** Sn^{4+} and Cr^{2+}

PQ-8. What is the standard electrode potential for a voltaic cell constructed in the appropriate way from these two half-cells?

Standard Reduction Potentials	$E°$
$Cr^{3+}(aq) + 3e^- \rightarrow Cr(s)$	–0.74 V
$Co^{2+}(aq) + 2e^- \rightarrow Co(s)$	–0.28 V

(A) –1.02 V **(B)** 0.46 V **(C)** 0.64 V **(D)** 1.02 V

PQ-9. What is the value of the missing standard reduction potential?
$Cu(s) + Pd^{2+}(aq) \rightarrow Cu^{2+}(aq) + Pd(s)$ $E° = 0.650$ V

Standard Reduction Potentials	$E°$
$Pd^{2+}(aq) + 2e^- \rightarrow Pd(s)$	?
$Cu^{2+}(aq) + 2e^- \rightarrow Cu(s)$	0.337 V

(A) 0.987 V **(B)** –0.987 V **(C)** 0.313 V **(D)** –0.313 V

PQ-10. The standard cell potential, $E°$, for this reaction is 0.79 V.

$$6I^-(aq) + Cr_2O_7^{2-}(aq) + 14H^+(aq) \rightarrow 3I_2(aq) + 2Cr^{3+}(aq) + 7H_2O(l)$$

What is the standard potential for $I_2(aq)$ being reduced to $I^-(aq)$ given that the standard reduction potential for $Cr_2O_7^{2-}(aq)$ changing to $Cr^{3+}(aq)$ is 1.33 V?

(A) 0.54 V **(B)** –0.54 V **(C)** 0.18 V **(D)** –0.18 V

Conceptual **PQ-11.** What is the standard cell potential, $E°$, for this reaction?
$Br_2(l) + 2Ce^{3+}(aq) \rightarrow 2Br^-(aq) + 2Ce^{4+}(aq)$

Standard Reduction Potentials	$E°$
$Ce^{4+}(aq) + e^- \rightarrow Ce^{3+}(aq)$	1.61 V
$Br_2(l) + 2e^- \rightarrow 2Br^-(aq)$	1.06 V

(A) –2.67 V **(B)** –2.16 V **(C)** –0.55 V **(D)** 2.67 V

Conceptual **PQ-12.** Which statement is true for an electrochemical cell built from an oxidation-reduction reaction if K for the reaction is greater than 1?

(A) $\Delta G°$ is negative, $E°$ is negative
(B) $\Delta G°$ is negative, $E°$ is positive
(C) $\Delta G°$ is positive, $E°$ is negative
(D) $\Delta G°$ is positive, $E°$ is positive

PQ-13. What is the equilibrium constant at 298 K for the spontaneous reaction that occurs when a $Cu^{2+}|Cu$ half-cell is connected to a $Ag^+|Ag$ half-cell?

Reduction Potentials	$E°$
$Cu^{2+}|Cu$	0.34 V
$Ag^+|Ag$	0.80 V

(A) 2.8×10^{-16} **(B)** 6.0×10^{7} **(C)** 3.6×10^{15} **(D)** 3.7×10^{38}

PQ-14. Consider the reaction at 298 K.

$$Sn^{2+}(aq) + 2Fe^{3+}(aq) \rightarrow Sn^{4+}(aq) + 2Fe^{2+}(aq) \qquad E° = 0.617\ V$$

What is the value of E when $[Sn^{2+}]$ and $[Fe^{3+}]$ are equal to 0.50 M and $[Sn^{4+}]$ and $[Fe^{2+}]$ are equal to 0.10 M?

(A) 0.069 V **(B)** 0.679 V **(C)** 0.658 V **(D)** 0.576 V

Conceptual **PQ-15.** Consider this reaction.

$$2Al(s) + 6HCl(aq) \rightleftharpoons 2AlCl_3(aq) + 3H_2(g) \quad E° = 1.66\ V$$

Which statement is true if the hydrogen ion concentration is initially at 1.0 M and the initial pressure of hydrogen gas is 1.0 atm?

(A) Addition of a base should result in a value of E which is less than 1.66 V.

(B) $n = 3$, because three moles of hydrogen is being produced.

(C) $E°$ is independent of the pH of the solution.

(D) $Q = \dfrac{[H_2]^3[AlCl_3]^2}{[Al]^2[HCl]^6}$

Conceptual **PQ-16.** According to the Nernst equation, when $E_{cell} = 0$ then

(A) $Q = K$ **(B)** $K = 1$ **(C)** $\Delta G° = 0$ **(D)** $E°_{cell} = 0$

PQ-17. What is E_{cell} at 25 °C if $[Ag^+] = 0.200$ M and $[Cu^{2+}] = 0.200$ M?

$$2Ag^+(aq) + Cu(s) \rightarrow Cu^{2+}(aq) + 2Ag(s) \qquad E°_{cell} = 0.460\ V$$

(A) 0.419 V **(B)** 0.439 V **(C)** 0.460 V **(D)** 0.481 V

Conceptual **PQ-18.** A voltmeter connected to the cell under standard state conditions reads +0.25 V, and the standard free energy ($\Delta G°$) is –44 kJ·mol^{-1}.

If the concentration of Ni^{2+} is increased to 2.0 M, the cell potential will _____ and ΔG will be ______ than $\Delta G°$.

$$Ni^{2+}(aq) + 2e^- \rightarrow Ni(s) \qquad E° = -0.25\ V$$

0.25 V
Salt Bridge
Ni(s)
$H_2(g)$ (1 atm)
Pt(s)
1 M $Ni^{2+}(aq)$
1 M $H^+(aq)$

(A) decrease, less negative **(B)** decrease, more negative

(C) increase, less negative **(D)** increase, more negative

PQ-19. What is the potential of a galvanic cell made from $Mn^{2+}|Mn$ and $Ag^+|Ag$ half-cells at 25 °C if $[Ag^+] = 0.10$ M and $[Mn^{2+}] = 0.10$ M?

Standard Reduction Potentials	$E°$
$Mn^{2+}(aq) + 2e^- \rightarrow Mn(s)$	–1.18 V
$Ag^+(aq) + e^- \rightarrow Ag(s)$	0.80 V

(A) 1.92 V **(B)** 1.95 V **(C)** 1.98 V **(D)** 2.01 V

PQ-20. What is the cell potential of the voltaic zinc-copper cell at 25 °C?

$$Zn(s)|Zn^{2+}(1.00\times10^{-5}\ M)||Cu^{2+}(1.00\times10^{-2})|Cu(s)$$

The standard cell potential of the cell is 1.10 V.

(A) 0.92 V **(B)** 1.28 V **(C)** 1.01 V **(D)** 1.19 V

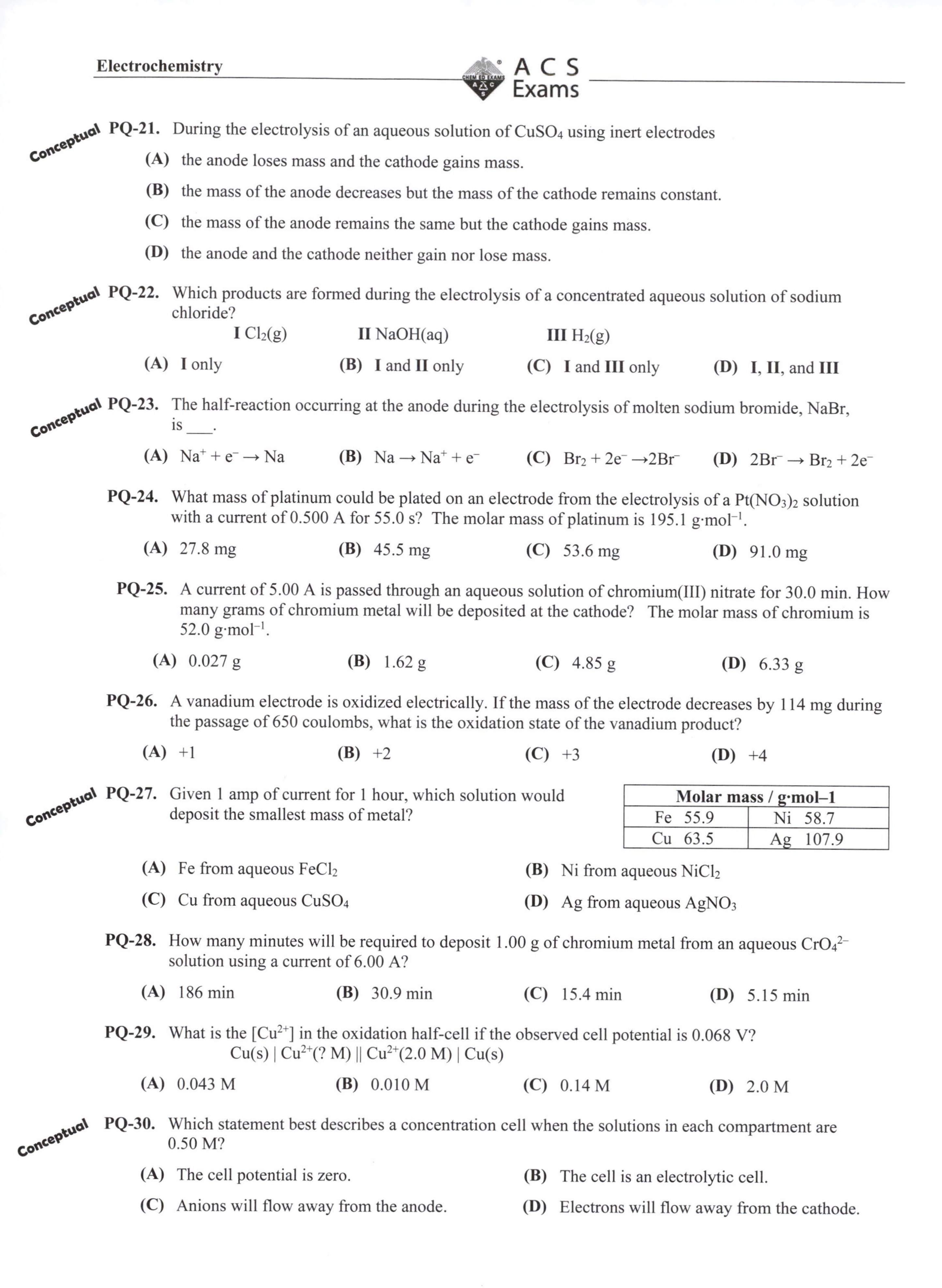

Conceptual **PQ-21.** During the electrolysis of an aqueous solution of $CuSO_4$ using inert electrodes

(A) the anode loses mass and the cathode gains mass.

(B) the mass of the anode decreases but the mass of the cathode remains constant.

(C) the mass of the anode remains the same but the cathode gains mass.

(D) the anode and the cathode neither gain nor lose mass.

Conceptual **PQ-22.** Which products are formed during the electrolysis of a concentrated aqueous solution of sodium chloride?

I $Cl_2(g)$ **II** $NaOH(aq)$ **III** $H_2(g)$

(A) **I** only **(B)** **I** and **II** only **(C)** **I** and **III** only **(D)** **I, II,** and **III**

Conceptual **PQ-23.** The half-reaction occurring at the anode during the electrolysis of molten sodium bromide, NaBr, is ___.

(A) $Na^+ + e^- \rightarrow Na$ **(B)** $Na \rightarrow Na^+ + e^-$ **(C)** $Br_2 + 2e^- \rightarrow 2Br^-$ **(D)** $2Br^- \rightarrow Br_2 + 2e^-$

PQ-24. What mass of platinum could be plated on an electrode from the electrolysis of a $Pt(NO_3)_2$ solution with a current of 0.500 A for 55.0 s? The molar mass of platinum is 195.1 $g{\cdot}mol^{-1}$.

(A) 27.8 mg **(B)** 45.5 mg **(C)** 53.6 mg **(D)** 91.0 mg

PQ-25. A current of 5.00 A is passed through an aqueous solution of chromium(III) nitrate for 30.0 min. How many grams of chromium metal will be deposited at the cathode? The molar mass of chromium is 52.0 $g{\cdot}mol^{-1}$.

(A) 0.027 g **(B)** 1.62 g **(C)** 4.85 g **(D)** 6.33 g

PQ-26. A vanadium electrode is oxidized electrically. If the mass of the electrode decreases by 114 mg during the passage of 650 coulombs, what is the oxidation state of the vanadium product?

(A) +1 **(B)** +2 **(C)** +3 **(D)** +4

Conceptual **PQ-27.** Given 1 amp of current for 1 hour, which solution would deposit the smallest mass of metal?

Molar mass / g·mol–1	
Fe 55.9	Ni 58.7
Cu 63.5	Ag 107.9

(A) Fe from aqueous $FeCl_2$ **(B)** Ni from aqueous $NiCl_2$

(C) Cu from aqueous $CuSO_4$ **(D)** Ag from aqueous $AgNO_3$

PQ-28. How many minutes will be required to deposit 1.00 g of chromium metal from an aqueous CrO_4^{2-} solution using a current of 6.00 A?

(A) 186 min **(B)** 30.9 min **(C)** 15.4 min **(D)** 5.15 min

PQ-29. What is the $[Cu^{2+}]$ in the oxidation half-cell if the observed cell potential is 0.068 V?

$Cu(s) \mid Cu^{2+}(?\ M) \parallel Cu^{2+}(2.0\ M) \mid Cu(s)$

(A) 0.043 M **(B)** 0.010 M **(C)** 0.14 M **(D)** 2.0 M

Conceptual **PQ-30.** Which statement best describes a concentration cell when the solutions in each compartment are 0.50 M?

(A) The cell potential is zero. **(B)** The cell is an electrolytic cell.

(C) Anions will flow away from the anode. **(D)** Electrons will flow away from the cathode.

Answers to Study Questions

1. C
2. C
3. B
4. B
5. B
6. A
7. B
8. B

Answers to Practice Questions

1. B
2. B
3. A
4. A
5. C
6. C
7. B
8. B
9. A
10. A
11. C
12. B
13. C
14. B
15. A
16. A
17. B
18. A
19. B
20. D
21. C
22. D
23. D
24. A
25. B
26. C
27. D
28. B
29. B
30. A

Chapter 16 – Nuclear Chemistry

Chapter Summary:

This chapter will focus on expressing and predicting nuclear reactions.

Specific topics covered in this chapter are:

- Radioactivity
- Radioactive decay
- Balancing nuclear reactions

Previous material that is relevant to your understanding of questions in this chapter include:

- Atomic Structure (***Chapter 1***)
- Electronic Structure (***Chapter 2***)
- Stoichiometry (***Chapter 4***)

Common representations used in questions related to this material:

Name	Example	Used in questions related to
Graphs	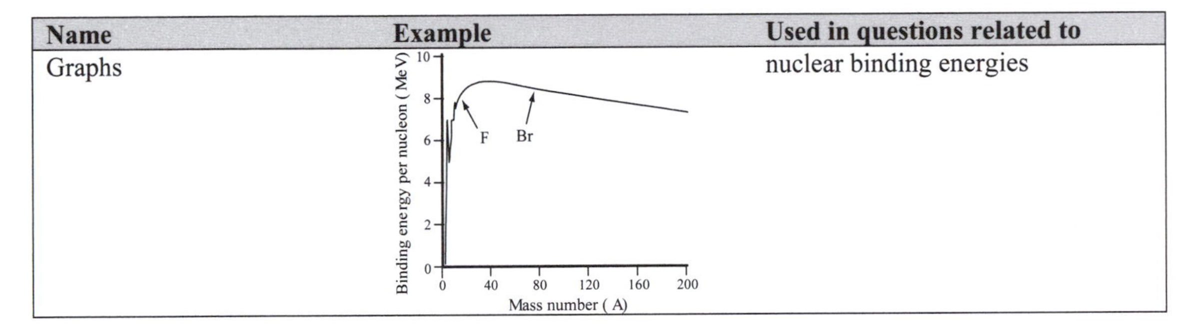	nuclear binding energies

Where to find this in your textbook:

The material in this chapter typically aligns to "Nuclear Chemistry" in your textbook. The name of your chapter(s) may vary.

Practice exam:

There are practice exam questions aligned to the material in this chapter. Because there are a limited number of questions on the practice exam, a review of the breadth of the material in this chapter is advised in preparation for your exam.

How this fits into the big picture:

The material in this chapter aligns to the Big Idea of Atomic Structure (1) as listed on page 12 of this study guide.

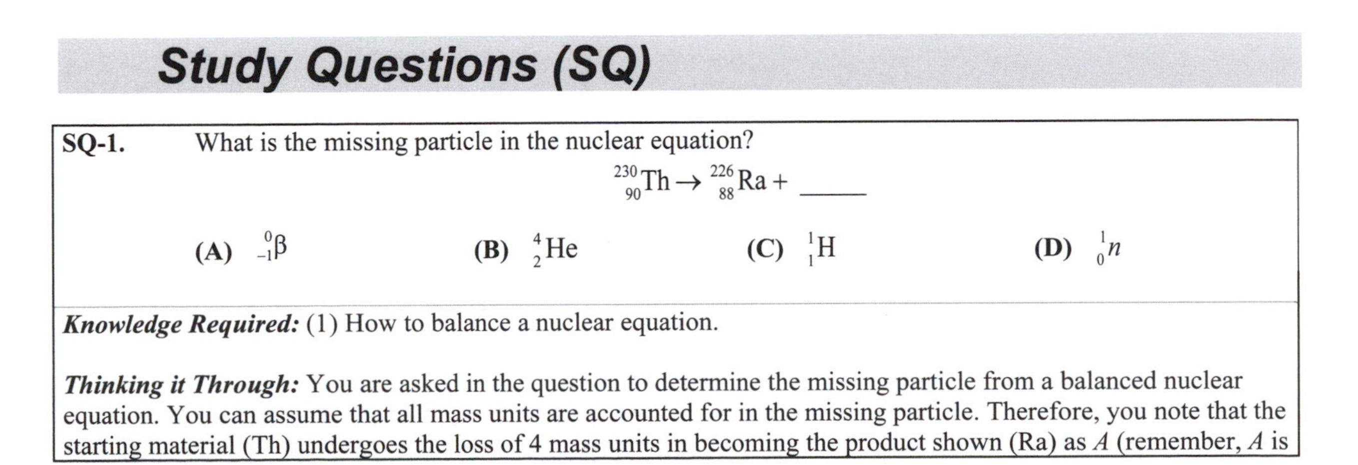

Study Questions (SQ)

SQ-1. What is the missing particle in the nuclear equation?

$$^{230}_{90}\text{Th} \rightarrow ^{226}_{88}\text{Ra} + \_\_\_\_$$

(A) $^{0}_{-1}\beta$ **(B)** $^{4}_{2}\text{He}$ **(C)** $^{1}_{1}\text{H}$ **(D)** $^{1}_{0}n$

Knowledge Required: (1) How to balance a nuclear equation.

Thinking it Through: You are asked in the question to determine the missing particle from a balanced nuclear equation. You can assume that all mass units are accounted for in the missing particle. Therefore, you note that the starting material (Th) undergoes the loss of 4 mass units in becoming the product shown (Ra) as *A* (remember, *A* is

mass number or the number of protons plus the number of neutrons) goes from 230 (^{230}Th) to 226 (^{226}Ra). The missing particle will have a mass number of 4 (^{4}X).

In addition, you note that the starting material (Th) is 2 atomic numbers (or two protons) more than the product (Ra). as Z goes from 90 ($_{90}$Th) to 88 ($_{88}$Ra). The missing particle will have an atomic number of 2 ($_{2}$X).

Four mass units with two of those as protons is a He particle. Therefore, you can conclude that this is an example of alpha decay and that $^{4}_{2}$He is the missing particle, choice **(B)**.

Choice **(A)** is not correct because it is the particle lost for a beta decay equation.
Choice **(C)** is not correct because it is a proton particle.
Choice **(D)** is not correct because it is a neutron particle.

Practice Questions Related to This: **PQ-1, PQ-2, PQ-3,** and **PQ-4**

SQ-2. A $^{235}_{92}$U atom undergoes nuclear fission by being bombarded with a neutron to produce a total of four neutrons, a $^{140}_{55}$Cs atom, and _____.

(A) $^{235}_{92}$U **(B)** $^{235}_{93}$Np

(C) $^{92}_{37}$Rb **(D)** $^{96}_{37}$Rb

Knowledge Required: (1) Definition of nuclear fission. (2). How to write and balance a nuclear equation.

Thinking it Through: You are asked in the question to determine the missing particle for a given nuclear fission reaction. The reaction is expressed in words; it is helpful in this situation to express the reaction as an equation:

$$^{235}_{92}\text{U} + {}^{1}_{0}\text{n} \rightarrow {}^{140}_{55}\text{Cs} + 4\,{}^{1}_{0}\text{n} + \_\_\_\_$$

Next, you balance the protons; there are 92 on the reactant side ($_{92}$U and $_{0}$n) and 55 on the product side ($_{55}$Cs and 4 × $_{0}$n) or 92–55 = 37 protons missing for the unknown atom.

You then balance the mass numbers; there are 235 + 1 on the reactant side (^{235}U and ^{1}n) and 140 + 4×1 on the product side (^{140}Cs and 4 × ^{1}n) or 236 – 144 = 92 mass numbers missing for the unknown atom. Thus, $^{92}_{37}X$ is missing in order to balance the reaction; this is the atom $^{92}_{37}$Rb, Choice **(C)**.

Choice **(A)** is not correct because it assumes no change in the starting material.
Choice **(B)** is not correct because it is the result of positron emission, i.e. a neutron becomes a proton.
Choice **(D)** is not correct because it does not account for the four neutrons that are produced in the nuclear fission reaction.

Practice Questions Related to This: **PQ-5**

SQ-3. An atom of the element of atomic number 53 and mass number 131 undergoes beta decay. The residual atom after this change has an atomic number of _____ and a mass number of _____.

(A) 54, 131 **(B)** 53, 131 **(C)** 53, 131 **(D)** 51, 127

Knowledge Required: (1) Radiation emitted when alpha decay, beta decay, gamma emission, and positron emission occur. (2) How to write and balance a nuclear reaction.

Thinking it Through: You are asked in the question to determine the atomic number and mass number for an atom after a particular radiation emission has occurred. In this instance, the atom undergoes beta decay (a $^{0}_{-1}\beta$ particle); this emission results in a net gain of one atomic number (or transformation such that the nucleus now has one additional proton). During beta decay, there is no overall change in mass number. You identify the element in the question as iodine, I (Z = 53).

$$^{131}_{53}\text{I} \rightarrow {}^{0}_{-1}\beta + {}^{131}_{54}\text{Xe}$$

Therefore, if a particle that is $^{131}_{53}I$ undergoes beta decay, the resulting atom is $^{131}_{54}Xe$ ($Z = 54$, $A = 131$), choice **(A)**.

Choice **(B)** is not correct because it would be the residual atom after a positron emission.
Choice **(C)** is not correct because it would be the residual atom after a gamma emission.
Choice **(D)** is not correct because it would be the residual atom after alpha decay.

Practice Questions Related to This: **PQ-6, PQ-7, PQ-8, PQ-9, PQ-10, PQ-11,** and **PQ-12**

Practice Questions (PQ)

PQ-1. Cobalt-60 undergoes beta decay. What is the balanced nuclear reaction for this process?

(A) $^{60}_{27}Co \longrightarrow ^{0}_{1}\beta + ^{60}_{26}Fe$ **(B)** $^{60}_{27}Co \longrightarrow ^{0}_{-1}\beta + ^{60}_{28}Ni$

(C) $^{60}_{27}Co \longrightarrow ^{4}_{2}\alpha + ^{56}_{25}Mn$ **(D)** $^{60}_{27}Co + ^{0}_{-1}e \longrightarrow ^{60}_{26}Fe$

PQ-2. What is the missing particle in the balanced nuclear equation?

$$^{23}_{12}Mg \rightarrow ^{23}_{11}Na + \_\_\_\_ + ^{0}_{0}\nu$$

(A) $^{0}_{-1}\beta$ **(B)** $^{0}_{1}\beta$ **(C)** $^{1}_{1}H$ **(D)** $^{1}_{0}n$

PQ-3. What is the missing product of this reaction?

$$^{32}_{15}P \rightarrow ^{32}_{16}S + \_\_\_\_$$

(A) $^{0}_{-1}\beta$ **(B)** $^{0}_{+1}\beta$ **(C)** $^{0}_{0}\gamma$ **(D)** $^{4}_{2}\alpha$

PQ-4. What is the missing particle in the balanced nuclear equation?

$$^{27}_{13}Al + ^{4}_{2}He \rightarrow ^{30}_{15}P + \_\_\_\_$$

(A) $^{0}_{-1}e$ **(B)** $^{0}_{1}e$ **(C)** $^{1}_{1}H$ **(D)** $^{1}_{0}n$

PQ-5. When $^{242}_{96}Cm$ captures an alpha particle, it produces a neutron and which isotope?

(A) $^{242}_{96}Cm$ **(B)** $^{245}_{98}Cf$ **(C)** $^{238}_{94}Pu$ **(D)** $^{239}_{94}Pu$

Conceptual

PQ-6. Which particle, if lost from the nucleus, will *not* result in a change in the atomic mass number?

(A) proton **(B)** alpha particle **(C)** beta particle **(D)** neutron

PQ-7. An atom of the element of atomic number 84 and mass number 199 emits an alpha particle. The residual atom after this change has an atomic number of _____ and a mass number of _____.

(A) 82, 195 **(B)** 84, 203 **(C)** 85, 195 **(D)** 86, 199

PQ-8. What is the product when $^{218}_{84}Po$ undergoes positron emission?

(A) $^{214}_{82}Pb$ **(B)** $^{222}_{86}Rn$ **(C)** $^{214}_{81}Tl$ **(D)** $^{218}_{83}Bi$

PQ-9. Which equation qualifies as electron capture?

(A) $^{6}_{3}Li + ^{1}_{0}n \rightarrow ^{3}_{1}H + ^{4}_{2}\alpha$ **(B)** $^{26}_{13}Mg + ^{1}_{1}p \rightarrow ^{23}_{11}Na + ^{4}_{2}\alpha$

(C) $^{37}_{18}Ar + ^{0}_{-1}e \rightarrow ^{37}_{17}Cl + h\nu$ **(D)** $^{38}_{19}K \rightarrow ^{38}_{20}Ca + ^{0}_{-1}\beta$

Conceptual **PQ-10.** Which nuclear process does not necessarily result in nuclear transmutation (the transformation of one element into another)?

(A) alpha (α) emission
(B) beta (β) emission
(C) gamma (γ) emission
(D) positron emission

PQ-11. Medical imaging is used when a patient ingests a radioactive emitter, called a tracer, so a doctor can view an internal system. Which tracer emission will penetrate through the body?

(A) alpha particle
(B) beta particle
(C) gamma particle
(D) helium nucleus

Conceptual **PQ-12.** Based on their binding energies, what nuclear transformations are predicted for F and Br?

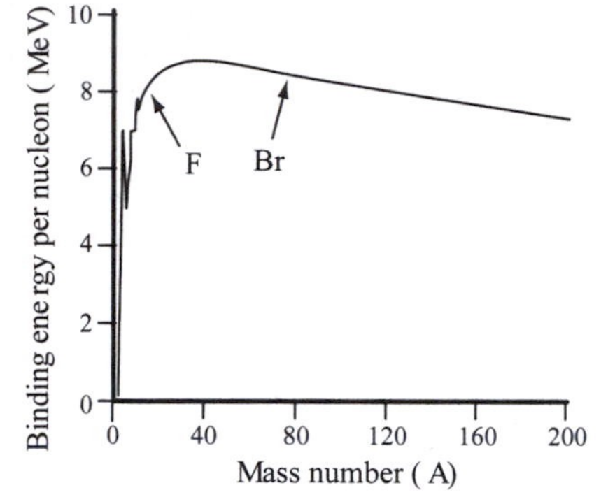

(A) Both could undergo nuclear fission.
(B) Both could undergo nuclear fusion.
(C) F could undergo nuclear fission and Br could undergo nuclear fusion.
(D) Br could undergo nuclear fission and F could undergo nuclear fusion.

Answers to Study Questions

1. B
2. C
3. A

Answers to Practice Questions

1. B
2. B
3. A
4. D
5. B
6. C
7. A
8. D
9. C
10. C
11. C
12. D